Lecture Notes in Computer Science 14478

Founding Editors

Gerhard Goos
Juris Hartmanis

Editorial Board Members

Elisa Bertino, *Purdue University, West Lafayette, IN, USA*
Wen Gao, *Peking University, Beijing, China*
Bernhard Steffen, *TU Dortmund University, Dortmund, Germany*
Moti Yung, *Columbia University, New York, NY, USA*

The series Lecture Notes in Computer Science (LNCS), including its subseries Lecture Notes in Artificial Intelligence (LNAI) and Lecture Notes in Bioinformatics (LNBI), has established itself as a medium for the publication of new developments in computer science and information technology research, teaching, and education.

LNCS enjoys close cooperation with the computer science R & D community, the series counts many renowned academics among its volume editors and paper authors, and collaborates with prestigious societies. Its mission is to serve this international community by providing an invaluable service, mainly focused on the publication of conference and workshop proceedings and postproceedings. LNCS commenced publication in 1973.

Yaroslav D. Sergeyev · Dmitri E. Kvasov ·
Annabella Astorino
Editors

Numerical Computations: Theory and Algorithms

4th International Conference, NUMTA 2023
Pizzo Calabro, Italy, June 14–20, 2023
Revised Selected Papers, Part III

Editors
Yaroslav D. Sergeyev
University of Calabria
Rende, Italy
Lobachevksy University of Nizhny Novgorod
Nizhny Novgorod, Russia

Dmitri E. Kvasov
University of Calabria
Rende, Italy

Annabella Astorino
University of Calabria
Rende, Italy

ISSN 0302-9743 ISSN 1611-3349 (electronic)
Lecture Notes in Computer Science
ISBN 978-3-031-81246-0 ISBN 978-3-031-81247-7 (eBook)
https://doi.org/10.1007/978-3-031-81247-7

© The Editor(s) (if applicable) and The Author(s), under exclusive license
to Springer Nature Switzerland AG 2025

This work is subject to copyright. All rights are solely and exclusively licensed by the Publisher, whether the whole or part of the material is concerned, specifically the rights of translation, reprinting, reuse of illustrations, recitation, broadcasting, reproduction on microfilms or in any other physical way, and transmission or information storage and retrieval, electronic adaptation, computer software, or by similar or dissimilar methodology now known or hereafter developed.
The use of general descriptive names, registered names, trademarks, service marks, etc. in this publication does not imply, even in the absence of a specific statement, that such names are exempt from the relevant protective laws and regulations and therefore free for general use.
The publisher, the authors and the editors are safe to assume that the advice and information in this book are believed to be true and accurate at the date of publication. Neither the publisher nor the authors or the editors give a warranty, expressed or implied, with respect to the material contained herein or for any errors or omissions that may have been made. The publisher remains neutral with regard to jurisdictional claims in published maps and institutional affiliations.

This Springer imprint is published by the registered company Springer Nature Switzerland AG
The registered company address is: Gewerbestrasse 11, 6330 Cham, Switzerland

If disposing of this product, please recycle the paper.

Preface

This book edited by Yaroslav D. Sergeyev, Dmitri E. Kvasov, and Annabella Astorino is part of three LNCS volumes containing selected peer-reviewed papers from the Fourth Triennial International Conference and Summer School NUMTA 2023 "Numerical Computations: Theory and Algorithms", held from 14 to 20 June 2023 at "TUI Magic Life Resort" in Calabria, Italy.

The Conference was organized by the Department of Computer Engineering, Modeling, Electronics and Systems Science of the University of Calabria, Italy in cooperation with the following departments of the same university: Department of Mathematics and Computer Science; Mechanical, Energy, and Management Engineering Department; and Civil Engineering Department. We are proud to inform you that, like the previous three editions, NUMTA 2023 was organized in cooperation with the Society for Industrial and Applied Mathematics (SIAM), USA. The previous editions of NUMTA took place in several beautiful places in Calabria in 2013, 2016, and 2019. This fourth edition was postponed from 2022 to 2023 due to the difficulties created by the COVID-19 pandemic.

The goal of all NUMTA Conferences is to create a multidisciplinary round table for an open discussion on numerical modeling using traditional and emerging computational paradigms. The NUMTA 2023 Conference discussed all aspects of numerical computations and modeling from foundations, philosophy, and teaching to advanced numerical techniques. New technological challenges and fundamental ideas from theoretical computer science, machine learning, linguistics, logic, set theory, and philosophy met requirements and new fresh applications from engineering, physics, chemistry, biology, economy, and teaching mathematics.

Researchers from both theoretical and applied sciences were invited to use this excellent possibility to exchange ideas with leading scientists from different research fields. Papers discussing new computational paradigms, relations with mathematical foundations, and their impact on natural sciences were particularly solicited. Special attention during the Conference was dedicated to numerical optimization techniques and a variety of issues related to theory and practice of the usage of infinities and infinitesimals in numerical computations. In particular, there was a substantial bunch of talks dedicated to a recent promising methodology allowing one to execute numerical computations with finite, infinite, and infinitesimal numbers on a new type of a supercomputer – the Infinity Computer – patented in several countries.

The Editors, together with all the organizers of the event, are proud to inform you that two hundred researchers from the following 31 countries participated in the Conference: Austria, Canada, China, Cyprus, Czech Republic, Finland, France, Germany, India, Iran, Italy, Japan, Kuwait, Latvia, Lebanon, Lithuania, Malta, Morocco, Norway, Philippines, Portugal, Romania, Russia, Serbia, Singapore, Spain, Thailand, Turkey, Ukraine, the UK, and the USA.

vi   Preface

The following plenary lecturers shared their achievements with the NUMTA 2023 participants:

- Luigi Brugnano, Florence, Italy: "Spectral solution in time of evolutionary problems";
- Louis D'Alotto, New York, USA: "Infinite computations and Büchi automata using Grossone";
- Renato De Leone, Camerino, Italy: "Use of Grossone in optimization, classification, and feature selection problems";
- Kalyanmoy Deb, East Lansing, USA: "Evolutionary multi-criterion optimization: An emerging computational problem-solving tool";
- Hoai An Le Thi, Metz, France: "DC Learning: Recent advances and ongoing developments";
- Francesca Mazzia, Bari, Italy: "Computing derivatives on the Infinity Computer";
- Panos Pardalos, Gainesville, USA : "Computational approaches for solving general systems of nonlinear equations in the cloud";
- Witold Pedrycz, Edmonton, Canada: "Society-oriented developments of machine learning: Challenges and opportunities";
- Nick Trefethen, Oxford, UK: "Smooth random functions and smooth random ODEs".

Moreover, the following tutorials were presented during the conference:

- Yaroslav Sergeyev, Rende, Italy: "Infinity computing: Foundations and practical computations with numerical infinities and infinitesimals".
- Vassili Toropov, London, UK: "Multidisciplinary topology optimisation: Where we are and where we aim to be".

Regular presentations were organized as the main stream and the following 18 special sessions (in alphabetical order):

1) Advanced numerical methods for global optimization; organized by Dmitri Kvasov and Maria Chiara Nasso (University of Calabria, Italy).
2) Advances in DC programming and DC Learning; organized by Hoai An Le Thi (University of Lorraine, France) and Tao Pham Dinh (National Institute of Applied Sciences, Rouen, France).
3) Application of computational systems, AI, machine learning, and classification methods in engineering; organized by Patrizia Piro, Michele Turco, Behrouz Pirouz, and Stefania Anna Palermo (University of Calabria, Italy).
4) Approximation theory and algorithms with applications to AI/ML; organized by Alessandra De Rossi and Elena Cordero (University of Turin, Italy).
5) Bioinformatics and health informatics: Methods and applications; organized by Mario Cannataro and Marianna Milano (University "Magna Graecia" of Catanzaro, Italy).
6) Computational aspects of dynamic geometry and applications; organized by Dorin Andrica (Babeş-Bolyai University, Romania) and Ovidiu Bagdasar (University of Derby, UK).
7) Computational tools and new trends in math and science education; dedicated to the memory of a young mathematics teacher, Giovanni Bozzolo, and organized by Annabella Astorino and Fabio Caldarola (University of Calabria, Italy).

8) Data-driven numerical analysis of complex and multiscale dynamical systems; organized by Constantinos Siettos (University of Naples Federico II, Italy) and Lucia Russo (IRC-CNR, Italy).
9) High-performance computing in modelling and simulation; organized by William Spataro, Donato D'Ambrosio, Rocco Rongo (University of Calabria, Italy), and Andrea Giordano (ICAR-CNR, Italy).
10) Local and global polynomial approximation and applications; organized by Francesco Dell'Accio (University of Calabria, Italy), Donatella Occorsio (University of Basilicata, Italy), and Woula Themistoclakis (IAC-CNR, Italy).
11) Numerical optimization and machine learning; organized by Annabella Astorino and Antonio Fuduli (University of Calabria, Italy).
12) Optimization for data-driven methods; organized by Giorgia Franchini (University of Modena and Reggio Emilia, Italy), Simone Rebegoldi (University of Florence, Italy), and Federica Porta (University of Modena and Reggio Emilia, Italy).
13) Optimization of marine prospects and sustainability of water resources under climate changes; organized by Fabio Caldarola, Manuela Carini, and Mario Maiolo (University of Calabria, Italy).
14) Pythagorean stream: From numbers to mathematical models, blockchain, and distributed AI; organized by Fabio Caldarola, Manuela Carini, and Gianfranco d'Atri (University of Calabria, Italy).
15) Quantum computing algorithms for science and industry; organized by Jacopo Settino (University of Calabria, Italy), Andrea Giordano, Carlo Mastroianni (ICAR-CNR, Italy), Francesco Plastina (University of Calabria, Italy), and Andrea Vinci (ICAR-CNR, Italy).
16) Recent advances in numerical computations for control theory; organized by Elizaveta Shmalko, Aleksandra Zhukova, and Askhat Diveev (FRC CSC of RAS, Russia).
17) Recent trends on ODEs methods, low-rank approximations, and image-processing techniques; organized by Antonella Falini and Francesca Mazzia (University of Bari Aldo Moro, Italy).
18) Variational analysis and optimization methods with applications in finance and economics; organized by David Barilla (University of Messina, Italy) and Tiziana Ciano (University of Valle d'Aosta, Italy).

During this edition of NUMTA, as happened in previous events, the NUMTA Research Awards for excellence in research and the Young Researcher Prize by Springer were awarded. The winners of the major prize were Francesca Mazzia (University of Bari, Italy) and Louis D'Alotto (City University of New York, USA), who delivered their awards-winning Plenary Lectures. The Young Researcher Prize by Springer was assigned to Davide Stocco (University of Trento, Italy).

The NUMTA 2023 proceedings collect 45 full papers and 60 short papers selected by members of the NUMTA 2023 Programming Committee and organizers of special sessions from 170 submissions to ensure high research standards of submissions for these volumes. The papers were accepted for publication after their thorough single-blind peer reviewing (which required up to three review rounds for some manuscripts) by PC members and independent reviewers.

Part I of the NUMTA 2023 proceedings contains 18 long and 17 short papers on numerical computations in: continuous and discrete single- and multi-objective problems, large-scale optimization, classification in machine learning, optimal control, and applications. The papers of Part I were selected from the NUMTA 2023 main stream and the above-mentioned special sessions 1, 2, 11, 12, 16, and 18.

Part II contains 11 long and 24 short peer-reviewed papers on various topics of computational and applied mathematics (such as approximation theory, computational geometry, computational fluid dynamics, dynamical systems and differential equations, numerical algebra, etc.) and applications in engineering and science. The papers of Part II were selected from the NUMTA 2023 main stream and special sessions 3, 4, 6, 10, 13, and 17. Part II also contains the paper of the winner of the Springer Young Researcher Prize for the best NUMTA 2023 presentation made by a young scientist (Davide Stocco, Trento, Italy). The support of the Springer LNCS editorial staff and Springer's sponsorship of the Young Researcher Prize are greatly appreciated.

Finally, Part III contains 16 long and 19 short peer-reviewed papers on numerical models and methods using traditional and emerging high-performance computational tools and paradigms and their application in artificial intelligence and data science, bioinformatics, engineering and technology, mathematical education, number theory and foundations of mathematics, etc. The papers of Part III were selected from the NUMTA 2023 main stream and special sessions 5, 7, 9, 14, and 15.

The Editors express their gratitude to the institutions that offered their generous support to the international conference NUMTA 2023. This support was essential to the success of the event:

- Department of Computer Engineering, Modeling, Electronics and Systems Science, University of Calabria, Italy;
- Department of Mathematics and Computer Science, University of Calabria, Italy;
- Mechanical, Energy, and Management Engineering Department, University of Calabria, Italy;
- Civil Engineering Department, University of Calabria, Italy;
- Institute of High Performance Computing and Networking of the National Research Council ICAR-CNR, Italy;
- Springer Nature Group;
- E4 Computer Engineering S.p.A., Italy;
- Bonomo Editore, Italy.

In conclusion, the Editors invite all participants of the previous NUMTA conferences and all the readers of these volumes to participate in the next edition of the International Conference and Summer School "Numerical Computations: Theory and Algorithms" which, hopefully, will be organized in 2026.

September 2024
Yaroslav D. Sergeyev
Dmitri E. Kvasov
Annabella Astorino

Organization

General Chair

Yaroslav Sergeyev   University of Calabria, Italy and "Lobachevsky" University of Nizhny Novgorod, Russia

Scientific Committee

Lidia Aceto	University of Eastern Piedmont "Amedeo Avogadro", Italy
Andrew Adamatzky	University of the West of England, UK
Francesco Archetti	University of Milano-Bicocca, Italy
Annabella Astorino	University of Calabria, Italy
Roberto Battiti	University of Trento, Italy
Giancarlo Bigi	University of Pisa, Italy
Luigi Brugnano	University of Florence, Italy
Sonia Cafieri	National School of Civil Aviation, France
Tianxin Cai	Zhejiang University, China
Fabio Caldarola	University of Calabria, Italy
Cristian Calude	University of Auckland, New Zealand
Antonio Candelieri	University of Milano-Bicocca, Italy
Mario Cannataro	University of Catanzaro "Magna Graecia", Italy
Francesco Carrabs	University of Salerno, Italy
Gianluca Caterina	Endicott College, USA
Carmine Cerrone	University of Genoa, Italy
Raffaele Cerulli	University of Salerno, Italy
Marco Cococcioni	University of Pisa, Italy
Salvatore Cuomo	University of Naples "Federico II", Italy
Louis D'Alotto	York College, City University of New York, USA
Renato De Leone	University of Camerino, Italy
Alessandra De Rossi	University of Turin, Italy
Kalyanmoy Deb	Michigan State University, USA
Suash Deb	"C.V. Raman" College of Engineering, India
Francesco Dell'Accio	University of Calabria, Italy
Branko Dragovich	Institute of Physics Belgrade, Serbia
Gintautas Dzemyda	Vilnius University, Lithuania
Adil Erzin	"Sobolev" Institute of Mathematics, Russia

Giovanni Fasano	University of Venice "Ca'Foscari", Italy
Şerife Faydaoğlu	Dokuz Eylül University, Turkey
Luca Formaggia	Polytechnic University of Milan, Italy
Elisa Francomano	University of Palermo, Italy
Roberto Gaudio	University of Calabria, Italy
Manlio Gaudioso	University of Calabria, Italy
Jonathan Gillard	Cardiff University, UK
Daniele Gregori	E4 Computer Engineering S.p.A., Italy
Vladimir Grishagin	"Lobachevsky" University of Nizhny Novgorod, Russia
Mario Guarracino	University of Cassino and Southern Lazio, Italy
Francesca Guerriero	University of Calabria, Italy
Jan Hesthaven	Karlsruhe Institute of Technology, Germany
Felice Iavernaro	University of Bari "Aldo Moro", Italy
Mikhail Khachay	"Krasovsky" Institute of Mathematics and Mechanics, Russia
Oleg Khamisov	"Melentiev" Energy Systems Institute, Russia
Timos Kipouros	University of Cambridge, UK
Yury Kochetov	"Sobolev" Institute of Mathematics, Russia
Vladik Kreinovich	University of Texas at El Paso, USA
Dmitri Kvasov	University of Calabria, Italy
Hoai An Le Thi	University of Lorraine, France
Wah June Leong	University of Putra Malaysia, Malaysia
Antonio Liotta	Free University of Bozen-Bolzano, Italy
Marco Locatelli	University of Parma, Italy
Stefano Lucidi	"Sapienza" University of Rome, Italy
Vladimir Mazalov	Institute of Applied Mathematical Research, Russia
Francesca Mazzia	University of Bari "Aldo Moro", Italy
Kaisa Miettinen	University of Jyväskylä, Finland
Edmondo Minisci	University of Strathclyde, UK
Ganesan Narayanasamy	Object Automation System Solutions Inc, USA
Ivo Nowak	Hamburg University of Applied Sciences, Germany
Donatella Occorsio	University of Basilicata, Italy
Panos Pardalos	University of Florida, USA
Hoang Xuan Phu	Institute of Mathematics, Vietnam
Remigijus Paulavičius	Vilnius University, Lithuania
Stefan Pickl	University of the Bundeswehr Munich, Germany
Patrizia Piro	University of Calabria, Italy
Raffaele Pisano	University of Lille, France
Mikhail Posypkin	Federal Research Center CSC RAS, Russia

Oleg Prokopyev	University of Zurich, Switzerland
Davide Rizza	University of East Anglia, UK
Massimo Roma	"Sapienza" University of Rome, Italy
Valeria Ruggiero	University of Ferrara, Italy
Maria Grazia Russo	University of Basilicata, Italy
Leonidas Sakalauskas	Vilnius University, Lithuania
Yaroslav Sergeyev (Chair)	University of Calabria, Italy and "Lobachevsky" University of Nizhny Novgorod, Russia
Khodr Shamseddine	University of Manitoba, Canada
Sameer Shende	University of Oregon, USA
Theodore Simos	Democritus University of Thrace, Greece
Vinai Singh	Motherhood University, India
Majid Soleimani-Damaneh	University of Tehran, Iran
William Spataro	University of Calabria, Italy
Maria Grazia Speranza	University of Brescia, Italy
Giandomenico Spezzano	ICAR-CNR, Italy
Rosa Maria Spitaleri	IAC-CNR, Italy
Alexander Strekalovskiy	Institute for System Dynamics and Control Theory, Russia
Gopal Tadepalli	Anna University, India
Tatiana Tchemisova	University of Aveiro, Portugal
Claudio Ternullo	Babeş-Bolyai University, Romania
Fernando Tohmé	National University of the South, Argentina
Gerardo Toraldo	University of Campania "Luigi Vanvitelli", Italy
Vassili Toropov	Queen Mary University of London, UK
Ivan Tyukin	King's College London, UK
Michael Vrahatis	University of Patras, Greece
Song Wang	Curtin University, Australia
Gerhard-Wilhelm Weber	Poznań University of Technology, Poland
Luca Zanni	University of Modena and Reggio Emilia, Italy
Anatoly Zhigljavsky	Cardiff University, UK
Antanas Žilinskas	Vilnius University, Lithuania
Julius Žilinskas	Vilnius University, Lithuania
Joseph Zyss	ENS Cachan, France

Organizing Committee

Annabella Astorino	University of Calabria, Italy
Francesco Dell'Accio (SIAM Representative)	University of Calabria, Italy
Alfredo Garro	University of Calabria, Italy

Francesca Guerriero — University of Calabria, Italy
Dmitri Kvasov (Chair) — University of Calabria, Italy
Maria Chiara Nasso — University of Calabria, Italy
Yaroslav Sergeyev — University of Calabria, Italy and "Lobachevsky" University of Nizhny Novgorod, Russia

Sponsors

UNIVERSITÀ DELLA CALABRIA
DIPARTIMENTO DI INGEGNERIA INFORMATICA, MODELLISTICA, ELETTRONICA E SISTEMISTICA
DIMES

UNIVERSITÀ DELLA CALABRIA
DIPARTIMENTO DI INGEGNERIA CIVILE

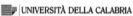
UNIVERSITÀ DELLA CALABRIA
DIPARTIMENTO DI INGEGNERIA MECCANICA, ENERGETICA E GESTIONALE

UNIVERSITÀ DELLA CALABRIA
DIPARTIMENTO DI MATEMATICA E INFORMATICA

Consiglio Nazionale delle Ricerche
Istituto di Calcolo e Reti ad Alte Prestazioni

E4 COMPUTER ENGINEERING

Springer

BONOMO EDITORE
COLLANA SCIENTIFICA

In cooperation with SIAM.

Contents - Part III

Long Papers

Circuit-Based Numerical Solutions of Transmission Lines: Application
to Korteweg-de Vries Equations 3
    *Giuseppe Alì, Francesca Bertacchini, Eleonora Bilotta,
Francesco Demarco, Pietro Pantano, and Stefano Vena*

Deep Learning Methods for fMRI Classification 15
    Luca Barillaro and Giuseppe Agapito

Deep Learning for Scoliosis Diagnosis: Methods and Databases 26
    Lorella Bottino, Marzia Settino, Luigi Promenzio, and Mario Cannataro

Understanding Spreading Dynamics of COVID-19 by Mining Human
Mobility Patterns 40
    Carmela Comito and Deborah Falcone

Variational Quantum Algorithms for Gibbs State Preparation 56
    Mirko Consiglio

Towards a Parallel Code for Cellular Behavior in Vitro Prediction 71
    Pasquale De Luca, Ardelio Galletti, and Livia Marcellino

Combinators as Observable Presheaves: A Characterization in the *Grossone*
Framework 84
    Rocco Gangle, Fernando Tohmé, and Gianluca Caterina

Applying Variational Quantum Classifier on Acceptability Judgements:
A QNLP Experiment 98
    *Raffaele Guarasci, Giuseppe Buonaiuto, Giuseppe De Pietro,
and Massimo Esposito*

Unimaginable Numbers: a Case Study as a Starting Point for an Educational
Experimentation 113
    Francesco Ingarozza, Gianfranco d'Atri, and Rosanna Iembo

Some Notes on a Continuous Class of Octagons 127
    Francesco Ingarozza and Aldo Piscitelli

Game Theory Presented to Italian High School Students in Connection
with Infinity Computing 139
   Corrado Mariano Marotta and Andrea Melicchio

Meta Discussion Pedagogical Model to Foster Mathematics Teacher's
Professional Development 154
   *Antonella Montone, Michele Giuliano Fiorentino,
and Giuditta Ricciardiello*

A Variational Quantum Soft Actor-Critic Algorithm for Continuous
Control Tasks 166
   *Antonio Policicchio, Alberto Acuto, Paola Barillà, Ludovico Bozzolo,
and Matteo Conterno*

Named Entity Recognition to Extract Knowledge from Clinical Texts 180
   Ileana Scarpino, Rosarina Vallelunga, and Francesco Luzza

Applied Mathematical Modelling in the Physics Problem-Solving
Classroom 193
   Annarosa Serpe

AIR SAFE: Leveraging IoT Sensors and AI Models to Foster Optimal
Indoor Conditions 207
   *Mariangela Viviani, Simone Colace, Daniele Germano, Sara Laurita,
Giuseppe Papuzzo, and Agostino Forestiero*

Short Papers

Unimaginable Numbers and Infinity Computing at School:
An Experimentation in Northern Italy 223
   Luigi Antoniotti, Annabella Astorino, and Fabio Caldarola

The Cantor-Vitali Function and Infinity Computing 232
   *Luigi Antoniotti, Corrado Mariano Marotta, Andrea Melicchio,
and Maria Anastasia Papaleo*

New Probabilistic Methods for Generating Risk Maps 240
   Arrigo Bertacchini, Pierpaolo Antonio Fusaro, and Massimo Zupi

How to Deal with Different Densities of Urban Spatial Data?
A Comparison of Clustering Approaches to Detect City Hotspots 248
   Eugenio Cesario, Paolo Lindia, and Andrea Vinci

The Impact of Vectorization on the Efficiency of a Parallel PIC Code
for Numerical Simulation of Plasma Dynamics in Open Trap 254
   Igor Chernykh, Igor Kulikov, Vitaly Vshivkov, Anna Efimova,
   Dmitry Weins, Ivan Chernoshtanov, and Marina Boronina

Algorithms for Design with CNC Machines: The Case Study of Wood
Furniture 262
   Francesco Demarco, Francesca Bertacchini, Eleonora Bilotta,
   Carmelo Scuro, and Pietro Pantano

PyGrossone: A Python Library for the Infinity Computer 270
   Alberto Falcone, Alfredo Garro, and Yaroslav D. Sergeyev

Towards Reproducible Research in Machine Learning via Blockchain 278
   Ernestas Filatovas, Linas Stripinis, Francisco Orts,
   and Remigijus Paulavičius

Dossier Classification to Support Workflow Management Optimization 286
   Simona Fioretto, Elio Masciari, and Enea Vincenzo Napolitano

Self-Sovereign Identification of IoT Devices by Using Physically
Unclonable Functions and Blockchain 293
   Gianluigi Folino, Agostino Forestiero, and Giuseppe Papuzzo

Introducing *Nondum*, A Mathematical Notation for Computation
with Approximations 301
   Francesco La Regina and Gianfranco d'Atri

Visualization of Multilayer Networks 309
   Ilaria Lazzaro and Marianna Milano

Legal Systems and Fractals, Towards Infinity Computing 316
   Maria Rita Maiolo and Mària Ivano

Exploit Innovative Computer Architectures with Molecular Dynamics 324
   Filippo Marchetti and Daniele Gregori

A Sentiment Analysis on Reviews of Italian Healthcare 332
   Maria Chiara Martinis and Chiara Zucco

A Numerical Approach to Basic Calculus 338
   Layla Nasr

Modelling Hyperentanglement for Quantum Information Processes 346
Luca Salatino, Luca Mariani, Carmine Attanasio, Sergio Pagano, and Roberta Citro

An Innovative Sentiment Analysis Model for COVID-19 Tweets 353
Areeba Umair and Elio Masciari

Exploring Hierarchical MPI Reduction Collective Algorithms Targeted to Multicore Node Clusters 362
Gladys Utrera, Marisa Gil, Xavier Martorell, William Spataro, and Andrea Giordano

Author Index 371

Long Papers

Circuit-Based Numerical Solutions of Transmission Lines: Application to Korteweg-de Vries Equations

Giuseppe Alì[1(✉)], Francesca Bertacchini[2], Eleonora Bilotta[1], Francesco Demarco[1(✉)], Pietro Pantano[1], and Stefano Vena[3]

[1] Department of Physics, University of Calabria, Rende, Italy
{giuseppe.ali,eleonora.bilotta,francesco.demarco,pietro.pantano}@unical.it
[2] Department of Mechanical, Energy and Management Engineering, University of Calabria, Rende, Italy
francesca.bertacchini@unical.it
[3] Altrama, Rende, Italy
stefano.vena@altrama.com

Abstract. Transmission lines, devices employed for the transmission of electrical signals, can be used for the approximation of non-linear partial differential equations (PDEs) also in the nonlinear case. To this end, transmission lines are used to spatially discretize PDEs allowing the problem to be solved numerically even for intricate boundary conditions - e.g. at intersection of many wires in the system. Using transmission lines to solve the Korteweg-de Vries Equation (KdV) allows for efficient and accurate numerical solutions, as it provides a method to investigate the propagation of wave-like information in nonlinear and dispersive media with multiple dimensions. In the present paper a software developed in C# is presented, the latter employs discretization by transmission lines and the Runge-Kutta 4–5 integration algorithm to numerically solve the KdV equation in the one-dimensional case. The implemented program, named "WireExplorer," is able to simulate the propagation of solitonic pulses on different types of circuits composed of one or more wires, even in the case of intersections. Such conditions are not canonically solvable by resolution of the KdV equations, while approximate numerical resolution allowed the evaluation of these intricate cases. In particular, the obtained results showed the subdivision of the wave into smaller components, at intersections between wires, which propagate in different directions showing also the formation of dispersive tails propagating in the direction opposite to that of the main wave.

Keywords: Trasmission Lines · KdV · Soliton

1 Introduction and Related Works

The nonlinear solitary wave problem, described by means of the Korteweg-De Vries (KdV) equation, has become increasingly important and widespread in recent decades. Solitons hold significant relevance in application-based research in many fields of science such as: oceanography, fluid mechanics, biology and numerous others [14,16,26,29,38]. Therefore, solving nonlinear partial differential equations (PDEs) has become a crucial problem in theoretical and applied research. Many approaches have been proposed over time to investigate the solving of solitonic equations, their evolution is also related to the computational capacity of the historical period in which they were implemented. We can distinguish more mature approaches such as the Painlevé analysis [34], the Darboux transform method [33], the inverse scattering transform method [4], the Backlund transform method [37], the Lie group and Lie algebra method [35]. Advances in machine learning have greatly improved the ability to address the challenges of scientific computing along with numerical analysis even in the field of solving PDEs. Deep learning, is currently most studied field of machine learning research, has played a key role in this regard. Differential equations and machine learning methods combined have been proven to be advantageous in solving soliton problems,one of the first case studies was implemented by Lee et al. [36]. They employed Hopfield neural networks to solve ordinary differential equation (ODE) models, to mention some of the more recent application Ying and Temuer [40] developed an artificial intelligence-based approach for solving Kdv [11,12,17,22,25]. Specifically, an algorithm based on Lie-group-based neural network has been implemnted. The main limitation of this method is related to the need to define large datasets for network training, as well as the need to employ a great deal of computing power precisely for the network training phase.

The approximation methods, then combined with numerical integration, do not require data sets and training steps allowing accurate numerical solutions to PDEs even for complex boundary conditions for which the exact solution cannot be computed. Significant applications of these approximation methods can be identified in the work of G. Borgese et al. [1,2,7,27]. They proposed a software with a graphical user interface (GUI) developed in MATLAB called SIMANSOUL that employs transmission lines for numerical solitonic wave solving. The solution method employed in SIMANSOUL is derived from the paradigm devised by Bilotta and Pantano [3,6,8,10,21,32]. In their work, they comprehensively defined the solving approach for KdV equations using Cellular Nonlinear Networks (CNNs) providing the theoretical basis for the work presented here as well [18,19,23,24,28,39].

The main advantage of the software presented in this paper is precisely its ability to provide accurate numerical solutions of the KdV equations for complex boundary conditions, a case that often occurs in applied research, for which the exact solution cannot be obtained. The accuracy of the solution is ensured by the approximation by transmission lines from the Cellular Neural Networks and the implementation of Runge-Kutta 4–5 (RK 4–5) integration. The software also features an intuitive GUI for defining system parameters, such as spatial domain length, discretization step and initial conditions. Next, the software exploits transmission lines and the RK 4–5 method to calculate the evolution of solitonic waves over time, returning a graphical representation of the approximate solutions.

2 CNN Trasmission Line

A cellular neural network consists of a discrete collection of nonlinear analog processors called cells, which are organized in a lattice or grid structure. Each cell is connected only to adjacent units, forming a restricted neighborhood or sphere of influence as showed in Fig. 1. This configuration also allows local nonlinear interactions, which enables CNNs to exhibit nonlinear dynamics.

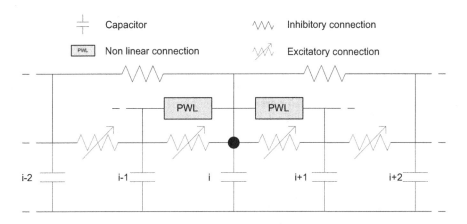

Fig. 1. In CNN each cell, consisting in a capacitor, interacts in both linear and nonlinear with their adjacent cells but only in a linear way with neighboring not adjacent cells.

In the case of the KdV equations, the transmission line is realized by employing a nonstandard one-dimensional CNN; the connections between the various cells are both linear and nonlinear. Each cell consists of a single capacitor C connected to neighboring cells by three different connections: The nearest cells are connected to each other by two distinct connections the first one, excitatory in nature, and the second one, nonlinear in nature; a third connection, inhibitory and linear in nature, connects the second neighbors. The nonlinear connection can be made by employing appropriate generators capable of modeling polynomial-type characteristic functions, as shown in Fig. 1. Let u_i be the state of the $i-th$ cell describing the voltage at the capacitor ends, R the factor governing the inhibitory connection, R' the factor governing the excitatory connection, and R'' the coefficient of the nonlinear term. The equation governing the state of the $i-th$ cell is as follows:

$$\frac{u_{i-2} - u_i}{R} - \frac{u_{i-1} - u_i}{R'} + \frac{u_{i-1}^2 - u_i^2}{R''} = \frac{u_{i+2} - u_i}{R} - \frac{u_{i+1} - u_i}{R'} + \frac{u_{i+1}^2 - u_i^2}{R''} + C\frac{du_i}{dt} \tag{1}$$

from which we obtain:

$$-C\frac{du_i}{dt} = \frac{u_{i+2} - u_{i-2}}{R} - \frac{u_{i+1} - u_{i-1}}{R'} + \frac{u_{i+1}^2 - u_{i-1}^2}{R''} \qquad (2)$$

By assuming $R' = \frac{R}{2}$ 2 becomes:

$$-\frac{du_i}{dt} = \frac{u_{i+2} - u_{i+1} + u_{i-1} - u_{i-2}}{RC} + \frac{u_{i+1}^2 - u_{i-1}^2}{R''C} \qquad (3)$$

Replacing $RC = 2h^2$ and $R''C = \frac{2h}{3}$, where h is a constant parameter thats depend on the number of cells and spatial interval, from 3 we obtain the following equation:

$$\frac{du_i}{dt} = \frac{1}{2h^3}(u_{i+2} - u_{i+1} + u_{i-1} - u_{i-2}) - \frac{2}{3h}(u_{i+1}^2 - u_{i-1}^2) \qquad (4)$$

Equation 4 is the spatial discretized version of 1D KdV equation and states that the dynamics of each cell is governed by the state of neighboring cells with a radius of influence $r = 2$. In particular, neighboring cells, interact with each other in both a linear and nonlinear way.

Assuming a large number of cells by setting $h \to 0$, the 4 approximates the KdV equation in 1D:

$$u_t + 6uu_x + uu_{xxx} \qquad (5)$$

CNN trasmission lines networks can be easily implemented in hardware Fig. 1, using single-cell capacitors and resistors for active and passive connections, respectively. Nonlinear connections can be modeled through Piecewise Linear (PWL) functions. The main advantage of this analog resolution method is based on the possibility of simulating, with appropriate modifications, the crossing between CNN transmission lines in 1D; analyzing the interaction between solitons belonging to different lines. Although the resolution of the canonical equations for these conditions is still far from being discovered.

Thus consider two transmission lines intersecting in a given cell (a,b), as shown in the Fig. 2, the state of each cell except for the intersecting cell is governed by Eq. 4. Cell (a,b) differs in that its neighborhood consists not only of the cells to the left and right but also of the cells below and above. the neighboring top and bottom cells have a linear excitatory connection, while the other connection is nonlinear. The intersection cell (a,b) has a different spatial discretized state equation than the other cells governed by 4. Specifically, the state equation is as follow:

$$\frac{du_{a,b}}{dt} = \frac{1}{2h^3}(u_{a-2,b} - u_{a+2,b} + u_{i-1} - 2u_{a-1,b} + 2u_{a+1,b})$$
$$-\frac{3}{2h}(u_{a,b-2}^2 - u_{a-1,b}^2) + \frac{1}{2h^3}(u_{a,b-2} - u_{a,b+2} - 2u_{a,b-1} + 2u_{a,b+1}) \qquad (6)$$
$$-\frac{3}{2h}(u_{a,b+1}^2 - u_{a,b-1}^2)$$

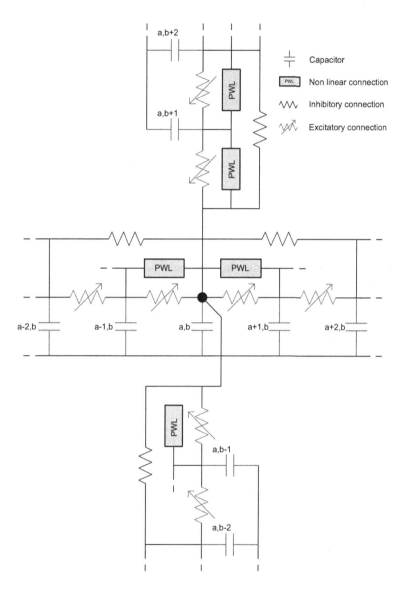

Fig. 2. CNN circuital representation of crossed trasmission lines

The parameter h serves as a normalization factor that can be adjusted to enhance the convergence and stability of the system model. The dynamics of cross cell (a,b) is governed by the states of its eight neighboring cells within a radius of 2. Neighboring cells within a radius of 1 interact with (a,b) in both linear and nonlinear manners.

The study of the interaction between solitons is examined in two distinct scenarios: the first involves the intersection of two transmission lines, both with a soliton present, while the second case involves the bifurcation of a single soliton on an intersection of two lines. These two case studies allow for the analysis of soliton dynamics; in particular, the intersection between transmission lines provides an investigation of typical dynamic behaviors of soliton interaction without complications due to topology.

To simulate the presence of a soliton on one of the wires it is necessary to consider the appropriate initial conditions, for this purpose a square hyperbolic secant function was considered:

$$s(x) = A sech^2 x \tag{7}$$

This function was chosen because it avoids divergence integration problems, thanks to its zero-tangent envelope for $x \to \pm\infty$. Boundary conditions were also imposed for both case studies in accordance to the definition by [13,41] the following was imposed:

$$u(x,0) = p(p+1) sech^2 x \quad \text{with} \quad p > 0 \tag{8}$$

from these conditions it is obtained that the eigenvalues of the Schrodinger equation spectrum are:

$$\lambda_n = (p-n)^2 \quad \text{with} \quad n = 0, 1, ..., N \tag{9}$$

If p is posed as an integer there will be p solitons from the initial conditions, otherwise there will be an integer number of solitons followed by a radiation tail.

The magnitude of solitons is

$$A_n = 2\lambda_n \tag{10}$$

The velocity of propagation is:

$$c_n = 4\lambda_n = 2A_n \tag{11}$$

Propagation speed, amplitude and width are correlated with each other. For example, amplitude is directly proportional to propagation velocity; solitons of amplitude 1 will always be used in the two case studies presented.

3 Wire Explorer

The Wire Explorer application was implemented with the specific purpose of being used for applied research in different fields, the assumed user audience therefore possesses scientific and computer science skills although not directly related to the mathematical field. To enable the widest possible use therefore, the software has been equipped with an intuitive Graphic User Interface (GUI) [20], shown in Fig. 3.

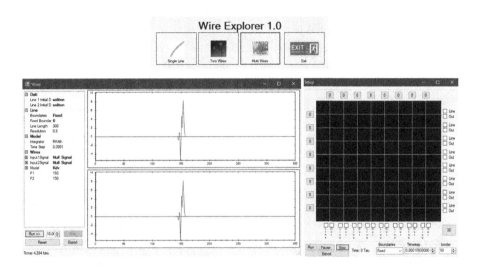

Fig. 3. Wire Explorer GUI software

With the goal of creating user-friendly software for the realization of numerical solving of KdV equations under unconventional conditions, the C# programming language in the integrated development environment (IDE) Visual Studio was chosen. Actually, the visual studio IDE allows both the back-end and GUI of the application to be built in a single platform, even using graphical tools [5,9,15,30]. The core of the software is the solving algorithm, in the literature there are many methods for solving KdV equations [31] and state ODEs, in the present work a 4th-order Runge-Kutta variable step method (RK4-5) was employed in order to obtain accurate results.

The GUI of wire explorer allows control, through different windows, of simulation parameters such as: boundary condition (periodic, zero and fixed flow), number of cells, iterations number, initial input function (Gaussian, hyperbolic secant, soliton), time step, spatial step and simulation scenario (crossed lines and network)

4 Setup Simulations in Wire Explorer

This section presents the results of simulations carried out using the Wire explorer tool. In both simulations, boundary conditions were set as shown in Table 1.

Table 1. Simulation parameters

Wire Explorer	Case 1	Case 2
Line 1 input	Soliton	Soliton
Line 2 input	Soliton	None
Number of cells (N)	300	300
Amplitude (A)	1	1
Time step (δt)	0.00001 s	0.00001 s
Spatial step (δx)	0.05	0.05

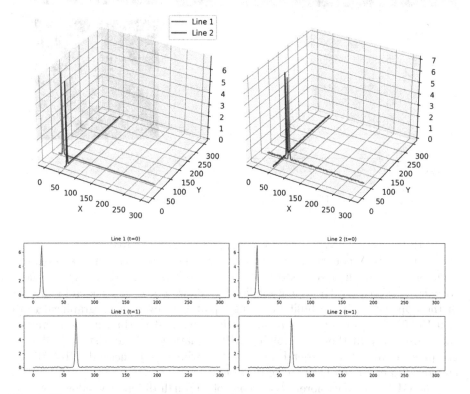

Fig. 4. The soliton on line 1 and line 2 before reaching the intersection cell and after the intersection; the solitons are virtually identical but have dispersion elements along the lines.

In Wire Explorer, the propagation of solitonic waves was analyzed in an unconventional condition, i.e., at the intersection of two transmission lines called "Line 1" and "Line 2" from now on. The numerical resolution tool enabled the study of soliton propagation before and after itting the intersection node; in addition to providing numerical results, the software also returns Space Time graphs for transmission lines 1 and 2.

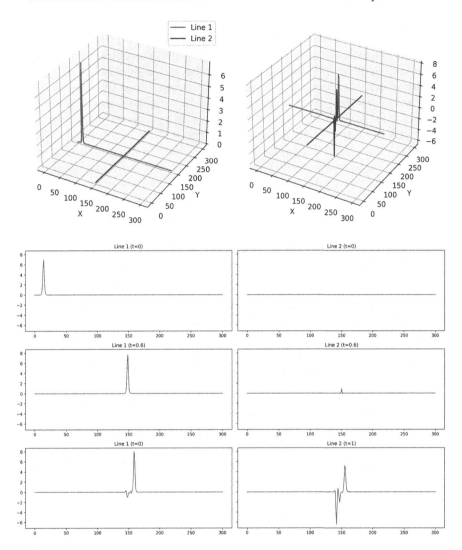

Fig. 5. The soliton on line 1 before reaching the intersection cell and two solitons with dispersive tails coming out of the intersection cell.

4.1 Case 1: One Soliton on Both Lines

As shown in Fig. 4, two solitons belonging to two different lines, in red line 1 and in blue line 2, meet at the point of intersection and separate again, the solitons maintain their individual characteristics as if the interaction had not occurred. Following the interaction, their distinctiveness remains preserved. The observed phenomenon primarily comprises a minor dispersive effect and the presence of reflected solitons, which can be attributed to the periodic boundary conditions rather than the interaction between the two solitons.

4.2 Case 2: One Soliton on Line 1

The second case involved the presence of a single soliton on line 1, in red, crossing the intersection of two transmission lines. As the soliton reaches the intersection point, its velocity decreases, resulting in disturbances along the two lines. Subsequent to the intersection, two distinct solitons emerge, each associated with its respective line. Additionally, two wave trains can be observed, with each traveling in the opposite direction compared to the solitons, as depicted in Fig. 5. Notably, there is a reduction in the magnitudes of both solitons. The magnitudes of the solitons are measured at 0.6 on line 1 and 0.5 on line 2. The larger soliton corresponds to the one residing on the same line as the incoming soliton (Line 1). Furthermore, the presence of dispersive tracks and reflected solitons can also be observed.

5 Conclusion

In the present work a new software, Wire Explorer, was proposed based on the use of discretized transmission lines using CNN. Upon establishing the fundamental equation governing the state of each cell, the existence and interaction of solitary waves have been identified. These waves engage in mutual interactions while preserving their individual characteristics. This behavior is expected, as in the continuous network case, the equation of state simplifies to the widely recognized Korteweg-de Vries (KdV) equation. Finally, simulations were conducted for two case studies concerning the intersection of two transmission lines using the presented software. The results obtained are promising, as many advantages were found in using CNNs to design transmission lines; in fact, they are easily implemented and provide an extremely flexible simulation environment for exploring nonlinear dynamics. The results obtained are promising and open the path for further development of Wire Explorer with special reference to solving multidimensional solitons.

Code and Data Availability. In support of the transparency and reproducibility of our research, the software "Wire explorer" presented in this study is available on GitHub in version 1.0. The repository contains the executable file as .exe and all support material for the working of the Wire Explorer software. Readers can access the code at the following URL: https://github.com/wetfire2k/wireExplorer.

References

1. Adamo, A., Bertacchini, P.A., Bilotta, E., Pantano, P., Tavernise, A.: Connecting art and science for education: learning through an advanced virtual theater with "talking heads". Leonardo **43**(5), 442–448 (2010)
2. Alì, G., Demarco, F., Scuro, C.: Propagation of elastic waves in homogeneous media: 2d numerical simulation for a concrete specimen. Mathematics **10**(15), 2673 (2022)

3. Alì, G., Torcicollo, I., Scuro, C.: Analysis of a nonlinear model arising in chemical aggression of marble. Int. J. Masonry Res. Innov. **8**(2–3), 263–274 (2023)
4. Arkadiev, V., Pogrebkov, A., Polivanov, M.: Inverse scattering transform method and soliton solutions for davey-stewartson ii equation. Physica D **36**(1–2), 189–197 (1989)
5. Bertacchini, F., Pantano, P., Bilotta, E.: Shaping the aesthetical landscape by using image statistics measures. Acta Physiol. (Oxf) **224**, 103530 (2022)
6. Bertacchini, F., Bilotta, E., Caldarola, F., Pantano, P.: The role of computer simulations in learning analytic mechanics towards chaos theory: a course experimentation. Int. J. Math. Educ. Sci. Technol. **50**(1), 100–120 (2019)
7. Bertacchini, F., Bilotta, E., Caldarola, F., Pantano, P., Bustamante, L.R.: Emergence of linguistic-like structures in one-dimensional cellular automata. In: AIP Conference Proceedings, vol. 1776, p. 090044. AIP Publishing LLC (2016)
8. Bertacchini, F., Bilotta, E., Carini, M., Gabriele, L., Pantano, P., Tavernise, A.: Learning in the smart city: a virtual and augmented museum devoted to chaos theory. In: Chiu, D.K.W., Wang, M., Popescu, E., Li, Q., Lau, R. (eds.) ICWL 2012. LNCS, vol. 7697, pp. 261–270. Springer, Heidelberg (2014). https://doi.org/10.1007/978-3-662-43454-3_27
9. Bertacchini, F., et al.: An emotional learning environment for subjects with autism spectrum disorder. In: 2013 International Conference on Interactive Collaborative Learning (ICL), pp. 653–659. IEEE (2013)
10. Bertacchini, F., Bilotta, E., Pantano, P.S.: On the temporal spreading of the sars-cov-2. PLoS ONE **15**(10), e0240777 (2020)
11. Bertacchini, F., Gabriele, L., Tavernise, A., et al.: Bridging educational technologies and school environment: implementations and findings from research studies. In: Educational Theory, pp. 63–82. Nova Science Publishers, Inc. (2011)
12. Bertacchini, F., Pantano, P.S., Bilotta, E.: Sars-cov-2 emerging complexity and global dynamics. Chaos Interdisc. J. Nonlinear Sci. **31**(12), 123110 (2021)
13. Bertacchini, F., Pantano, P.S., Bilotta, E.: Jewels from chaos: a fascinating journey from abstract forms to physical objects. Chaos Interdisc. J. Nonlinear Sci. **33**(1) (2023)
14. Bertacchini, F., Scuro, C., Pantano, P., Bilotta, E.: Modelling brain dynamics by boolean networks. Sci. Rep. **12**(1), 16543 (2022)
15. Bertacchini, F., Scuro, C., Pantano, P., Bilotta, E.: A project based learning approach for improving students' computational thinking skills. Front. Rob. AI **9** (2022)
16. Bilotta, E., Bossio, E., Pantano, P.: Chaos at school: Chua's circuit for students in junior and senior high school. Int. J. Bifurcat. Chaos **20**(01), 1–28 (2010)
17. Bilotta, E., Di Blasi, G., Stranges, F., Pantano, P.: A gallery of chua attractors part vi. Int. J. Bifurcat. Chaos **17**(06), 1801–1910 (2007)
18. Bilotta, E., Gabriele, L., Servidio, R., Tavernise, A.: Edutainment robotics as learning tool. In: Transactions on Edutainment III, pp. 25–35 (2009)
19. Bilotta, E., Lafusa, A., Pantano, P.: Life-like self-reproducers. Complexity **9**(1), 38–55 (2003)
20. Bilotta, E., Lafusa, A., Pantano, P.: Searching for complex ca rules with gas. Complexity **8**(3), 56–67 (2003)
21. Bilotta, E., Pantano, P.: Cellular nonlinear networks meet kdv equation: a new paradigm. Int. J. Bifurcat. Chaos **23**(01), 1330003 (2013)
22. Bilotta, E., Pantano, P., Vena, S.: Artificial micro-worlds part i: a new approach for studying life-like phenomena. Int. J. Bifurcat. Chaos **21**(02), 373–398 (2011)

23. Bilotta, E., Pantano, P., Vena, S.: Speeding up cellular neural network processing ability by embodying memristors. IEEE Trans. Neural Netw. Learn. Syst. **28**(5), 1228–1232 (2016)
24. Bilotta, E., et al.: Imaginationtools (tm)-a 3d environment for learning and playing music. In: Eurographics Italian Chapter Conference, vol. 1, pp. 139–144 (2007)
25. Bilotta, E., Stranges, F., Pantano, P.: A gallery of chua attractors: part iii. Int. J. Bifurcat. Chaos **17**(03), 657–734 (2007)
26. Bonnemain, T., Doyon, B., El, G.: Generalized hydrodynamics of the kdv soliton gas. J. Phys. A: Math. Theor. **55**(37), 374004 (2022)
27. Borgese, G., Pace, C., Pantano, P., Bilotta, E.: FPGA-based distributed computing microarchitecture for complex physical dynamics investigation. IEEE Trans. Neural Netw. Learn. Syst. **24**(9), 1390–1399 (2013)
28. Gabriele, L., Bertacchini, F., Tavernise, A., Vaca-Cárdenas, L., Pantano, P., Bilotta, E.: Lesson planning by computational thinking skills in Italian pre-service teachers. Inf. Educ. **18**(1), 69–104 (2019)
29. Gabriele, L., Marocco, D., Bertacchini, F., Pantano, P., Bilotta, E.: An educational robotics lab to investigate cognitive strategies and to foster learning in an arts and humanities course degree. Int. J. Online Eng. **13**(4) (2017)
30. Gabriele, L., Tavernise, A., Bertacchini, F.: Active learning in a robotics laboratory with university students. In: Increasing Student Engagement and Retention Using Immersive Interfaces: Virtual Worlds, Gaming, and Simulation, vol. 6, pp. 315–339. Emerald Group Publishing Limited (2012)
31. Gardner, C.S., Greene, J.M., Kruskal, M.D., Miura, R.M.: Method for solving the korteweg-devries equation. Phys. Rev. Lett. **19**(19), 1095 (1967)
32. Giglio, S., Bertacchini, F., Bilotta, E., Pantano, P.: Machine learning and points of interest: typical tourist Italian cities. Curr. Issue Tour. **23**(13), 1646–1658 (2020)
33. Huang, D.J., Li, D.S., Zhang, H.Q.: Explicit n-fold darboux transformation and multi-soliton solutions for the $(1+1)$-dimensional higher-order broer–kaup system. Chaos Solitons Fractals **33**(5), 1677–1685 (2007)
34. Khater, A., Callebaut, D., Shamardan, A., Ibrahim, R.: Bäcklund transformations and painlevé analysis: exact soliton solutions for strongly rarefied relativistic cold plasma. Phys. Plasmas **4**(11), 3910–3922 (1997)
35. Kumar, S., Kumar, D.: Solitary wave solutions of $(3+1)$-dimensional extended zakharov-kuznetsov equation by lie symmetry approach. Comput. Math. Appl. **77**(8), 2096–2113 (2019)
36. Lee, H., Kang, I.S.: Neural algorithm for solving differential equations. J. Comput. Phys. **91**(1), 110–131 (1990)
37. Nimmo, J., Freeman, N.: The use of backlund transformations in obtaining n-soliton solutions in Wronskian form. J. Phys. A: Math. Gen. **17**(7), 1415 (1984)
38. Sahoo, H., Das, C., Chandra, S., Ghosh, B., Mondal, K.K.: Quantum and relativistic effects on the kdv and envelope solitons in ion-plasma waves. IEEE Trans. Plasma Sci. **50**(6), 1610–1623 (2021)
39. Vaca-Cárdenas, L.A., et al.: Coding with scratch: the design of an educational setting for elementary pre-service teachers. In: 2015 International Conference on Interactive Collaborative Learning (ICL), pp. 1171–1177. IEEE (2015)
40. Wen, Y., Chaolu, T.: Learning the nonlinear solitary wave solution of the korteweg-de vries equation with novel neural network algorithm. Entropy **25**(5), 704 (2023)
41. Zabusky, N.J., Kruskal, M.D.: Interaction of "solitons" in a collisionless plasma and the recurrence of initial states. Phys. Rev. Lett. **15**(6), 240 (1965)

Deep Learning Methods for fMRI Classification

Luca Barillaro[1](✉) and Giuseppe Agapito[2]

[1] Data Analytics Research Center, Department of Medical and Surgical Sciences, University "Magna Græcia" of Catanzaro, 88100 Catanzaro, Italy
luca.barillaro@unicz.it
[2] Department of Law, Economics and Social Sciences, University "Magna Græcia" of Catanzaro, 88100 Catanzaro, Italy
agapito@unicz.it

Abstract. This paper aims to review some deep learning-based methods for the classification of functional Magnetic Resonance Imaging (fMRI) brain images. fMRI is one of the MRI modalities which provides information about the activity of the neurons in the brain. In particular, the fMRI signal is sensitive to blood dynamic changes, which drive the neuron firing. This relationship is known as the Blood Oxygenation Level Dependent (BOLD) effect. Two main applications are possible: a task-specific activity and a resting state. The main difference is that in the first case, a patient is asked to perform a specific task, and the fMRI shows the brain response, while in the resting state, no particular action is required.

The study of brain activity and response represents one of the significant research challenges to understand diseases and potential damages better. Using automated techniques based on artificial intelligence, such as machine learning or deep learning, allow the analysis of large fMRI datasets. The classification task is essential because it enables the possibility to support the decision-making process by healthcare experts by identifying particular diseases from images.

This paper reviews deep learning methods for fMRI classification, their corresponding accuracies, and the required computational resources for massive fMRI datasets.

Keywords: fMRI · deep learning · medical imaging

1 Introduction

Functional Magnetic Resonance Imaging is one of the MRI modalities which provides information about the activity of the neurons in the brain. fMRI [5] signal is sensitive to blood dynamic changes which drive the neuron firing. This relationship is known as the Blood Oxygenation Level Dependent (BOLD) effect.

Two main applications are possible: a task specific activity and a resting state one. The main difference is that in the first case a patient is asked to perform a specific task and the fMRI shows the brain response, while in the resting state no particular action is required.

This kind of MRI was introduced in 1990 and since that it opened the way to several deeper analysis of the human brain. In particular it had a big impact on neuroscience and several other medical fields.

From a technical perspective we have to recall that in this kind of imaging we are dealing with volumes instead of common images. In addition acquisitions are not taken in a single instant but involves an acquisition time. This implies that even a small number of patients produces a huge amount of data to be analyzed.

To perform analysis on such large dataset, automated techniques are very useful. In addition there is the possibility to discover hidden information or properties by using these approaches. Deep learning approaches are particularly useful for these purpose since they combine a high power of data representation with high accuracy. In our paper we focused on classification tasks on fMRI data since is useful with respect to supporting decision in therapies.

The structure of this paper is the following: in Sect. 2 we make a description of functional magnetic resonance imaging; in Sect. 3 we first recall some deep learning key concepts in Subsect. 3.1 followed in Subsect. 3.2 by some examples of classification task of fMRI using deep learning approaches. Finally in Sect. 4 we conclude the paper.

2 Functional Magnetic Resonance Imaging

In this section, our aim is to give a description of the functional Magnetic Resonance (fMRI) technique and its applications.

2.1 MRI Key Concepts

We briefly recall that Magnetic Resonance Imaging (MRI) is a powerful and widely used medical imaging technique that has transformed the field of diagnostic imaging. It was a major breakthrough in medical imaging thanks to the separate discovers by Paul C. Lauterbur [13] and Peter Mansfield [14], for which they were awarded of the (shared) 2003 Nobel Prize in Physiology or Medicine [26].

MRI relies on powerful magnetic fields and radio waves, providing detailed and accurate images of the internal structures of the human body. Unlike other imaging modalities, such as X-rays or CT scans, MRI does not involve ionizing radiation, making it a safe and versatile tool for diagnosing a wide range of conditions [8].

2.2 fMRI

By utilizing the principles of Magnetic Resonance Imaging (MRI), the Functional Magnetic Resonance Imaging (fMRI) is a specialized technique that has the goal to measure and map brain activity. It was introduced by S. Ogawa et all in 1990 [16].

It is a non-invasive neuroimaging method that provides insights into the functional organization of the brain by detecting changes in blood oxygenation levels associated with neural activity.

While conventional MRI primarily focuses on capturing structural information, fMRI goes beyond by revealing the dynamic patterns of brain activation, allowing researchers and clinicians to investigate cognitive processes, localize functional brain regions, and explore the neural basis of various mental functions and disorders.

By combining the spatial resolution of MRI with the ability to assess brain function, fMRI has become an invaluable tool in the field of neuroscience and has opened up new avenues for understanding the complexities of the human brain.

Key Concepts. At its core, fMRI relies on the magnetic properties of blood, specifically the oxygenation levels of hemoglobin [12]. When neurons in the brain are active, they consume more oxygen, leading to an increase in regional blood flow to the active areas.

This phenomenon, known as the hemodynamic response [21], forms the basis of fMRI. By using powerful magnetic fields and radio waves, fMRI scanners can detect these subtle changes in blood oxygenation and create detailed maps of brain activity.

One of the key advantages of fMRI is its ability to provide both spatial and temporal information about brain function. Spatially, fMRI can pinpoint the areas of the brain that are activated during specific tasks or cognitive processes. These activations are represented as statistical maps known as activation maps, which highlight the regions of the brain that show a significant increase or decrease in neural activity. By overlaying these maps onto high-resolution anatomical images, researchers can precisely localize brain activations.

Temporally, fMRI has a relatively high temporal resolution compared to other imaging techniques like positron emission tomography (PET). It can capture changes in neural activity with a time resolution of a few seconds, allowing researchers to study dynamic brain processes. This capability has been instrumental in investigating various cognitive functions, such as attention, memory, language processing, and emotion regulation.

By manipulating experimental conditions and observing corresponding changes in brain activity, researchers can uncover the neural underpinnings of complex cognitive processes.

Applications. The applications of fMRI span across multiple domains, including cognitive neuroscience, psychology, and clinical research. In cognitive neuroscience [18], fMRI has been instrumental in unraveling the neural mechanisms underlying perception, decision-making, and social cognition. It has also shed light on the functional connectivity between different brain regions, revealing intricate networks that support various cognitive functions.

In psychiatry, fMRI has provided [29] valuable insights into mental disorders such as schizophrenia, depression, and anxiety. By comparing brain activity patterns in healthy individuals and patients, researchers can identify aberrant neural circuits associated with these disorders, leading to better understanding and potential therapeutic interventions.

In clinical research, fMRI has proven to be a valuable tool for pre-surgical mapping in patients with brain tumors or epilepsy. By identifying the critical brain regions involved in language, motor function, or sensory processing, surgeons can plan safer and more precise interventions, minimizing the risk of post-operative deficits.

Typologies. Functional Magnetic Resonance Imaging (fMRI) can be categorized into various typologies based on the experimental design and data analysis techniques employed. Here are two common typologies of fMRI:

1. Task-Based fMRI: In task-based fMRI, participants are presented with specific tasks or stimuli while their brain activity is measured using fMRI [23,24]. The goal is to identify brain regions that are selectively activated during the task performance. This approach allows researchers to investigate the neural correlates of cognitive processes, such as attention, memory, language processing, and motor control. By comparing brain activity during different task conditions or between groups of individuals, insights into the functional organization of the brain can be gained.
2. Resting-State fMRI: Resting-state fMRI involves measuring spontaneous brain activity in the absence of any specific task or stimulation. Participants are instructed to relax and keep their minds at rest while fMRI scans are conducted. The analysis focuses on identifying intrinsic patterns of brain connectivity, known as resting-state networks (RSNs), which represent synchronized activity between different brain regions. Resting-state fMRI provides valuable information about the functional connectivity and communication between brain regions, offering insights into the brain's intrinsic organization and potential biomarkers for neurological and psychiatric disorders.

These typologies of fMRI reflect different experimental designs and data analysis approaches, each offering unique insights into the functional organization of the brain. By combining these approaches, researchers can gain a comprehensive understanding of brain function and its relevance to various cognitive processes and clinical conditions.

Drawbacks. fMRI's indirect measure of neural activity and relies on the assumption that changes in blood flow accurately reflect underlying neuronal events. Additionally, fMRI is limited by its relatively low spatial resolution, as it cannot capture activity at the level of individual neurons. Techniques such as multivariate pattern analysis (MVPA) have been developed to overcome these limitations and extract more detailed information from fMRI data.

To summarize, functional magnetic resonance imaging (fMRI) is a powerful technique that has revolutionized our understanding of the human brain. By measuring changes in blood oxygenation levels, fMRI allows researchers to investigate brain function non-invasively. Its spatial and temporal resolution have enabled breakthroughs in cognitive neuroscience, psychology, and clinical research. While fMRI has its limitations, its contribution to the field of neuroscience cannot be overstated. As technology advances and data analysis techniques improve, fMRI will continue to be at the forefront of brain research, unraveling the complexities of the human mind.

3 Deep Learning Methods for fMRI

In this section we first recall some deep learning key concepts. Then we will provide a description of several deep learning based method for fMRI classification.

3.1 Deep Learning Key Concepts

Basics. Deep learning is a subclass of machine learning approaches that are part of the artificial intelligence class. The key innovation with respect to classical machine learning is the way of representing and using input data. It relies on an extended (deep) layer stratification.

The founding stone is the neural network, which is trained to understand the input by creating a coherent characteristic representation. This approach lead to several advantages. One of the most important is a reduced preprocessing phase because it is done inside layers stratification, implying a reduced human factor.

The key data structure in deep learning models are tensors (multidimensional arrays) which are manipulated with calculations that are known mathematical operations such as geometric transformations. The goal of these modifications is to minimize the loss function as a measure of the model's error in representing actual data.

These calculations are made on high dimensional matrices (tensors), resulting in high computation costs, even on benchmark examples. In addition, the loss function evaluation itself is computationally expensive because it is based on a training loop that updates the weights according to what the network has learned from the training data.

Deep learning gained popularity and feasibility due to the increased computational power mainly lead by performing them on Graphical Processing Unit (GPU).

There are several frameworks to obtain efficient deep learning applications. The most popular are TensorFlow [1] with its front-end API Keras [3], and Pytorch [17].

Common Neural Networks for Classification. We briefly describe common approaches to classification using deep learning. We have to say that there are several ways to perform classification task depending on data type.

Convolutional Neural Networks. For data of biomedical imaging such as Computed Tomography (CT) a popular approach is to use a Convolutional Neural Network (CNN) [25,28]. This kind of network has been developed for computer vision purposes and performs well at detecting patterns and extracting features from raw pixel data.

Their structure is made of multiple layers, including convolutional, pooling, and fully connected layers, each performing a specific operation on the input data. The convolutional layers apply filters to the input image, capturing local features and preserving spatial relationships. Pooling layers downsample the feature maps, reducing their dimensionality while retaining the most salient information. Finally, fully connected layers process the extracted features and generate predictions or classifications.

Whereas these networks performs well in image recognition it must be noted that they could fail in particular scenarios such as same images taken by different angle. It represents a potential drawback in the medical context, where there are lot of source of noise (e.g. the movement induced by breathing). In this case a solution could be the application of image augmentation [22] during preprocessing. This technique takes input images and applies some operation such as rotation or flip to enhance the ability of the model to adapt to noisy images.

Recurrent Neural Networks. Another common network suitable for medical applications is the Recurrent Neural Network (RNN) [2] [4]. They are designed to process sequential data, such as time series, speech, and text. They provide a feedback connection, allowing them to maintain a hidden state that captures information from previous steps in the sequence.

This recurrent nature enables RNNs to model temporal dependencies and capture context, making them well-suited for tasks like language translation, speech recognition, and sentiment analysis.

The key component of an RNN is the recurrent layer, which processes each input along with the previous hidden state and generates an output and a new hidden state. This hidden state acts as a memory, retaining information about previous inputs and influencing future predictions.

However, one of the major drawback is that RNNs suffer from the vanishing gradient problem, where gradients become extremely small or vanish over long sequences, hindering their ability to capture long-term dependencies. To address this, variations like Long Short-Term Memory (LSTM) [9] and Gated Recurrent Unit (GRU) were introduced, incorporating gating mechanisms to better preserve and update information over time.

Since some biomedical data, such as fMRIs, involve sequential data analysis, the major advantage of RNNs is that they are powerful tools for modeling and predicting sequential patterns and suitable also for classification purposes.

Convolutional Recurrent Neural Networks. A third typology comes from the combinations of CNNs and RNNs. The Convolutional Recurrent Neural

Network (convRNN), is a hybrid architecture that combines the strengths of both convolutional neural networks (CNNs) and recurrent neural networks (RNNs).
It is designed to process sequential data with spatial dependencies, such as videos or time series with spatial dimensions. The ConvRNN architecture typically consists of convolutional layers to capture spatial features and extract local patterns, followed by recurrent layers to model temporal dependencies and capture long-term context.
By incorporating convolutional operations within the recurrent framework, ConvRNNs can efficiently process and analyze both spatial and temporal information simultaneously, and this feature could be useful in some medical application [31]. ConvRNNs have shown [11] promising results in various computer vision and video processing applications, leveraging the strengths of both CNNs and RNNs to achieve better performance in tasks involving sequential and spatial data.

3.2 Methods for fMRI Classification

Here we review some deep learning based approach to fMRI classification. Since the wide usages of fMRI, we faced several domain of applications in terms of medical conditions. Other differences could be the type of fMRI (resting state or task-based), the used framework and the neural network. We focused our attention on resting state fMRI and we now report a description for every paper which includes these relevant data and the classification score. We then summarize these approaches in Table 1.

First of all, we analyzed a paper from Wang et al. [27] which deals with the classification of resting state fMRI data by applying a ConvRNN network. The goal is the individual identification, giving the possibility to identify a specific subject from a large group [7]. This work was based on data from the Human Connectoma Project and involved 100 subjects. On every subject (patient) four fMRI session were performed, with 1200 volumes per session. For each fMRI session, fMRI data with 1200 volumes was divided into twelve 100-frame clips as inputs of ConvRNN. Data from day 1 was used as the training dataset. The two sessions from Day 2 were used as validation and testing datasets, respectively. The best model was decided based on the validation dataset and the final performance was assessed on the testing dataset. They performed also some preprocessing steps on data, such as motion correction and denoising. The model was implemented using the Keras/Tensorflow framework. By using the convRNN model they achieved an accuracy of 98.50%. They also made some comparisons with previous work where two different RNN configurations where used. These resulted in an accuracy of 94.43% and 95.33%.

While a paper from Santana et al. [19] deals with resting state fMRI for chronic pain condition classification. They performed binary classification task on a dataset of patients divided into an healthy group (control group), and a group of chronic pain condition: fibromyalgia and chronic back pain. The entire dataset, after some preprocessing phases and data cleaning is composed by 140 subjects. A total of 240 whole-brain echo-planar images were acquired in a time of 10 min. In this case, authors made a comparison among four classifiers, three of which used convolution neural network approaches. The first was the classifier BrainNetCNN, defined by Kawahara et al. [10] and it proposes three new conventional filters. The second was a modified version of BrainNetCNN. Since it was added of batch normalization layers between the BrainNetCNN layers, the proposed version is called BrainNetCNNBatch. The third classifier was based on the work of Meszlnyi et al. [15] and had a sequence of two one-dimensional convolutional layers followed by a fully connected layer and a softmax layer with two outputs. While the fourth classifier was made with an automated machine learning toolkit called TPOT4[1]. This toolkit was used to test different classical machine learning models and feature engineering processes with a reduced computational cost. Along as the classifiers, they analyzed how different brain parcelations and connectivity measures affect the classification performance. They used four different parcelations including ROI and group-ICA based parcelations, and two different measures of functional brain connectivity, such as correlations and Dynamic Time Warping. Summarizing all of these resulted in a the best model using the Ann4brain (the first classifier) architecture and MSDL (Multi-Subject Dictionary Learning) parcelation which reached a balanced accuracy of 86.8% and concluded that while the parcelation does not have a relevant impact on classification, the use of Dynamic Time Warping significantly increase the classifier performance.

In a paper from Sarraf and Tofighi [20] the classification of fMRI data for Alzheimer subjects from normal controls was performed. In this case they applied a Convolutional Neural Network (CNN) using the popular LeNet-5 architecture. The dataset was composed by 15 control subjects and 28 diagnosed with Alzheimer. After various preprocessing stages and application of the LeNet neural network they were able to achieve a 96.86% accuracy, which was relevant with respect to their reviewed paper on same scenarios but using a classical machine learning approach (Supporting Vector Machine).

An example of CNN application can be found in a paper from Zheng et al. [30] deals with the psychiatric disease of schizophrenia. This work used a slightly modified version of common convolutional neural network model, the VGG16 net, and the transfer learning methodology to perform a binary classification between healthy patients and affected by the disease. The dataset, composed by 102 health patients and 98 affected by schizophrenia comes from the public data set of the Center for Biomedical Research Excellence (COBRE). In this work authors achieved a 87.85% accuracy on the selected dataset by using first the VGG16 for migration learning, then extracts the features of fMRI by designing the convolution structure of the neural network, and finally uses the fully connected layer for training and continuous optimization to obtain the optimal weight parameters.

In a paper from Dvornek et al. [6] an experiment of classification of autism spectrum disorder by using resting state fMRI is conducted. Their effort was to

---

[1] http://epistasislab.github.io/tpot/.

combine a RNN, according to literature, since it is claimed to perform well on rs-fMRI time series as input and phenotypic features (like sex or age) often available. They built and tested several deep learning models on a popular dataset from Autism Brain Imaging Data Exchange, ABIDE, which resulted in 529 autism subjects and 571 case controls. To establish a baseline for classification performance, authors first used a LSTM model. They then explored several ways to combine phenotypic and fMRI data into a single network. The best in term of mean classification accuracy was the one that combines phenotypic data directly with score from rsfMRI input. In this case phenotypic data and the final output of the LSTM model are input into dense layers, with a single node in the last layer. This model achieved a mean classification score of 70.1%.

Table 1. Relevant data for analyzed papers.

Paper	Medical domain	DL Network Type	Best Accuracy
Wang et al.	Individual identification	convRNN	98.5%
Santana et al.	Cronic pain condition	CNN-BrainNetCNN	86.8%
Sarraf and Tofighi	Alzheimer disease	CNN-LeNet-5	96.86%
Zheng et al.	Schizophrenia	CNN-VGG16	87.85%
Dvornek et al.	Autism Spectrum Disorder	RNN-LSTM	70.1%

4 Conclusions

In this paper our aim was to review some deep learning based approaches on functional magnetic resonance imaging classification.

We introduced the imaging technique by giving some details on its importance, applications and key concepts. We also made a brief recall on deep learning key concepts with a focus on common classification models.

Then we reviewed some application of deep learning based classification on fMRI data. We selected some different applications to show the wide usage of fMRI and the power of deep learning techniques.

This paper is intended as a base for further experiences with fMRI classification, possibly by building our own model and by proposing novel approaches to problems already studied in literature.

References

1. Abadi, M., et al.: TensorFlow: large-scale machine learning on heterogeneous distributed systems (2016). arXiv:1603.04467

2. Al-Askar, H., Radi, N., MacDermott, A.: Chapter 7 - recurrent neural networks in medical data analysis and classifications. In: Al-Jumeily, D., Hussain, A., Mallucci, C., Oliver, C. (eds.) Applied Computing in Medicine and Health, pp. 147–165. Emerging Topics in Computer Science and Applied Computing, Morgan Kaufmann (2016). https://doi.org/10.1016/B978-0-12-803468-2.00007-2. https://www.sciencedirect.com/science/article/pii/B9780128034682000072
3. Chollet, F., et al.: Keras (2015). https://github.com/fchollet/keras
4. Cui, R., Liu, M.: RNN-based longitudinal analysis for diagnosis of alzheimer's disease. Comput. Med. Imaging Graph. **73**, 1–10 (2019). https://doi.org/10.1016/j.compmedimag.2019.01.005. https://www.sciencedirect.com/science/article/pii/S0895611118303987
5. DeYoe, E.A., Bandettini, P., Neitz, J., Miller, D., Winans, P.: Functional magnetic resonance imaging (FMRI) of the human brain. J. Neurosci. Methods **54**(2), 171–187 (1994). https://doi.org/10.1016/0165-0270(94)90191-0
6. Dvornek, N.C., Ventola, P., Duncan, J.S.: Combining phenotypic and resting-state FMRI data for autism classification with recurrent neural networks. In: Proceedings, IEEE International Symposium on Biomedical Imaging, vol. 2018, pp. 725–728 (2018). https://doi.org/10.1109/ISBI.2018.8363676
7. Finn, E.S., et al.: Functional connectome fingerprinting: identifying individuals using patterns of brain connectivity. Nat. Neurosci. **18**(11), 1664–1671 (2015). https://doi.org/10.1038/nn.4135
8. Grover, V.P., Tognarelli, J.M., Crossey, M.M., Cox, I.J., Taylor-Robinson, S.D., McPhail, M.J.: Magnetic resonance imaging: principles and techniques: lessons for clinicians. J. Clin. Exp. Hepatol. **5**(3), 246–255 (2015). https://doi.org/10.1016/j.jceh.2015.08.001
9. Hochreiter, S., Schmidhuber, J.: Long short-term memory. Neural Comput. **9**(8), 1735–1780 (1997). https://doi.org/10.1162/neco.1997.9.8.1735
10. Kawahara, J., et al.: BrainNetCNN: convolutional neural networks for brain networks; towards predicting neurodevelopment. NeuroImage **146**, 1038–1049 (2017). https://doi.org/10.1016/j.neuroimage.2016.09.046
11. Keren, G., Schuller, B.: Convolutional RNN: an enhanced model for extracting features from sequential data (2017)
12. Kim, S.G., Bandettini, P.A.: Principles of functional MRI. In: Faro, S.H., Mohamed, F.B. (eds.) BOLD fMRI: A Guide to Functional Imaging for Neuroscientists, pp. 3–22. Springer, Heidelberg (2010)
13. Lauterbur, P.C.: Image formation by induced local interactions: examples employing nuclear magnetic resonance. Nature **242**(5394), 190–191 (1973). https://doi.org/10.1038/242190a0
14. Mansfield, P., Grannell, P.K.: NMR 'diffraction' in solids? J. Phys. C: Solid State Phys. **6**(22), L422–L426 (1973). https://doi.org/10.1088/0022-3719/6/22/007
15. Meszlényi, R.J., Buza, K., Vidnyánszky, Z.: Resting state fMRI functional connectivity-based classification using a convolutional neural network architecture. Front. Neuroinf. **11**, 61 (2017). https://doi.org/10.3389/fninf.2017.00061
16. Ogawa, S., Lee, T.M., Kay, A.R., Tank, D.W.: Brain magnetic resonance imaging with contrast dependent on blood oxygenation. Proc. Natl. Acad. Sci. U.S.A. **87**(24), 9868–9872 (1990). https://doi.org/10.1073/pnas.87.24.9868
17. Paszke, A., et al.: PyTorch: an imperative style, high-performance deep learning library. arXiv:1912.01703 (2019)
18. Poldrack, R.A.: The role of fMRI in cognitive neuroscience: where do we stand? Curr. Opin. Neurobiol. **18**(2), 223–227 (2008). https://doi.org/10.1016/j.conb.2008.07.006

19. Santana, A.N., Cifre, I., de Santana, C.N., Montoya, P.: Using deep learning and resting-state fMRI to classify chronic pain conditions. Front. Neurosci. **13**, 1313 (2019). https://doi.org/10.3389/fnins.2019.01313
20. Sarraf, S., Tofighi, G.: Deep learning-based pipeline to recognize Alzheimer's disease using fMRI data. In: 2016 Future Technologies Conference (FTC). pp. 816–820. IEEE, San Francisco (2016). https://doi.org/10.1109/FTC.2016.7821697
21. Shibasaki, H.: Human brain mapping: hemodynamic response and electrophysiology. Clin. Neurophysiol. **119**(4), 731–743 (2008). https://doi.org/10.1016/j.clinph.2007.10.026
22. Shorten, C., Khoshgoftaar, T.M.: A survey on image data augmentation for deep learning. J. Big Data **6**(1), 60 (2019). https://doi.org/10.1186/s40537-019-0197-0
23. Silva, M.A., See, A.P., Essayed, W.I., Golby, A.J., Tie, Y.: Challenges and techniques for presurgical brain mapping with functional MRI. NeuroImage. Clin. **17**, 794–803 (2018). https://doi.org/10.1016/j.nicl.2017.12.008
24. Soares, J.F.: Task-based functional MRI challenges in clinical neuroscience: choice of the best head motion correction approach in multiple sclerosis. Front. Neurosci. **16**, 1017211 (2022). https://doi.org/10.3389/fnins.2022.1017211
25. de Sousa, P.M., et al.: A new model for classification of medical CT images using CNN: a COVID-19 case study. Multimedia Tools Appl. **82**(16), 25327–25355 (2023). https://doi.org/10.1007/s11042-022-14316-7
26. The Nobel Prize in Physiology or Medicine 2003. https://www.nobelprize.org/prizes/medicine/2003/summary/
27. Wang, L., Li, K., Chen, X., Hu, X.P.: Application of convolutional recurrent neural network for individual recognition based on resting state fMRI data. Front. Neurosci. **13**, 434 (2019). https://doi.org/10.3389/fnins.2019.00434
28. Xu, M., et al.: Segmentation of lung parenchyma in CT images using CNN trained with the clustering algorithm generated dataset. Biomed. Eng. Online **18**(1), 2 (2019). https://doi.org/10.1186/s12938-018-0619-9
29. Zhan, X., Yu, R.: A window into the brain: advances in psychiatric fMRI. BioMed Res. Int. **2015**, 542467 (2015). https://doi.org/10.1155/2015/542467
30. Zheng, J., Wei, X., Wang, J., Lin, H., Pan, H., Shi, Y.: Diagnosis of schizophrenia based on deep learning using fMRI. Comput. Math. Methods Med. **2021**, 1–7 (2021). https://doi.org/10.1155/2021/8437260
31. Zreik, M., et al.: A recurrent CNN for automatic detection and classification of coronary artery plaque and stenosis in coronary CT angiography. IEEE Trans. Med. Imaging **38**(7), 1588–1598 (2019). https://doi.org/10.1109/TMI.2018.2883807

Deep Learning for Scoliosis Diagnosis: Methods and Databases

Lorella Bottino[1]([⊠]), Marzia Settino[1], Luigi Promenzio[2], and Mario Cannataro[1]

[1] Data Analytics Research Center, Department of Medical and Surgical Sciences, University Magna Graecia of Catanzaro, 88100 Catanzaro, Italy
lorella.bottino@unicz.it
[2] Chief of Pediatric Orthopaedics Department, Villa Serena for Children, 88100 Catanzaro, Italy

Abstract. The advent of data-driven science and artificial intelligence (AI) has provided a deeper knowledge about data that has driven the clinical research to unprecedented change. AI has shown great potential in the scoliosis diagnosis for which the current widely adopted standard of evaluation is the manual measurement on the X-ray radiographs of the Cobb angle to quantify the magnitude of spinal deformities in scoliosis. The reliability of the Cobb angle measurement mainly depends on the subjective experience of the operators and it is time-consuming. Machine learning (ML) and Deep Learning (DL) methods can help surgeons to avoid misjudgment about scoliosis screening, diagnosis and classification by providing a powerful solution for saving time and effort in the Cobb angle measurement. The contribution of this work is twofold. Primarily it aims to provide an overview of the main ML approaches, with special focus on DL, that can be used in the spine field to make physicians and researchers aware of the benefits of the ML approaches for scoliosis diagnosis and treatment with respect to the traditional methods. Furthermore, because the reliability of the all ML approaches depends strongly on the training data and often it is difficult to obtain a large amount of representative data, we survey the main databases containing spinal data with the purposes of providing the stakeholders with a knowledge that can substantially help them to choose the best dataset in the spine field for training their ML models. Furthermore, with this work we intend to lay the foundations for development of a ML-based tool for supporting the physicians in the diagnosis, classification, screening and prognosis prediction of scoliosis.

Keywords: Scoliosis · Machine Learning · Deep Learning · Convolutional neural network · CNN · Cobb angle · Spine

1 Introduction

Scoliosis is a spinal curvature in the frontal plane. Adolescent idiopathic scoliosis (AIS) is the most common type of scoliosis; the most widely known risk factors for adolescent idiopathic scoliosis are gender, skeletal maturity and heredity. Menarche, i.e. the transition into adulthood in females, greatly influences the course of the curve. In fact girls who show even mild scoliosis in childhood may experience severe scoliotic curve worsening in the 3 months before and after menarche. Once menarche is reached, progression of scoliotic curve becomes more predictable [1,2]. Skeletal maturity is an other important factor risk as AIS curves progress during the phase of rapid growth of the patient and for this reason need to be evaluated carefully.

The diagnosis of scoliosis is usually performed on the basis of the physical examination which includes the evaluation of some parameters which take into account geometric and morphological variations in the trunk and rib cage. Landmarks located on specific points of the patient's body can be used for the scoliosis severity evaluation. Landmarks-based analysis provides information about the symmetry or asymmetry of the body.

In those cases where scoliosis is suspected, a standing posterior-anterior radiograph should be obtained [3].

Artificial intelligence (AI) is gradually changing medical practice including the procedures to address the issues related to the scoliosis.

In particular, machine learning (ML) and deep learning (DL) are assuming a crucial role in supporting scoliosis diagnosis and classification, screening and prognosis prediction [4,5].

This work aims to provide an overview of the recent AI approaches that are revolutionizing the clinical practice of scoliosis. We hope that our work can contribute to improve the existing methods and/or to develop new ones beside of encouraging clinicians to apply even more AI methods to the clinical practice related to scoliosis. Furthermore, since ML is based on a data-driven approach, we survey the most popular databases containing spinal images.

The rest of the paper is organized as follows: conventional approaches for calculating the Cobb angle in spine X-Ray are discussed in Sect. 2. Machine learning approaches, with a particular focus on deep learning methods, for automated extraction of Cobb angle measurements are discussed in Sect. 3. Finally, Sect. 4 concludes the paper and underlines future work.

2 Traditional Methods of Measuring the Degree of Spinal Curvature

Spinal radiography represents the standard of imaging for the scoliosis evaluation. The Cobb angle is the most widely used measurement to quantify the degree of spinal deformities from X-ray images. It is defined as the angle between the two tangents of the upper and lower endplates of the upper and lower end vertebra, respectively. On the basis of the value of the Cobb angle, the severity of the scoliosis deformation can be determined as shown in Table 1.

In the clinical setting, a protractor is employed in estimating manually the angle of spinal curvature.

Table 1. Cobb angle size with the related scoliosis severity

Cobb angle size	Scoliosis severity
10°–20°	Mild
20°–40°	Moderate
> 40°	Severe

When using a traditional measurement method, significant interobserver and intraobserver variations can be found.

An alternative to protractor is represented by goniometer method, an manual measurement method that relies on the use of a particular instrument that follows the line of the spine while the patient stands in a forward bending position, and measures the angle of trunk rotation (ATR). One limb of the device is placed on the upper margin of the upper end-vertebra and the other limb on the lower margin of the lower end vertebra; the sustained angle is read off and recorded.

Recently the use of smartphone apps are widely used to measure spine curve in simple, accurate way and independently by the clinician intervention, bringing a series of advantages both on the doctor and on the patient side [6].

3 Machine Learning Approaches for Scoliosis Diagnosis and Evaluation

With the large increase in available digitized data and the progress made in computer science, AI applied to the biomedical field produced medical insights previously unavailable beside of providing a valid support to the clinical decision-making process for personalized medical treatments.

Machine learning is a sub-field of the AI that achieved excellent results in several areas including the medical field. It provides great opportunities in the spine domain demonstrating its value in the scoliosis diagnosis, screening and classification [7–9].

An accurate scoliosis classification has an high impact on the scoliosis evaluation and treatment and it is a very critical task for the clinicians because it can be affected by the observer bias and/or influenced by the medical image quality. Furthermore, anatomical variability, low tissue contrast and image artifacts due to surgical implants can make difficult to identify the vertebrae for an accurate spinal curvature evaluation [10]. Computer-aided methods can successfully overcome these issues by reducing or eliminating bias and error to achieve more accurate diagnoses and enable more personalized treatments.

ML-based approach for the scoliosis classification by using X-rays images are proposed in several works with satisfactory results [11,12]. With respect to reliability, Cobb angle measurements based on ML-based approach show an high reliability compared with a manual approach (error of cobb angles measurements can reach 0.5° or less) [13].

3.1 ML Methods

ML algorithms are generally categorised into the following categories:

3.1.1 Supervised Methods

The goal of the supervised algorithms is to learn by example: they use a training set (typically a labeled dataset) to learn the model and to make predictions. There are two main subcategory of supervised learning (SL) problems that are *classification* that consists in predicting a class label and *regression* that consists in predicting a numerical value. For instance, classification algorithms can be well-trained to automatically detect and diagnose cancer by identifying features on histopathological images (e.g.; through alterations of the color in image areas) that could indicate a pathological condition [14]. Support Vector Machines (SVMs) are supervised learning models for classification and regression problems. They can be used to predict the scoliosis curve type through the analysis of the surface of the trunk [15].

3.1.2 Unsupervised Methods

Unlike supervised learning, algorithms based on unsupervised learning assess the data to identify patterns from unlabeled datasets. They aim to group samples based on their features only, i.e.; samples that are similar between them belong to the same group while those that are dissimilar belong to another group. The unsupervised learning is adopted to classify patient affected by the AIS versus healthy one using rasterstereography data [16]. Clustering is an application of unsupervised learning that has been used to assign to different groups patients with osteoporotic vertebral fractures (the assignment is based on the increasing pain) [17].

3.1.3 Semi-Supervised Methods

Algorithms based on semi-supervised learning use a small amount of labeled data and a large amount of unlabeled data to train a predictive model. A semi-supervised learning approach has been used in [15] for the scoliosis curve type prediction on the basis of the trunk back surface analysis.

3.1.4 Reinforcement Learning Methods

Reinforcement learning is a feedback-based technique based on rewarding desired behaviors and/or punishing undesired ones, i.e., for a good action, an agent gets positive feedback, and for a bad action, the agent gets negative feedback or penalty. There is no labeled datasets therefore the agent learns by its experience only.

Among the ML techniques, Deep Learning methods have been successfully applied to several scoliosis images such as X-ray, computed tomography (CT), or magnetic resonance (MRI) thanks to their high recognition accuracy and feature extraction ability when they process images [11,18–20].

3.2 Scoliosis Diagnosis and Classification Using Deep Learning

3.2.1 Structure of Deep Neural Network

Deep learning is a subset of the ML and it is based on the concept of the Artificial Neural Networks (ANNs) [21] that is widely applied in various areas such as healthcare, visual recognition, cybersecurity, etc. for its great ability to find complex patterns in huge datasets [22]. The adjective "deep" in deep learning refers to the use of multiple layers in the network, i.e., neural networks with more than three layers (including input and output).

Figure 1 shows the structure of a DL network that consists of a collection of interconnected processing nodes (i.e., neurons) organized into layers that work together. Circles represent the neurons while the arrows represent the weighted connections between neurons. The input layers receive data as input and it passes them to the next intermediate layers, i.e., hidden layers. Input layers do not apply any operations to the input data.

The multiple hidden layers contain several neurons which apply non-linear transformations to the data to learn complex patterns. Weights represent the strength of the connection between two neurons and they establish how much influence the input will have on the output. The output layer produces the final result. Training a DL model consists in learning (i.e., calculate) the optimal values for all the weights from labeled examples [23].

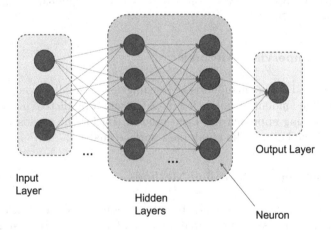

Fig. 1. Structure of a deep neural network

3.2.2 Deep Learning: Strengths, Weaknesses and Applications in Spine Field

Thanks to its great potential in the image processing, DL has became rapidly the natural methodological approach to address the spine-related diseases using radiographic images [4,24].

In general, DL-based methods offer several advantages over traditional ML methods. Feature extraction is the process of extracting features (e.g.; characteristics, properties and attributes) from raw data. Feature extraction plays a critical role in improving the accuracy and generalization of the ML models and it can be performed manually or automatically. Primarily, DL has the greater ability to learn directly from raw data. This means that preprocessing phase is limited because the single neural network is trained for complex tasks by using directly the raw data as input, without any manual feature extraction. Unlike ML, DL performs well when it deals with large volumes of unstructured and complex datasets that require high-level abstraction and representation, high accuracy and performance.

On the other hands, DL is computationally expensive due to several reasons. This is because DL algorithms typically use multiple layers that process the input data through several computations beside of requiring large amount of data to be trained to achieve a significant accuracy. Moreover, DL algorithms are often trained using iterative methods which require significant amount of computational power. This implies that graphics processing units (GPUs) and central processing units (CPUs) with high performance and price should be used for DL.

Interpretability in ML is the degree to which the model behaviours can be understood by a human. Interpretability is a key factor when comparing DL with ML. DL-based models generally behave as "black-boxes" meaning that they provide results without explanation of how they are achieved. This implies the lack of interpretability of the DL models that limits their wider adoption in healthcare [52]. A common problem in ML is the overfitting that occurs when a model performs significantly better for training data with respect to new data. It leads inevitably to a low performance of the ML models. Overfitting can occur due to several reasons but primarily it happens because the size of the training dataset is too small and it does not contain enough data to accurately represent all possible input data values. With respect to ML models, DL models are more prone to overfitting and they require large amounts of data to avoid it.

Nevertheless, thanks to the advantage of not being dependent on the human intervention to learn and thanks to its capacity to deal with unstructured data, in recent years DL has become the preferred technique for the image analysis tasks in the spine field. Table 2 shows the main Strengths and Weaknesses of ML and DL models. The automatic *localization* and *segmentation* in the medical images are crucial steps for the computer-aided detection and classification of the spine deformity [25].

Image localization (IL) is a computer vision task used to identify the location of one object in an image, i.e.; to find the position of the object and draw a bounding box around it [26]. IL is used for example to localize the spinal vertebrae.

Image segmentation (IS) is a computer vision task that consists in labelling each image pixel based on its belonging to a specific area (i.e., anatomic region). IS allows to identify regions and objects that compose the image to obtain an image representation easier to be further analyzed [27]. For instance, vertebra segmentation is used to identify fractures and spine deformities such as kyphosis and scoliosis [28].

Manually performed IL and IS are time-consuming, tedious and they are prone to error (e.g.; due to the subjectivity of the clinician) [29]. Furthermore, despite the high resolution of the computed tomography (CT) imaging, vertebra segmentation remains a difficult task due to the wide range of shapes and pathological alterations that can be found in images [30].

Several DL-based methods have been proposed for improving localization and segmentation reliability and efficiency for the spinal curvature assessment. For instance, DL has been used for locating and identifying vertebrae from Computed tomography (CT) and Magnetic Resonance (MR) images respectively in the works [31,32] and it has been employed for the vertebral segmentation spinal misalignment classification in [33].

In addition to the application of DL techniques to analyze radiographic images for the estimation of the Cobb angle, another concrete application of these techniques is the analysis of the photos of the patient's back for the Trunk Rotation Angle (ATR) measurement.

ATR is used as parameter for the scoliosis diagnosis and it represents a valid alternative to the usage of the instruments based on ionizing radiation whose use may have harmful consequences for the patient's health. However, the measurement of this parameter is strongly dependent on the visual observation of the orthopedic professional and on his experience.

For this reason, in the context of an early diagnosis of scoliosis, automatizing the ATR measurement process has a whole series of advantages, including more accurate and repeatable measurements.

Table 2. Strengths and Weaknesses in ML and DL models

	Strengths	Weaknesses
ML	– Low computational cost	– Prone to overfitting
	– Models can be interpretable	– Low performance
	– Low dependence on dataset size	– Manual Features extraction
DL	– Low complexity of preprocessing	– Prone to high overfitting
	– High performance	– High computational cost
	– Automatic Features extraction	– Lack of interpretability
		– High dependence on dataset size

3.2.3 Convolutional Neural Networks: Concept and Applications

Convolutional Neural Networks (CNNs) are a class of ANNs that have at least one convolution layer. The convolution layer applies a filter to the input to create a feature map that summarizes the detected features. CNNs have been used to automatize the IL and IS procedures on which relies the Cobb angle measurement process.

Indeed, CNNs are specialized in processing data that have a grid-like topology such as images and therefore they are well-suited to be applied in spine domain [34–39]. Like the traditional ANNs, the CNNs structure consists of neurons, layers, and weights. However, CNNs are distinguished from the ANNs because CNNs have three types of layers: convolution, pooling, and fully connected layers [40]. Each layer performs a specific task, such as detecting edges or patterns in the input image. The convolution layer applies a mathematical operation called convolution to the input and then it passes the result to the next layer. CNNs demonstrated widely their superiority in performance in object recognition and classification from images with respect to the traditional ANNs thanks to their ability of detecting different features of an input image through back-propagation and multiple layers [41–45].

Despite the undoubted success of the CNNs in the medical field, the large number of images required for their training and testing as well as the difficulty in collecting high-quality medical images that often must be annotated and labeled by clinicians, represent barriers that hinder the full application of CNNs in the clinical practice [46,47]. Due to the difficulty in obtaining large amount of representative data, because for smaller datasets classical ML algorithms often outperform DL algorithms, the first are preferred with respect to the latter. Fortunately, nowdays several large databases of annotated medical images including spine-focused images are emerging to address this obstacle.

3.2.4 Deep Learning-Based Framework for Cobb Angle Estimation

Generally, DL-based methods are widely applied to measure the Cobb angle by x-rays images. Figure 2 shows the DL-based framework usually adopted for Cobb angle estimation. The framework consists of three main steps that are:

- *Data Preprocessing*: a preliminary step for removing noise and distortion in the image is required. This step impacts significantly on the outcome.
- *Localization and segmentation*: The IS process distinguishes the constituent parts or objects in the image and it is stopped when the region of interest (ROI) is detected. It is mainly used to distinguish the objects from the background. This step consists in localizing anatomical structures in the image. The segmentation of each vertebra from the spine image is performed in this step. Despite the improvements in the recent years, the automatic segmentation of vertebrae continues to be a challenging task due to complexity of the anatomical structure in which vertebrae are in close proximity to the corresponding ribs, blood vessels and other structures.

– *Severity estimation*: The premise of measuring Cobb angle is the spinal segmentation and the vertebral corner recognition. Once concluded these tasks, the Cobb angle measurement can be performed. Several DL-based methods have been proposed to measure automatically the Cobb angle size in this step. For example, Cobb angle is calculated on the vertebral slopes estimation in [48]. Cobb angle is automatically measured through the tangent line of spine curve in [49].

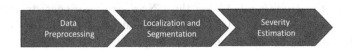

Fig. 2. DL-based framework usually adopted for Cobb angle estimation

3.3 Spine Imaging Databases

Spine imaging is an essential tool for assessing spine curvature. Nevertheless, errors in the manual selection of the vertebrae can affect the Cobb angle calculation. Therefore, even experienced radiologists can provide different evaluations for the same clinical case despite having adopted the standard diagnostic scheme. Therefore, considering the weaknesses in the process of manual measurement of Cobb angle in terms of efforts, time to spend and low accuracy, there is an urgent need of advanced Computer-aided methods that support the physician in diagnosis, treatment, planning and monitoring of the spine curvature evolution. Machine learning and deep learning algorithms can automate some complex medical tasks and reduce the physicians' workload. Data collection and preprocessing are key steps in ML because data quality and quantity strongly impact on the reliability of the ML model. Therefore, the availability of large databases of radiographic images of the spine can contribute successfully to improving the ML-based techniques or developing new ones for an automated assessment of the spinal curvature.

It is important that the model is trained on high-quality data to accurately represent the target population, i.e., scoliosis patients [1,50].

In this section we provide a survey of the most popular databases containing spinal radiographic images that can be used, for instance, to train and test ML models developed for computer-aided measurement of Cobb angle. Because the reliability of the all ML approaches depends strongly on the data (i.e.; size and quality) that are used for training the algorithms and often it is difficult to obtain a large amount of representative data that are also of acceptable quality, the quality of data sources plays a critical role in the reliable training of algorithms and in the widespread adoption of ML in clinical practice. For this reason, public databases in spine field are fundamental starting points for developing novel ML models able to: i) classify patients as healthy or affected by scoliosis, for a first skimming; ii) stratify patients based on the level of severity of scoliosis (mild, moderate and severe); iii) predict, based on some features, how a patient's scoliosis will evolve over time.

An overview of the main features of the most popular Databases in the spine field are shown in Table 3. Each database is described through the evaluation of the following characteristics:

- **Content types**. Content types represent content category of data collected on database. In general we can have collections of x-rays, photo, multiple choice questionnaires, cases or articles.
- **Data collection center**. Data collection center is the main source where the data comes from. Once collected, the data can be used to seek answers to scientific questions and evaluate outcomes.
- **Database size**. Database size is the total size of used storage from database or, in other words, the space the files physically consume on disk.
- **Healthy subjects**. Healthy subjects indicates if there are healthy subjects among subjects recruited for the study.
- **Number of subjects**. Number of subjects indicates the total number of subjects which have been recruited to collect the data. Often the number of spine images does not correspond with the number of subjects recruited because we may have more than one scan for the same subject.
- **Categories of subjects**. Categories of subject indicates to which group of pathology or absence of pathology the recruited subjects belong.
- **Type of access**. Access to the data can be free or require registration or require the payment of a fee.
- **Licence**. A licence (a official permission) may be required to use database. There are two basic categories of licenses: Commercial and Academic (Research) Use or Open Source. Research licenses permit research institutions or individual researchers to use DB free of charge or at a research use rate. Such research use licenses also leave open the possibility of future traditional commercial licensing. Open Source licenses allow free and less restricted distribution of DB.
- **References**. References are the scientific publications where it is possible to find the studies that have been done on the database in question.

Table 3. Overview of the main database in the spine field. Acronyms: DC, Data Collection; N.A., Not Available; N.C., Non-commercial use; qtns, questionaries; subjs, subjects; Ref., references; Ann., Annotations, JUST, Jordan University of Science and Technology; UW, University of Washington.

Database	Content types	DC center	Dataset size	Healthy Subjs	#N subjs / images	Subjs categories	Access type	License	Ref.	Models	Ann.
Mendeley	- X-ray	JUST[a]	81 MG	Yes	- 338 subjs	- scoliosis - spondylolisthesis - normal vertebrae	Free	N.C.	[53]	DL	No
Biomedia	- CT	UW[b]	18,4 GB	No	- 125 subjs - 242 scans types	various types of pathologies	Free	N.C.	[54]	N.A.	Yes
Radiopaedia	- articles - cases - qtns	N.A.	N.A.	Yes	N.A.	various types of pathologies	Free	N.C.	[55]	N.A.	Yes

[a] https://www.just.edu.jo/Pages/Default.aspx.
[b] https://rad.washington.edu.

- **Models.** Models are machine learning models that have been created by training algorithms with either labeled or unlabeled data, or a mix of both. Training, validation and testing are the three phases in which the initial dataset is randomly partitioned. The training dataset is the largest group, because the model needs as many examples as possible to learn the subject domain.
- **Annotations.** Data Annotation (sometimes called "Data Labeling") refers to the active labeling to help machines understand what exactly is in it and what is important. Annotations field can have a positive or negative value.

4 Conclusions and Future Work

Applying Machine Learning algorithms to address the spine-related issues such as scoliosis can contribute meaningfully to improve the accuracy in diagnosis and the patient outcomes. In particular, Deep Learning techniques such as Convolutional Neural Networks are specifically designed for tasks of object recognition and classification from images. Because medical images such as X-ray, CT and MRI play a crucial role in the diagnosis, monitoring, and management of scoliosis, CNNs represent the natural approach to address spine-related issues by images in the clinical practice. On the other hand, several issues hinder their successfully application in the spine field mainly related to the large amount of annotated medical images required by CNNs that are difficult to collect. Fortunately, several large databases of spine-focused images are emerging to address this obstacle. However, ML-based systems poses further challenges in the medical field due to the need of understanding the working mechanisms underlying the decision-making process of the ML algorithms.

Explainable Artificial Intelligence (XAI) is an emerging research topic of ML that can be used to address these challenges by revealing the black-box for creating trustworthy ML-based systems in the medical field [51]. XAI plays a key role in the widespread adoption of ML in clinical practice because the lack of trust in the ML-based systems led inevitably to the lack of adoption of that system in the medical applications. Therefore, we believe that the next advancements in XAI can boost significantly the adoption of AI in the medical domain.

This work aims primarily to provide a overview of the main ML approaches, with special focus on DL, that can be used in the spine field. Furthermore it provides an overview of the main databases containing spinal data for training ML-based models for scoliosis treatment. With this work we intend to lay the foundations for development of a ML-based tool for supporting the physicians in the diagnosis, classification, screening and prognosis prediction of scoliosis.

References

1. Taher, H., Grasso, V., Tawfik, S., Gumbs, A.: The challenges of deep learning in artificial intelligence and autonomous actions in surgery: a literature review. Artif, Intell. Surg. **2**(3), 144–158 (2022)

2. Yochum, T.R., Rowe, L.J.: Essentials of skeletal radiology (1987)
3. Janicki, J.A., Alman, B.: Scoliosis: review of diagnosis and treatment. Paediatrics Child Health **12**(9), 771–776 (2007)
4. Yang, J., et al.: Development and validation of deep learning algorithms for scoliosis screening using back images. Commun. Biol. **2**, 10 (2019)
5. Kadoury, S., Mandel, W., Roy-Beaudry, M., Nault, M.-L., Parent, S.: 3-d morphology prediction of progressive spinal deformities from probabilistic modeling of discriminant manifolds. IEEE Trans. Med. Imaging **36**(5), 1194–1204 (2017)
6. Bottino, L., Settino, M., Promenzio, L., Cannataro, M.: Scoliosis management through apps and software tools. Int. J. Environ. Res. Public Health **20**(8), 5520 (2023)
7. Obermeyer, Z., Emanuel, E.J.: Predicting the future - big data, machine learning, and clinical medicine. N. Engl. J. Med. **375**(13), 1216–1219 (2016). PMID: 27682033
8. Haenssle, H., et al.: Man against machine: diagnostic performance of a deep learning convolutional neural network for dermoscopic melanoma recognition in comparison to 58 dermatologists". Ann. Oncol. **29**(8), 1836–1842 (2018)
9. Yahara, Y., et al.: A deep convolutional neural network to predict the curve progression of adolescent idiopathic scoliosis: a pilot study. BMC Musculoskelet. Disord. **23**, 610 (2022)
10. Malfair, D., et al.: Radiographic evaluation of scoliosis: Review. AJR. Am. J. Roentgenol. **194**, s8-22 (2010)
11. Rothstock, S., Weiss, H.R., Krueger, D., Paul, L.: Clinical classification of scoliosis patients using machine learning and markerless 3D surface trunk data. Med. Biol. Eng. Comput. **58**, 2953–2962 (2020)
12. Galbusera, F., Casaroli, G., Bassani, T.: Artificial intelligence and machine learning in spine research. JOR Spine **2**(1), e1044 (2019)
13. Li, J., Li, S., Yang, Z., Wu, T., Hu, Y.: An automatic scoliosis diagnosis platform based on deep learning approach. In: 2022 4th Asia Pacific Information Technology Conference, APIT 2022, pp. 215–223, Association for Computing Machinery, New York (2022)
14. Saravi, B., et al.: Artificial intelligence-driven prediction modeling and decision making in spine surgery using hybrid machine learning models. J. Pers. Med. **12**, 03 (2022)
15. Seoud, L., Adankon, M., Labelle, H., Dansereau, J., Cheriet, F.: Prediction of scoliosis curve type based on the analysis of trunk surface topography, pp. 408–411 (2010)
16. Colombo, T., et al.: Supervised and unsupervised learning to classify scoliosis and healthy subjects based on non-invasive rasterstereography analysis. PLoS ONE **16**(12), e0261511 (2021)
17. Galbusera, F., Casaroli, G., Bassani, T.: Artificial intelligence and machine learning in spine research. JOR SPINE **2**, e1044 (2019)
18. Galbusera, F., et al.: Fully automated radiological analysis of spinal disorders and deformities: a deep learning approach. Eur. Spine J. **28**, 951–960 (2019)
19. Alharbi, R.H., Alshaye, M.B., Alkanhal, M.M., Alharbi, N.M., Alzahrani, M.A., Alrehaili, O.A.: Deep learning based algorithm for automatic scoliosis angle measurement. In: 2020 3rd International Conference on Computer Applications & Information Security (ICCAIS), pp. 1–5 (2020)
20. Tan, Z., et al.: An automatic scoliosis diagnosis and measurement system based on deep learning. In: 2018 IEEE International Conference on Robotics and Biomimetics (ROBIO), pp. 439–443. IEEE Press (2018)

21. Hornung, A., et al.: Artificial intelligence in spine care: current applications and future utility. Eur. Spine J. **31**, 08 (2022)
22. Tajdari, M., et al.: Image-based modelling for adolescent idiopathic scoliosis: mechanistic machine learning analysis and prediction. Comput. Methods Appl. Mech. Eng. **374**, 113590 (2021)
23. Sze, V., Chen, Y.-H., Yang, T.-J., Emer, J.: Efficient processing of deep neural networks: a tutorial and survey. Proc. IEEE **105**, 03 (2017)
24. Fraiwan, M., Audat, Z., Fraiwan, L., Manasreh, T.: Using deep transfer learning to detect scoliosis and spondylolisthesis from x-ray images. PLOS ONE **17**, 1–21 (2022)
25. Spiller, J.,, Marwala, T.: Medical image segmentation and localization using deformable templates, vol. 14, pp. 2292–2295 (2007)
26. Dasiopoulou, S., Mezaris, V., Kompatsiaris, I., Papastathis, V., Strintzis, M.: Knowledge-assisted semantic video object detection. IEEE Trans. Circ. Syst. Video Technol. **15**, 1210–1224 (2005)
27. Gonzalez, R., Faisal, Z.: Digital Image Processing, 2nd edn. (2019)
28. Qadri, S.F., et al.: Ct-based automatic spine segmentation using patch-based deep learning. Int. J. Intell. Syst. **2023**, 1–14 (2023)
29. Vania, M., Mureja, D., Lee, D.: Automatic spine segmentation from ct images using convolutional neural network via redundant generation of class labels. J. Comput. Des. Eng. **6**(2), 224–232 (2019)
30. Courbot, J.B., Rust, E., Monfrini, E., Collet, C.: Vertebra segmentation based on two-step refinement. J. Comput. Surg. **4**, 1 (2016)
31. Chen, H., et al.: Automatic localization and identification of vertebrae in spine CT via a joint learning model with deep neural networks. In: Navab, N., Hornegger, J., Wells, W.M., Frangi, A.F. (eds.) MICCAI 2015. LNCS, vol. 9349, pp. 515–522. Springer, Cham (2015). https://doi.org/10.1007/978-3-319-24553-9_63
32. Suzani, A., Rasoulian, A., Seitel, A., Fels, S., Rohling, R., Abolmaesumi, P.: Deep learning for automatic localization, identification, and segmentation of vertebral bodies in volumetric mr images, vol. 9415 (2015)
33. Masood, R.F., Taj, I.A., Khan, M.B., Qureshi, M.A., Hassan, T.: Deep learning based vertebral body segmentation with extraction of spinal measurements and disorder disease classification. Biomed. Signal Process. Control **71**, 103230 (2022)
34. Yamashita, R., Nishio, M., Do, R., Togashi, K.: Convolutional neural networks: an overview and application in radiology. Insights Imaging **9**, 06 (2018)
35. Renganathan, G., Manaswi, N., Ghionea, I., Cukovic, S.: Automatic vertebrae localization and spine centerline extraction in radiographs of patients with adolescent idiopathic scoliosis. Stud. Health Technol. Inform. **281**, 288–292 (2021)
36. Huang, X., et al.: The comparison of convolutional neural networks and the manual measurement of cobb angle in adolescent idiopathic scoliosis. Global Spine J. 21925682221098672 (2022)
37. Caesarendra, W., Rahmaniar, W., Mathew, J., Thien, A.: Automated cobb angle measurement for adolescent idiopathic scoliosis using convolutional neural network. Diagnostics **12**(2), 396 (2022)
38. Choi, R., et al.: Cnn-based spine and cobb angle estimator using moire images (2017)
39. Alharbi, R.H., et al.: Deep learning based algorithm for automatic scoliosis angle measurement. In: 2020 3rd International Conference on Computer Applications & Information Security (ICCAIS), pp. 1–5 (2020)
40. Yamashita, R., Nishio, M., Do, R.K.G., Togashi, K.: Convolutional neural networks: an overview and application in radiology. Insights Imaging **9**, 611–629 (2018)

41. Kokabu, T., et al.: An algorithm for using deep learning convolutional neural networks with three dimensional depth sensor imaging in scoliosis detection. Spine J. **21**(6), 980–987 (2021)
42. Horng, M.-H., Kuok, C.-P., Fu, M.-J., Lin, C.-J., Sun, Y.-N.: Cobb angle measurement of spine from x-ray images using convolutional neural network. Comput. Math. Methods Med. **2019**, 1–18 (2019)
43. Cheng, P., Yang, Y., Yu, H., He, Y.: Automatic vertebrae localization and segmentation in CT with a two-stage Dense-U-Net. Sci. Rep. **11**, 22156 (2021)
44. Chen, H., Dou, Q., Wang, X., Qin, J., Cheng, J.C.Y., Heng, P.-A.: 3D fully convolutional networks for intervertebral disc localization and segmentation. In: Zheng, G., Liao, H., Jannin, P., Cattin, P., Lee, S.-L. (eds.) MIAR 2016. LNCS, vol. 9805, pp. 375–382. Springer, Cham (2016). https://doi.org/10.1007/978-3-319-43775-0_34
45. Wang, Z., et al.: Accurate scoliosis vertebral landmark localization on x-ray images via shape-constrained multi-stage cascaded cnns (2022)
46. Zhou, Z., Siddiquee, M.M.R., Tajbakhsh, N., Liang, J.: Unet++: redesigning skip connections to exploit multiscale features in image segmentation. IEEE Trans. Med. Imaging **39**(6), 1856–1867 (2019)
47. Fatima, J., Akram, M., Jameel, A., Syed, A.: Spinal vertebrae localization and analysis on disproportionality in curvature using radiography-a comprehensive review. EURASIP J. Image Video Process. **2021**, 06 (2021)
48. Zhang, J., Li, H., Lv, L., Zhang, Y.: Computer-aided cobb measurement based on automatic detection of vertebral slopes using deep neural network. Int. J. Biomed. Imaging **2017**, 9083916 (2017)
49. Tu, Y., Wang, N., Tong, F., Chen, H.: Automatic measurement algorithm of scoliosis cobb angle based on deep learning. J. Phys. Conf. Ser. **1187**, 042100 (2019)
50. Ong, W., et al.: Application of artificial intelligence methods for imaging of spinal metastasis. Cancers **14**(16), 4025 (2022)
51. Yang, G., Ye, Q., Xia, J.: Unbox the black-box for the medical explainable AI via multi-modal and multi-centre data fusion: a mini-review, two showcases and beyond. Inf. Fusion **77**, 29–52 (2022)
52. Sarker, I.H.: Deep learning: a comprehensive overview on techniques, taxonomy, applications and research directions. SN Comput. Sci. **2**(6), 420 (2021)
53. Russo, G.L., Spolveri, F., Ciancio, F., Mori, A.: Mendeley: an easy way to manage, share, and synchronize papers and citations. Plast. Reconstr. Surg. **131**(6), 946e–947e (2013)
54. Glocker, B., Zikic, D., Konukoglu, E., Haynor, D.R., Criminisi, A.: Vertebrae localization in pathological spine CT via dense classification from sparse annotations. In: Mori, K., Sakuma, I., Sato, Y., Barillot, C., Navab, N. (eds.) MICCAI 2013. LNCS, vol. 8150, pp. 262–270. Springer, Heidelberg (2013). https://doi.org/10.1007/978-3-642-40763-5_33
55. Glocker, B., Feulner, J., Criminisi, A., Haynor, D.R., Konukoglu, E.: Automatic localization and identification of vertebrae in arbitrary field-of-view CT scans. In: Ayache, N., Delingette, H., Golland, P., Mori, K. (eds.) MICCAI 2012. LNCS, vol. 7512, pp. 590–598. Springer, Heidelberg (2012). https://doi.org/10.1007/978-3-642-33454-2_73

Understanding Spreading Dynamics of COVID-19 by Mining Human Mobility Patterns

Carmela Comito[✉][iD] and Deborah Falcone[iD]

ICAR-CNR, Via P. Bucci 8-9C, 87036 Rende, CS, Italy
{carmela.comito,deborah.falcone}@icar.cnr.it

Abstract. The combination of health social data analytics and machine learning led to the so called big geo-social data. However, analyzing such data brings significant challenges. Therefore, new methods and high computing tools are required to effectively analyze the big data from social media platforms. In this direction, we propose an approach for gathering health-related information from social networks and combine such data with geographical, social and temporal data normally embedded in social media. Our research may offer insights into diseases, uncover social dynamics that lead to broad phenomena and forecast disease outbreaks. Social media played also a key role in the management of COVID-19 especially in tracking disease spreading as users self-report their health-related issues. In this paper we present a case study of the proposed methodology to monitor COVID-19 spreading characteristics. Our aim is to understand spreading dynamics of the virus in US, through the analysis of people movements between US states by exploiting official surveillance data and geo-tagged tweets related to COVID-19. In the analysis we investigated human mobility patterns by considering COVID-19 related tweets posted in regions exhibiting high correlation with the official number of COVID-19 cases. As first step of the methodology, we mine users trajectories from the collected geo-tagged posts and build a mobility map including the most frequent movements. After that we extracted a set of different spatial-temporal features characterizing the trajectories, including the frequency of visit in a specific locations, the direction of movements among locations, the frequency of movements. The approach gives us the possibility of monitoring the spread of the epidemics and detecting outbreak locations.

Keywords: Social Media · COVID-19 · Mobility Mining

1 Introduction

The use of social media offers extensive opportunities for exchanging information and for social connection. They have a significant impact on our lives and we rely on them to stay updated on friends, family, and global events. Easy access to information is a key benefit of these platforms, therefore many individuals are now more likely to choose social media news over more conventional media sources. Their effects have obviously had an impact on medicine, creating opportunities to keep people safe, informed and connected. Social media can act as support systems for health since they allow users to access information about health topics they feel significant. Many people look at social networks for information and direction related to health issues. The interaction between patients and healthcare organizations as well as the sharing of medical expertise and knowledge is made easier for healthcare workers, researchers, and patients. These tools can quickly disseminate crucial new information, pertinent new scientific findings, share diagnostic, treatment, and follow-up protocols, as well as compare various approaches globally, erasing geographic boundaries for the first time in history. Other two revolutionary aspects are the real-time and location-aware nature of social contents. Spatial-temporal information associated with posts allows to comprehend human mobility. These distinguishing characteristics point to a potential use for this technology in disease surveillance, early public health warning, early outbreak detection, and disease spread monitoring. According to research, social media may be an important part of the disease surveillance toolset, which will help public health experts to identify disease outbreaks more quickly and to improve outbreak response. The literature review provided by [3], shows the value of social media in promoting and enhancing public health as well as in identifying intervention target audiences. The review's main recommendation is to find ways for public health experts to use social media analytics into disease surveillance and outbreak management practice. Recent studies have looked into the possible advantages of using social networking sites to capture different health issues and for public health monitoring, such as for recording the mental health condition of impacted population groups in the event of a disaster and tracking their long-term recovery [7,10], for influenza monitoring [2,6], for HIV surveillance and prevention [14]. Monitoring of infectious diseases has been the subject of other social media studies. Twitter, for instance, has been used to monitor dengue infection, cholera, E.coli, and ebola.

Large-scale spatial-transmission models of infectious diseases depend heavily on human movements, which can now easily and quickly gleaned from social networks. Accurate modeling and measurement of people mobility is essential for improving epidemic control, however this process may be complicated by challenging social data. Advancements in automated data processing, machine learning, and natural language processing (NLP) open up the prospect of using these enormous data sources for public health monitoring and surveillance.

In this vein, we offer a method to map the spread of an infectious disease and control its dissemination. The basic concept behind this work is the use of social media posts as sensors for keeping track on public health. We base our assumption on the findings of earlier studies in the field. Recently, [9] used Twitter data to derive spatio-temporal behavioral patterns to track the sites of flu outbreaks. The authors demonstrated how closely connected flu-related traffic on social media is to actual flu outbreaks. In [4] a detailed analysis of Twitter data to inspect how information about the COVID-19 epidemics spread in US is presented. The research study examined the relationship between Twitter data and COVID-19 confirmed cases, along with an assessment of the viability of using Twitter to predict the spread of an outbreak. The findings show a strong correlation between tweets and actual COVID-19 data, demonstrating that Twitter can be used as a reliable indicator of an epidemic's spread and that user activity data on social media is increasingly important for identifying and comprehending epidemic outbreaks.

Considering these outcomes, we propose a prediction model that allows us to forecast the potential outbreak locations and the areas that warrant closer attention because a disease may spread there more in the upcoming time period. If early outbreaks detection can be made, effective interventions can be taken to contain the epidemics. To solve that challenge, we followed a step-by-step approach. At the beginning, by analyzing the spatio-temporal information of social posts, we reconstruct the flow of user movements in a geographical area. Then, by extracting rules and regularities in moving trajectories and performing a spatial-temporal characterization of patterns, we compute a prediction algorithm. The algorithm predicts the geographical area where disease outbreaks are most likely to occur at present and future time.

We examined the COVID-19 spreading features as case study, in order to evaluate the prediction model's accuracy. Social media platforms have been widely used during the COVID-19 pandemic, when they were a crucial components for the dissemination of information. Numerous studies have been conducted to determine the impact of social media on the COVID-19 pandemic. In [12], authors discussed the contribution of population movement to the spread of COVID-19. They assert that the geographic shifts in movement, the diversity of jobs held by migrants, the rapid expansion of travel for business and pleasure, and the greater distances traveled for family reunions are what distinguish the spread of COVID-19 from that of SARS. The purpose of the study in [13] was to investigate the possibility of detecting the COVID-19 outbreak in 2019 using the Chinese social media WeChat. Differently from us, they evaluate the WeChat Index, a data service that illustrates how frequently a given keyword appears in posts. They demonstrated the possibility of forecasting outbreaks two weeks before the epidemic declaration by plotting daily WeChat Index results for terms related to coronavirus.

We make use of the CoronaVis Twitter dataset [11], a cleaned-up and processed dataset containing coronavirus-related tweets. Guided by [4,9], we assume that the movements revealed from the CoronaVis dataset provide us with information regarding the mobility of people affected by the COVID-19 in the USA. Our findings confirm what is stated in [11], in which the authors showed the correlation between people mobility and number of infection using the twitter data. They observe the number of cases and user movement between multiple states in every week, and increment a mobility count if a user moves between two or more states within 14 days. We have enriched this result providing: (i) the most frequent movements among the American states; (ii) the potential outbreak locations; (iii) the locations where pay more attention because there could be further virus spread in the near future.

The main contributions of the paper can be summarized as follows:

- we propose an effective approach for detecting relevant users movements from geo-tagged posts;
- we formulate a prediction algorithm that, by exploiting the spatio-temporal characterization of the user's travel patterns, identifies social movements that generate widespread phenomenon and predicts disease outbreaks;
- we analyze spreading dynamics of COVID-19 as a case study, by exploring a real-world dataset of tweets posted within the 50 American states and the District of Columbia. As result of the analysis, useful insights in terms of outbreak locations are provided with a remarkable accuracy.

The rest of the paper is organized as follows. Section 2 formalizes the outbreaks prediction problem. Section 3 describes the proposed model, detailing the spatial-temporal features and the prediction algorithm. Section 4 presents the COVID-19 case study and the evaluation results of prediction approach. Finally, Sect. 5 concludes the paper.

2 Problem Formulation

Understanding human movement patterns yield insight into a variety of important societal issues, such as urban planning, public transport management, location based advertisement, and opens the door to powerful studies on health monitoring.

The main idea of this work is to use social media posts as sensors for public health monitoring. Posts on social networks provide us with information about users' mobility in addition to providing us with hints regarding peoples' health. This is a unique chance to comprehend the dynamics of how infectious diseases spread. Moreover, social media posts give us real-time information. The prospect of using such real-time data, to provide early public health warning, has led to a new wave of studies probing ways to use social media in medicine. For instance, studies have shown that if early detection is possible, effective steps to contain the outbreaks can be taken. The Centers for Disease Control and Prevention (CDC) in the US traditionally gathers information about influenza-like illness (ILI) from "sentinel" medical practices. According to [1], there is typically a 1–2 week wait between the time a patient is diagnosed and when that data point appears in aggregate ILI reports. In [9], authors discovered that online posts about the flu tend to precede clinical flu encounters, and identify several public locations from which a majority of posts initiated. Social data is therefore an important step in order to develop epidemic models endowed with realism. In this work, user social profiles are thought of as virtual passive sensors. They are referred to as passive because they require data mining in order to yield useful information about users' health. They all exhibit different characteristics, some are quite active and others are not, furthermore their signals are noisy when compared to data from regular physical sensors. To extract health-related information from social media and analyze movements of people to discover mobility patterns in a geo-spatial region, we adopt the methodology and the formal modeling proposed in [5]. Figure 1 highlights the main phases of the methodology, which are:

- *sentinel detection*, that detects posts with illness mentions, named *sentinel posts*;
- *disease monitoring*, that collects and processes sentinel posts related to different time windows and users, located in a geographical area of reference, with the option of selecting various degrees of geographic resolution (e.g., Country, Region, Site).

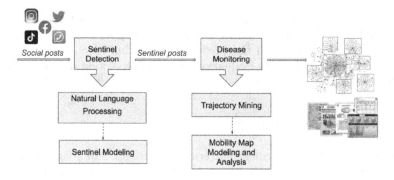

Fig. 1. Methodology for mining and retrieving health-related data from social media.

In this work we collect and process sentinel posts and historical data of users who move in a target geographic area, in order to extract their trajectories traveled during periods of potential illness.

Formally, a sentinel post sp is defined as a tuple $sp = <id, u, t, l, m, S, \bar{D}>$ where: id is the identification of the post; u is the identification of user who post it; t is the timestamp at which it has been posted; l represents the location from where it has been posted; m is the user mood; S is a set of symptoms it refers to; $\bar{D}$ is a set of diseases it refers to.

The historical posts (HP_u) of a user u represent the set of posts in the time window $W = [t_{sp-ub}, t_{sp}]$, namely those that u shared from the beginning of the incubation period (t_{sp-ub}) up to the timestamp of sentinel post (t_{sp}). The incubation period depends on the disease and determines how long W is. It is formally defined as: $HP_u = \{<id_i, u, t_i, l_i, m_i> \mid t_{sp-ub} \leq t_i < t_{sp} \forall i \in HP\}$.

Tracking user mobility patters from geo-tagged posts may allows us to early detect a disease outbreak and monitor the spread of a disease or contagions. In other word, it's critical to recognize who interacts with whom. The likelihood of infection can vary depending on the number of contacts. Contagious diseases can spread to other people by: direct physical contact with the ill person; unintentional contact with their fluids or objects; or airborne routes. Each disease has a different level of contagiousness, and those with long incubation times may be more contagious. These illnesses include the common cold and flu as well as more serious ones like SARS, cholera, and COVID-19.

In light of the arguments above, we formalize the outbreaks prediction problem as follows:

Given a disease d, a related set of sentinel posts SP and historical data HP shared by a set of users U and located in a set of locations L, by exploiting spatial-temporal mobility patterns mined from the set $SP \cup HP$, we aim to predict:

- *the set of outbreak locations O_{curr} at current time t_{curr},*
- *the set of locations O_{succ} where d could potentially spread at future time t_{succ}.*

We defined a step-by-step prediction model to deal with the outbreaks prediction problem:

1. a collection of trajectories is initially retrieved from sentinel posts and historical data;
2. a mobility map is created based on the users trajectories;
3. mobility patterns are detected from the mobility map;
4. visited locations and mobility patterns are characterized through a set of spatio-temporal features;
5. a prediction algorithm is finally employed to predict the outbreak locations of a target disease.

3 Prediction Model

In this section we define a prediction model in order to monitor the transmission of a disease by tracking user mobility patterns from geo-tagged posts. The goal is to identify illness outbreaks before they become widespread. The first step in the suggested prediction model is the extraction of a collection of user trajectories that go through a target geographic area during likely disease periods. Given a target disease d, a set of historical posts HP_u and a sentinel post sp, movements made by user u during the time window $W = [t_{sp-ub}, t_{sp}]$ are extracted. First of all, social posts are grouped on the basis of a given temporal granularity w, that must be less than or equal to the duration of W. As an illustration, posts can be grouped daily (w =1), weekly (w =7), bimonthly (w =14), or monthly (w =30). Users movements are modeled as temporally ordered sequences of visited location, and are referred to as trajectories. Formally:

A trajectory is a spatio-temporal sequence of locations visited by the user u according to temporal order during the time period $w \in W$:

$tr_{u,w} = l_{0,T_0} \rightarrow l_{1,T_1} \rightarrow \cdots \rightarrow l_{n,T_n}$

where, l_{i,T_i} is the i-th location in which u posts contents at the i-th time range during the time period w. The time range $T_i = (t_{ifirst} \ldots t_{ilast})$ is a sequence of consecutive timestamps relative to posts that u publishes in l_i before moving to the successive location l_{i+1}, with $t_{ilast} < t_{i+1first}$ $\forall \, 0 < i < n$.

Trajectories are first extracted for all users $u_i \in U$ who post related contents to the disease d until current time t_{curr}, then, they are merged to build the mobility map of d. Formally:

A *mobility map of a disease d* is a set of trajectories traveled by users who post contents related to d.

$M_d = \{TR_{u_0}, ..., TR_{u_m}\} \quad \forall u_i \in U, \quad 0 \leq i \leq m$

where $TR_{u_i} = \{tr_{u_i,w_0}, ..., tr_{u_i,w_k}\} \quad \forall w_j \in W, \quad 0 \leq j \leq k$ is the set of trajectories of a user u_i, traveled during the considered time window W, related to d.

The mobility map is examined by an algorithm to extract user mobility patterns. In particular, we propose a tailored version of the famous frequent sequential pattern mining algorithm PrefixSpan [8], named *MPSpan*, that will find all the sequential patterns whose frequencies are no smaller than a minimum support. A pattern's support is a measure of how many trajectories contain it. Mobility patterns describe frequent behaviors, in terms of both space and time. A mobility pattern is formally defined in a similar way to a trajectory but according to its support:

A *mobility pattern* is a trajectory consisting of a sequence of locations frequently visited together, with a frequency no smaller than a minimum support s, that can be expressed in the form:

$mp = l_{0,T_0} \rightarrow l_{1,T_1} \rightarrow \cdots \rightarrow l_{n,T_n}(s)$

where, l_{i,T_i} is the i-th location in which users post at the i-th time range $T_i = (t_{ifirst} \ldots t_{ilast})$, and s is the number of trajectories that contain mp.

3.1 Spatio-Temporal Prediction Features

Here is a description of how to obtain useful highlights on the mined users trajectories, to predict possible disease epicenters and understand spreading dynamics. We define a collection of spatio-temporal features that take advantage of various information dimensions regarding users' travels, such as their pathways of travel, the places they visit, and the movements among those sites. The most relevant features are listed below both for locations and mobility patterns.

Given a target disease d, we denote the set of users as U, the set of locations as L, the set of trajectories as TR, and the set of mobility patterns as MP.

Support. The support of a location $l \in L$ represents the number of times that users visit the location along trajectories, and it can be defined as follows:

$$s(l) = |\{tr_i \in TR : l \in tr_i\}|. \tag{1}$$

Given a user u, we define $s(l, u)$ as the number of times that u visits the location l along its trajectories.

The support of a mobility pattern $mp \in MP$ represents the number of times that users move along the trajectory, and it can be defined as follows:

$$s(mp) = |\{tr_i \in TR : mp \subseteq tr_i\}|. \tag{2}$$

Given a user u, we define $s(mp, u)$ as the number of times that u moves along the mobility pattern mp.

Popularity. The popularity of a location $l \in L$ represents the number of distinct users visiting the location along trajectories, and it can be defined as follows:

$$p(l) = |\{u \in U : s(l, u) \neq 0\}| \quad (3)$$

The popularity of a mobility pattern $mp \in MP$ represents the number of distinct users traveling along the trajectory, and it can be defined as follows:

$$p(mp) = |\{u \in U : s(mp, u) \neq 0\}| \quad (4)$$

Entropy. Entropy describes the distribution of movements among users and explains how a location is visited, how a trajectory is traveled, and if people tend to visit a location or follow a trajectory frequently or if they just pass through them occasionally. We utilized Shannon Entropy for this purpose:

$$e(X) = \sum_{u=1}^{n} P(X) \, log \, P(X). \quad (5)$$

For a location $l \in L$, the probability $P(X) = f(l, u)$. It depicts the user's percentage of visits to the place l and is described as:

$$f(l, u) = s(l, u)/s(l) \quad (6)$$

For a mobility pattern $mp \in MP$, the probability $P(X) = f(mp, u)$. It depicts the user's percentage of trips along the mobility pattern mp and is defined as:

$$f(mp, u) = s(mp, u)/s(mp) \quad (7)$$

3.2 Prediction Algorithm

Mobility patterns and spatio-temporal features are used to define the prediction algorithm showed in Fig. 2. Given a disease d, we attempt to predict locations where disease outbreaks are most likely to occur and/or where disease might spread in near future.

To predict outbreaks at current and future time, locations are selected from a list of candidate places ranked according to the prediction features. The algorithm accepts as an input parameter the metric ρ that combine the spatio-temporal features, on the basis of which it will rank locations and mobility patterns.

The method *computeSTfeatures* extracts features for each location and each mobility pattern.

Based on the metric ρ, the method *rank* ranks locations and mobility patterns.

Algorithm Outbreaks Prediction

Input: a set of users $U \neq \emptyset$
a set of locations $L \neq \emptyset$
a set of trajectories $TR \neq \emptyset$
a set of mobility patterns $MP \neq \emptyset$
timestamps $t_{curr}, t_{succ} : t_{succ} > t_{curr}$
ranking metrics ρ

Output: O_{curr}, O_{succ}

---

for $l \in L$ do
    $computeSTfeatures(l, U, TR, MP);$
end for
$rankL = rank(L, \rho)$
$O_{curr} = predict(t_{curr}, rankL)$

for $mp \in MP$ do
    $computeSTfeatures(mp, U, TR, MP)$
end for
$rankMP = rank(MP, \rho)$
$O_{succ} = predict(t_{succ}, O_{curr}, rankMP)$

Fig. 2. Disease outbreaks prediction algorithm.

The *predict* method is the one that deals with making the prediction. When called to make a prediction at current time t_{curr}, the method returns the set of locations O_{curr}, that are at the top of the ranking and that represent potential disease outbreaks. When the method is invoked to predict at future time t_{succ}, the method analyzes the mobility patterns. Among them, it considers only those whose starting location is an outbreak, i.e. locations in the set O_{curr}. For these mobility patterns, it extracts the locations visited during the trajectories (except for the starting points). The locations, that are at the top of the ranking, are returned as output in O_{succ} and represent the places to pay more attention to since the disease is more likely to spread than in the others. The ranking of the mobility patterns affects how the locations are ranked. If a place belongs to more than one mobility pattern, it will adhere to the ranking of the highest-ranked mobility pattern.

4 Evaluation Study: Understanding Spreading Dynamics of COVID-19

In order to prove the accuracy of the proposed prediction model, we applied the methodology to the COVID-19 scenario. We used a cleaned and processed dataset named CoronaVis Twitter dataset, available to the research community at https://github.com/mykabir/COVID19. The aim is to understand spreading dynamics of the virus in the geographic area of United States of America and the District of Columbia, through the analysis of people movements between states. The COVID-19 tweets data repository, is continuously updated since March 5,

2020 and will keep fetching the tweets using Twitter Streaming API. Data have US states as geographical resolution. In this work we considered the data up to April 27, 2020, and only data with the geo-location information. In Table 2 the statistics of the dataset are shown (Table 1).

Table 1. Summary statistics of CoronaVis dataset.

Statistic	Total
Number of tweets	7229508
Number of distinct users	1918834
Number of locations	51

As demonstrated in [4], considering a certain geographical area, there is a correlation between real cases of COVID-19 and the number of tweets that talk about it, therefore we consider that the routes depicted in the CoronaVis dataset give us indication about the mobility of people affected by coronavirus in the USA, leading to the possibility of monitoring the spread contagions and detecting outbreak locations.

Table 2. Trajectories and mobility patterns statistics w.r.t. temporal granularity.

w	1		7		14		30	
#tr	201		842		1279		2019	
#mp	26	6	1819	22	165	71	268	1221
|mp|	2	3	2	3	56 2	36	2	3 45

We extracted users trajectories from the collected geo-tagged posts, considering different temporal granularity w. In particular we extracted daily ($w = 1$), weekly ($w = 7$), biweekly ($w = 14$) and monthly ($w = 30$) trajectories. We considered the trajectories with at least a movement, that is with at least two distinct visited locations. We set *MPSpan* algorithm in order to find the mobility patterns with a minimum support $s = 2$. Table 2 summarizes the trajectories and mobility patterns statistics for each temporal granularity w, reporting the number of extracted trajectories ($\#tr$) and the number of mobility patterns ($\#mp$) with a specific length ($|mp|$).

We considered biweekly trajectories, as for COVID-19 is the time that passes between exposure to the virus (the moment of infection) and the onset of symptoms, therefore we focus on the statistical characterization of the 173 mobility patterns for $w = 14$. To this aim we exploited the features defined in the previous section. Most of patterns found are 2-length trajectories. The overall analysis of traffic flows across time and space confirms that COVID-19 spread fast with

Table 3. Top-20 locations ranking based on popularity.

Location name	USPS code	s	p	e
California	CA	285	269	8.02
New York	NY	225	199	7.55
Texas	TX	194	186	7.52
Washington	WA	181	157	7.12
Florida	FL	149	139	7.06
Georgia	GA	124	119	6.87
Illinois	IL	91	89	6.46
Massachusetts	MA	94	81	6.22
Nevada	NV	80	74	6.15
New Jersey	NJ	79	69	5.86
Virginia	VA	68	66	6.03
North Carolina	NC	67	64	5.97
Pennsylvania	PA	65	61	5.90
Ohio	OH	62	61	5.92
Maryland	MD	66	60	5.85
Arizona	AZ	59	58	5.85
Louisiana	LA	54	53	5.72
Colorado	CO	53	52	5.69
Tennessee	TN	57	48	5.49
Michigan	MI	48	48	5.58

Table 4. Top-20 mobility patterns ranking based on popularity.

Mobility pattern	s	p	e
NY → CA	20	20	4.32
CA → NY	17	17	4.09
CA → NV	16	16	4.00
CA → TX	15	15	3.91
NJ → NY	14	14	3.81
NY → NJ	14	14	3.81
WA → DC	15	13	3.64
WA → CA	13	13	3.70
MA → NY	12	12	3.58
DC → WA	13	11	3.39
CA → GA	11	11	3.46
FL → NY	11	10	3.28
TX → LA	10	10	3.32
CA → AZ	10	10	3.32
FL → TX	9	9	3.17
NY → FL	9	9	3.17
TX → FL	9	9	3.17
AZ → CA	9	9	3.17
SC → WA	9	9	3.17
MD → WA	9	9	3.17

the movements of the people. As multiple contacts play the most important role in the probability of infection, our prediction algorithm ranked locations and mobility patterns according to their popularity, that indicates the number of distinct users visiting a location or traveling along a trajectory.

The support (s), popularity (p) and entropy (e) values for the top-20 locations and top-20 mobility patterns are presented in Table 3 and 4, respectively. Looking at the Table 3, it can be seen that the locations visited by a large number of distinct users are those that have the highest number of visits, that is very high support. Moreover, as indicated from high entropy values in these places people tend to have a less stable attitude. In other words, many users visit such locations but in an occasionally way, not regularly or permanently. Table 4 lists the top-20 mobility patterns. As expected most of them involve the most popular locations. As with US states, the top-20 popular trajectories also have high support and entropy.

4.1 Prediction Accuracy

The following illustrates the accuracy evaluation of our findings using data and statistics obtained from the Centers for Disease Control and Prevention (CDC) of the United States. The aim is to demonstrate that our prediction model give the correct insights to determine possible outbreaks of a disease and to monitor its spreading.

Table 5. Location rankings on the basis of popularity and COVID-19 total cases.

Location	Popularity on 4/27/2020	Location	Tot. cases on 4/27/2020	Location	Tot. cases on 5/11/2020
CA	269	NY	131507	NY	151698
NY	199	NJ	111188	NJ	140037
TX	186	CA	56574	CA	84127
WA	157	MA	56462	IL	79007
FL	139	MI	50129	MA	78462
GA	119	IL	45883	MI	69456
IL	89	PA	42050	PA	57154
MA	81	FL	31290	FL	40982
NV	74	LA	27111	TX	39869
NJ	69	CT	25997	MD	34061
VA	66	TX	25297	CT	33765
NC	64	GA	23268	GA	32005
PA	61	MD	20113	LA	31881
OH	61	OH	16325	VA	25800
MD	60	IN	15961	OH	24777
AZ	58	VA	14339	IN	24627
LA	53	CO	13798	CO	19735
CO	52	WA	13745	WA	17266
TN	48	TN	10303	TN	16341
MI	48	NC	10064	NC	16183

By setting the measure ρ to correspond with the popularity feature, we ranked locations and mobility patterns. Table 5 lists: (i) the top-20 locations ranked by ρ on April 27, 2020; (ii) the top-20 locations ranked by CDC COVID-19 total cases on April 27, 2020; (iii) the top-20 locations ranked by CDC COVID-19 total cases on May 11, 2020.

We compute two metrics, using different lists of locations with length $X \in {5, 10, 15, 20}$:

- the accuracy $Acc@X$, that is the average accuracy of prediction. A location l is successfully ranked if it is in the top-X positions of the comparison list;
- the mean squared error $MSE@X$, that is the average squared difference between the estimated rank positions of locations and its actual positions in the comparison list.

The top-20 locations ranked by ρ on April 27, 2020 (t_{curr}) in Table 5 represent the set of outbreak locations O_{curr} predicted by our algorithm. In order to assess the prediction at current time, we compared them against the US states ranked by the CDC based on the total number of coronavirus cases in the same date. Figures 3(a) and (b) confirm that USs most popular states record a high number of COVID-19 cases. This is evidence of a correlation between people mobility and number of infections. The findings demonstrate that the states of California and New York, which represent the two probable main outbreaks, are highly ranked and record a significant number of actual cases of the virus. As expected, the accuracy (Fig. 3(a)) of the locations improves by considering lists of increasing size, reaching 90% of accuracy when 20 states are considered. The correlation between mobility and disease spread is strengthened by the $MSE@X$ values which give us information on the precision of the ranking. Inversely to accuracy, the mean squared error decreases as the size of locations list increase (Figs. 3(b)).

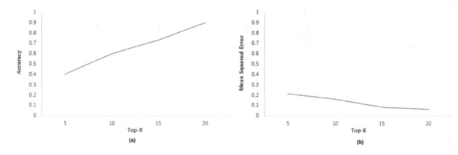

Fig. 3. Prediction accuracy ($Acc@X$) and mean squared error ($MSE@X$) w.r.t. $X \in 5, 10, 15, 20$, at current time.

To evaluate the accuracy of the prediction at future time, we considered the mobility patterns that start from outbreak locations in O_{curr}. Figure 4 displays how mobility patterns are distributed across the top-20 outbreak locations, considered as starting point of the trajectories. From mobility patterns we extracted the set of visited locations $\dot{L}$, ignoring the starting ones. For example, given the mobility pattern $NY \rightarrow NC \rightarrow CA$, we consider only NC and CA as locations belonging to the set $\dot{L}$. We assume that the starting locations are actual outbreaks and we want to show that destination states may have a greater spread of the infection recording a significant increase in COVID-19 cases. There are 140 mobility patterns taken into consideration and 37 possible outbreaks locations.

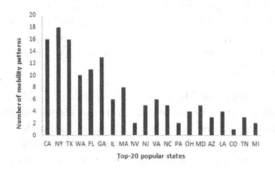

Fig. 4. Distribution of mobility patterns among Top-20 popular locations.

Also in this case, we compute the accuracy $Acc@X$ and the mean squared error $MSE@X$ with $X \in 5, 10, 15, 20$. Given that a person with a coronavirus may take up to 14 days before experiencing symptoms, we took the ranking relative to the 14th day after the target date. Due to this, we compare the set of locations $\dot{L}$ reached up to April 27, 2020 (t_{curr}) with the top-X locations ranked based on the CDC total number of cases on May 11, 2020 (t_{succ}).

Figure 5(a) reports the accuracy of the top-20 possible outbreak states extracted by our prediction algorithm. This set of experiments give evidence of the spread of COVID-19 due to the movements of people. In fact after 14 days of moving, locations visited along mobility patterns, that take hold from highly infected states, appear to have more contagions. The accuracy of the results is 70% in the case of the Top-10 locations and reaches 90% when the locations considered are 20. Even in this case, the mean squared error in Fig. 5(b) remains low and decreases with the locations lists length, with a mean value of about 18%.

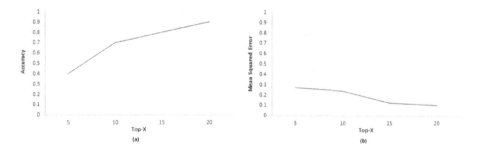

Fig. 5. Prediction accuracy ($Acc@X$) and mean squared error ($MSE@X$) w.r.t. $X \in 5, 10, 15, 20$, at future time.

5 Conclusion

Geo-tagging and real-time information widely shared by general population on social media are two revolutionary aspects of these platforms. This technology finds application also in the medical field and expands its frontiers. Social media data have particular strengths over traditional health monitoring. This includes the capacity to rapidly assess data on entire or particular populations in order to ascertain health issues, find people who are around a disaster, or guide clinical decision-making. Additionally, social media data can help with the real-time updating of health assessments, making it possible to recognize and react to the changing needs of the population or to the onset of new illness phases. In this work, we analyzed health-related social data in order to anticipate the areas where outbreaks of a particular disease are most likely to happen and/or where illness might spread in the near future. We suggest a prediction model which uses a machine learning algorithm to identify relevant user movements from posts with space-time information. Then, on the basis of a set of spatio-temporal features of user's travel patterns, it identifies social movements that lead to broad phenomena and forecast disease outbreaks.

Social media played a key role in the management of COVID-19 especially in tracking disease spreading as users self-report their health-related issues. For that reason, we evaluate the proposed prediction model considering COVID-19 as case study. The research yielded insightful information on outbreak locations inside the 50 American states and the District of Columbia. Our results demonstrate the relationship between people's mobility and the number of infections. Additionally, the prediction algorithm identifies possible outbreaks and locations that need to be monitored for potential spread of the epidemic in the near future.

References

1. Achrekar, H., Gandhe, A., Lazarus, R., Yu, S.H., Liu, B.: Predicting flu trends using twitter data. In: 2011 IEEE Conference on Computer Communications Workshops (INFOCOM WKSHPS), pp. 702–707 (2011). https://doi.org/10.1109/INFCOMW.2011.5928903
2. Aramaki, E., Maskawa, S., Morita, M.: Twitter catches the flu: detecting influenza epidemics using twitter. In: Proceedings of the 2011 Conference on Empirical Methods in Natural Language Processing, pp. 1568–1576 (2011)
3. Charles-Smith, L.E., et al.: Using social media for actionable disease surveillance and outbreak management: a systematic literature review. PLoS ONE **10**(10), e0139701 (2015)
4. Comito, C.: How covid-19 information spread in U.S.? the role of twitter as early indicator of epidemics. IEEE Trans. Serv. Comput. **15**(3) (2022)
5. Comito, C., Falcone, D., Forestiero, A.: Integrating iot and social media for smart health monitoring. In: Proceedings of the Web Intelligence and Intelligent Agent Technology (2020)
6. Culotta, A.: Towards detecting influenza epidemics by analyzing twitter messages. In: Proceedings of the First Workshop on Social Media Analytics, pp. 115–122 (2010)
7. Gruebner, O., et al.: Mental health surveillance after the terrorist attacks in paris. The Lancet **387**(10034), 2195–2196 (2016)
8. Han, J., et al.: Prefixspan: mining sequential patterns efficiently by prefix-projected pattern growth. In: proceedings of the 17th International Conference on Data Engineering, pp. 215–224. Citeseer (2001)
9. Hassan Zadeh, A., Zolbanin, H.M., Sharda, R., Delen, D.: Social media for now-casting flu activity: spatio-temporal big data analysis. Inf. Syst. Front. **21**, 743–760 (2019)
10. Jones, N.M., Thompson, R.R., Dunkel Schetter, C., Silver, R.C.: Distress and rumor exposure on social media during a campus lockdown. Proc. Natl. Acad. Sci. **114**(44), 11663–11668 (2017)
11. Kabir, M.Y., Madria, S.: Coronavis: a real-time covid-19 tweets data analyzer and data repository (2020)

12. Shi, Q., Dorling, D., Cao, G., Liu, T.: Changes in population movement make covid-19 spread differently from sars. Social Sci. Med. **255**, 113036 (2020). https://doi.org/10.1016/j.socscimed.2020.113036. https://www.sciencedirect.com/science/article/pii/S0277953620302550
13. Wang, W., Wang, Y., Zhang, X., Jia, X., Li, Y., Dang, S.: Using wechat, a chinese social media app, for early detection of the covid-19 outbreak in december 2019: retrospective study. JMIR Mhealth Uhealth **8**(10), e19589 (2020)
14. Young, S.D., Rivers, C., Lewis, B.: Methods of using real-time social media technologies for detection and remote monitoring of hiv outcomes. Prev. Med. **63**, 112–115 (2014)

Variational Quantum Algorithms for Gibbs State Preparation

Mirko Consiglio[✉][iD]

Department of Physics, University of Malta, Msida MSD 2080, Malta
mirko.consiglio@um.edu.mt

Abstract. Preparing the Gibbs state of an interacting quantum many-body system on noisy intermediate-scale quantum (NISQ) devices is a crucial task for exploring the thermodynamic properties in the quantum regime. It encompasses understanding protocols such as thermalization and out-of-equilibrium thermodynamics, as well as sampling from faithfully prepared Gibbs states could pave the way to providing useful resources for quantum algorithms. Variational quantum algorithms (VQAs) show the most promise in efficiently preparing Gibbs states, however, there are many different approaches that could be applied to effectively determine and prepare Gibbs states on a NISQ computer. In this paper, we provide a concise overview of the algorithms capable of preparing Gibbs states, including joint Hamiltonian evolution of a system–environment coupling, quantum imaginary time evolution, and modern VQAs utilizing the Helmholtz free energy as a cost function, among others. Furthermore, we perform a benchmark of one of the latest variational Gibbs state preparation algorithms, developed by Consiglio et al. [16], by applying it to the spin-1/2 one-dimensional XY model.

Keywords: Quantum Computing · Quantum Algorithms · Quantum Thermodynamics

1 Introduction

Gibbs states (also known as thermal states) can be used for quantum simulation [12], quantum machine learning [6,28], quantum optimization [49], and the study of open quantum systems [38]. Moreover, semi-definite programming [9], combinatorial optimization problems [49], and training quantum Boltzmann machines [28], can be tackled by sampling from well-prepared Gibbs states. Nevertheless, the preparation of an arbitrary quantum state on a quantum computer, is a QMA-hard problem [59]. Preparing Gibbs states, specifically at low temperatures, could be as hard as finding the ground-state of that Hamiltonian [2].

From a physical point of view, A Gibbs state is the quantum state that is at thermal equilibrium with the surrounding environment, in the canonical ensemble. Let's consider a Hamiltonian $\mathcal{H}$, describing n interacting qubits. The

Gibbs state at inverse temperature $\beta \equiv 1/(k_B T)$, where k_B Boltzmann constant and T is the temperature, is defined as

$$\rho(\beta, \mathcal{H}) = \frac{e^{-\beta \mathcal{H}}}{\mathcal{Z}(\beta, \mathcal{H})}, \quad \mathcal{Z}(\beta, \mathcal{H}) = \text{Tr}\{e^{-\beta \mathcal{H}}\} = \sum_{i=0}^{d-1} e^{-\beta E_i}, \tag{1}$$

where $\mathcal{Z}(\beta, \mathcal{H})$ is called the partition function. Here the dimension $d = 2^n$, while $\{E_i\}$ are the eigenenergies of $\mathcal{H}$ (with $\{|E_i\rangle\}$ denoting the corresponding eigenstates), i.e. $\mathcal{H}|E_i\rangle = E_i|E_i\rangle$.

The objective of this study is twofold: firstly, to offer a comprehensive overview of Gibbs state preparation algorithms in Sect. 2, and to explore a noteworthy application of Gibbs states in quantum Boltzmann machines. Secondly, to present the recent variational Gibbs state preparation algorithm by Consiglio et al. [16] in Sect. 3. In Sect. 4, we apply the variational quantum algorithm (VQA) to prepare Gibbs states of the XY model, while also qualitatively investigating the scalability of the algorithm. Finally in Sect. 5, we draw our conclusions.

2 Overview of Gibbs State Preparation Algorithms

The first algorithms for preparing Gibbs states on a quantum computer were based on the idea of coupling the system to a register of ancillary qubits, and letting the system and environment evolve under a joint Hamiltonian, simulating the process of thermalization [35,45,53]. In the limits of small time-steps or weak system-bath interactions, the dynamics are described by a Lindblad master equation that thermalizes the system in the long-time limit. Nevertheless, implementing time-evolution on a noisy intermediate-scale quantum (NISQ) devices is currently impractical, since it requires long coherence times, and a significant number of ancilla qubits and/or precise qubit reset capabilities. One could resort to implementing the Lindblad generator directly on the quantum computer, however this would once again require applying quantum phase estimation [41] or other quantum subroutines related to the quantum Fourier transform [10]. Refs. [5,27] studied thermalization and Lindblad generators from the perspective of mathematical physics, showing rapid convergence to the Gibbs state and Gibbs distribution, respectively, under certain conditions. Ref. [8] devised a method for preparing the Gibbs state of a local Hamiltonian, given some assumptions. Ref. [64] also developed a quantum algorithm, based on a local quantum Markov process that can be used to sample from the Gibbs distribution.

Temme et al. [52] proposed the quantum Metropolis algorithm, that is capable of preparing Gibbs states through random walks, which is inspired from the Metropolis–Hastings algorithm. A quadratic speedup is achieved over the direct implementation of the Metropolis–Hastings algorithm by using Grover's algorithm or related quantum algorithmic techniques [11,38]. Using modern quantum algorithmic approaches, further polynomial speedups could be achieved, such as using dimension reduction to sample from Gibbs states [7], which requires the use of a precise quantum phase estimation subroutine, among others [14,37,60,63].

A similar approach is the one in Ref. [42], which requires the use of a circuit very similar to that used in Shor's algorithm [46], which can only be feasibly implemented on fault-tolerant quantum devices. Ref. [20] showed how to adapt "Coupling from the Past"-algorithms proposed by Ref. [40], to sample from the exact Gibbs distribution, rather than an approximation, using the quantum Metropolis algorithm. Nevertheless, compared with the local updates of the classical Metropolis algorithm, all of these algorithms require the implementation of large, global quantum circuits across the whole system, making them infeasible for NISQ devices. A recent review of current quantum Gibbs-sampling algorithms can be found in Ref. [10].

A promising approach for NISQ devices is VQAs, where a hybrid quantum–classical approach of minimizing an objective function, using a parametrized quantum circuit (PQC) as a variational ansatz, leads to the preparation of the Gibbs state. Some approaches utilize a physically-inspired objective function, the Helmholtz free energy, such in Refs. [13, 16, 19, 33, 44, 61]. Others employ an engineered cost function [39, 43], while others usequantum approximate optimization algorithm (QAOA)-based approaches [61, 65]. Alternative variational approaches consist of using: truncated Taylor series to evaluate an approximation of the free energy [57]; adaptive derivative assembled problem-tailored (ADAPT)-VQA applied to a similar cost function to the free energy [58]; and a VQA based on McLachlan's variational principle to initialize and evolve a thermofield double state [29, 33, 39, 43, 61, 65].

Alternative algorithms prepare thermal states via quantum imaginary time evolution (QITE), which typically require, either starting from a maximally mixed state [21, 34, 66], or a maximally entangled state [62], and carrying out imaginary time evolution of time τ, preparing the Gibbs state at inverse temperature β, which is proportional to τ. Refs. [22, 36, 50] utilize minimally entangled typical thermal states (METTS), in combination with QITE or measurement-based variational QITE to compute thermal averages and correlation functions. Ref. [51] developed a linear scaling QITE algorithm, called Fast QITE, that can also compute thermal averages. Ref. [56] also employed variational QITE, that is inspired from the measurement-based approach [36], however, it is adapted to be ansatz-based QITE. Ref. [47] utilizes random quantum circuits with intermediate measurements to impose QITE, while Ref. [48] proposed a fragmented QITE algorithm, for sampling from the Gibbs distribution.

Variational ansätze based on multi-scale entanglement renormalization [44] and product spectrum ansatz [33] have also been proposed for preparing Gibbs states. Other VQA approaches rely on variational quantum simulation [23, 55] and variational autoregressive networks [31]. The hybrid quantum–classical algorithms proposed in Ref. [32] differ from VQA, in that they compute microcanonical and canonical properties of many-body systems, using: filtering operators, similar to quantum phase estimation; and Monte Carlo simulations. Ref. [15] employs Green's functions to sample from the Gibbs distribution. Finally, Ref. [24] used quantum-assisted simulation to prepare thermal states, which contrary to VQAs, does not require a hybrid quantum–classical framework.

2.1 Quantum Boltzmann Machines

One pertinent application of Gibbs states is in Gibbs states is in quantum Boltzmann machines (QBMs) [3], which are a type of quantum machine learning model that are based on the principles of classical Boltzmann machines (CBMs) [1], but incorporate quantum effects such as entanglement and superposition. CBMs are variants of neural networks embedded in an undirected graph, where the weights and biases of the network represent the information encoded within it. Typically, the network consists of two groups of nodes: the visible nodes, that determine the input and output of the network; and the hidden nodes; which act as latent variables [66]. The purpose of training a Boltzmann machine is to learn the sets of weights and biases such that the resulting network is capable of approximating a target probability distribution, whereby the model learns by feeding in training data. The Boltzmann machine can then be used for both discriminative and generative learning [3].

A Boltzmann machine is represented by an undirected graph, where V is the set of vertices (nodes) and E is the set of edges. The states of the CBM are defined by $Z = \{U, H\}$, where U represents the state of the visible nodes, and H represents the state of the hidden nodes. This defines an energy function

$$E(Z) = -\sum_{i \in V} b_i z_i - \sum_{\{i,j\} \in E} w_{ij} z_i z_j, \qquad (2)$$

with $w_{ij} \in \boldsymbol{w}, b_i \in \boldsymbol{b}$, denoting the weights and biases (parameters) of the model, respectively. z_i is a binary unit, and to remain consistent with quantum mechanics, we define $z_i \in \{-1, 1\}$. Thus, the probability to observe a particular configuration U of visible nodes is defined as

$$p_U = \frac{1}{\mathcal{Z}(\beta, Z)} \sum_H e^{-\beta E(Z)}, \quad \mathcal{Z}(\beta, Z) = \sum_Z e^{-\beta E(Z)}. \qquad (3)$$

The Boltzmann distribution summed over the hidden variables in Eq. (3) is called the marginal distribution. The goal of a CBM is to determine the weights and biases of (2), such that p_U approximates well p_U^{data} defined by the training data. Typically, the cost function employed in the optimization procedure of a CBM is the Kullback-Leibler divergence, defined as

$$\mathcal{S}(p_U^{\text{data}} \| p_U) = \sum_U p_U^{\text{data}} \log p_U^{\text{data}} - \sum_U p_U^{\text{data}} \log p_U. \qquad (4)$$

The main difference between a CBM and a QBM, is that the latter employs nodes that are defined by Pauli matrices rather than binary units, and therefore constructs a network defined by a parameterized Hamiltonian:

$$\mathcal{H}(\boldsymbol{b}, \boldsymbol{w}) = -\sum_{i \in V} b_i \sigma_i^z - \sum_{\{i,j\} \in E} w_{ij} \sigma_i^z \sigma_j^z. \qquad (5)$$

In this case, the QBM is defined as the Gibbs state

$$\rho(\beta, \mathcal{H}) = \frac{e^{-\beta \mathcal{H}(\boldsymbol{b}, \boldsymbol{w})}}{\mathcal{Z}(\beta, \mathcal{H})}, \quad \mathcal{Z}(\beta, \mathcal{H}) = \text{Tr}\{e^{-\beta \mathcal{H}(\boldsymbol{b}, \boldsymbol{w})}\}. \qquad (6)$$

The cost function is thus the generalized Kullback-Leibler divergence for density matrices, which corresponds to the quantum relative entropy:

$$\mathcal{S}(\sigma\|\rho(\beta,\mathcal{H})) = \text{Tr}\{\sigma \ln \sigma\} - \text{Tr}\{\sigma \ln \rho(\beta,\mathcal{H})\}, \tag{7}$$

where σ is the target density matrix representing the data embedded into a mixed state. The goal of QBM training is to find a set of weights and biases of the Hamiltonian, such that the Gibbs state approximates the target density matrix. There are various methods of tackling the training of QBMs [18,21,25,26,66], and an analogous problem to QBM training is Hamiltonian learning [4].

3 Variational Gibbs State Preparation

In this section we will describe how one can prepare a Gibbs state using a VQA with the free energy as a cost function, specifically discussing the algorithm presented in Ref. [16]. Sections 3.1 and 3.2 are intended to serve as a qualitative description of the estimation of the von Neumann entropy and the free energy, respectively. Consequently, in Sect. 3.3, a quantitative description of the framework of the VQA is carried out, mathematically linking the von Neumann entropy and energy expectation, with the estimation of the free energy.

Suppose one chooses a Hamiltonian $\mathcal{H}$ and inverse temperature β. For a general state ρ, one can define a generalized Helmholtz free energy as

$$\mathcal{F}(\rho) = \text{Tr}\{\mathcal{H}\rho\} - \beta^{-1}\mathcal{S}(\rho), \tag{8}$$

where the von Neumann entropy $\mathcal{S}(\rho)$ can be expressed in terms of the eigenvalues, p_i, of ρ,

$$\mathcal{S}(\rho) = -\sum_{i=0}^{d-1} p_i \ln p_i. \tag{9}$$

Since the Gibbs state is the unique state that minimizes the free energy, a variational form can be put forward that takes Eq. (8) as an objective function, such that

$$\rho(\beta,\mathcal{H}) = \arg\min_{\rho} \mathcal{F}(\rho). \tag{10}$$

In this case, $p_i = \exp(-\beta E_i)/\mathcal{Z}(\beta,\mathcal{H})$ is the probability of getting the eigenstate $|E_i\rangle$ from the ensemble $\rho(\beta,\mathcal{H})$.

3.1 Computing the von Neumann Entropy

The difficulty in measuring the von Neumann entropy, defined by Eq. (9), of a quantum state on a NISQ device is typically the challenging part of variational Gibbs state preparation algorithms, since $\mathcal{S}(\rho)$ is not an observable. To exactly compute the von Neumann entropy on a quantum device, one would have to perform full state tomography on the Gibbs state. As a consequence, different works have employed various techniques to compute an approximation for the von Neumann entropy, such as using: a Fourier series approximation to the von Neumann entropy [13]; a thermal multi-scale entanglement renormalization ansatz [44]; sums of Rényi entropies [19]; among others. On the other hand, Ref. [16] devised a method of computing the von Neumann exactly, via post-processing of measurement results acting on ancillary register.

When preparing an n-qubit state, the unitary gates used in quantum computers ensure that the final quantum state of the entire register, starting from the input state $|0\rangle^{\otimes n}$, remains pure. As a result, in order to prepare an n-qubit Gibbs state on the system register, an $m \leq n$-qubit ancillary register is required. For example, in the case of the infinite-temperature Gibbs state, which is a fully mixed state, $m = n$ qubits are needed in the ancillary register to achieve maximal von Neumann entropy. To accurately evaluate the von Neumann entropy without any approximations, the complete Boltzmann distribution is prepared on the ancillary register, thus, $m = n$ is set regardless of the temperature.

3.2 Computing the Free Energy

Following the prescription of Ref. [16], we shall denote the ancillary register as A, while the preparation of the Gibbs state will be carried out on the system register S. The purpose of the VQA is to effectively create the Boltzmann distribution on A, which is then imposed on S, via intermediary CNOT gates, to generate a diagonal mixed state in the computational basis. In the ancillary register we can choose a unitary ansatz capable of preparing such a probability distribution. Thus, the ancillary qubits are responsible for classically mixing in the probabilities of the thermal state, while also being able to access these probabilities via measurements in the computational basis. On the other hand, the system register will host the preparation of the Gibbs state, as well as the measurement of the expectation value of our desired Hamiltonian during the optimization routine.

The specific design of the PQC allows classical post-processing of simple measurement results, carried out on ancillary qubits in the computational basis, to determine the von Neumann entropy. A diagrammatic representation of the structure of the PQC is shown in Fig. 1. Note that while the PQC of the algorithm has to have a particular structure—a unitary acting on the ancillae and a unitary acting on the system, connected by intermediary CNOT gates—it is not dependent on the choice of Hamiltonian $\mathcal{H}$, inverse temperature β, or the variational ansätze, U_A and U_S, employed within. This is in addition to enjoying a sub-exponential scaling in the number of shots needed to precisely compute the Boltzmann probabilities.

3.3 Modular Structure of the PQC

The PQC, as shown in Fig. 1 for the VQA, is composed of a unitary gate U_A acting on the ancillary qubits, and a unitary gate U_S acting on the system qubits, with CNOT gates in between. Note that the circuit notation we are using here means that there are n qubits for both the system and the ancillae, as well as n CNOT gates that act in parallel, $\text{CNOT}_{AS} \equiv \bigotimes_{i=0}^{n-1} \text{CNOT}_{A_i S_i}$. The parametrized unitary U_A acting on the ancillae, followed by CNOT gates between the ancillary and system qubits, is responsible for preparing a probability distribution on the system. The parametrized unitary U_S is then applied on the system qubits to transform the computational basis states into the eigenstates of the Hamiltonian.

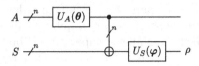

Fig. 1. PQC for Gibbs state preparation, with systems A and S each carrying n qubits. CNOT gates act between each qubit A_i and corresponding S_i.

Throughout the paper we will describe unitary operations and density matrices in the computational basis: $|0\rangle \equiv (1\ 0)^\top$ and $|1\rangle \equiv (0\ 1)^\top$, which is spanned by one qubit (and products of the basis states in the case of multiple qubits). We will also denote the n-fold tensor product of a state $|\psi\rangle$, of an n-qubit register R, as $|\psi\rangle_R^{\otimes n} \equiv \bigotimes_{i=0}^{n-1} |\psi\rangle_i$, where $i \in R$. Denote a general unitary gate of dimension $d = 2^n$, as $U_A = (u_{i,j})_{0 \leq i,j \leq d-1}$. Starting with the initial state of the $2n$-qubit register, $|0\rangle_{AS}^{\otimes 2n}$, we apply the unitary gate U_A on the ancillae to get a quantum state $|\psi\rangle_A$, such that $(U_A \otimes I_S)|0\rangle_{AS}^{\otimes 2n} = |\psi\rangle_A \otimes |0\rangle_S^{\otimes n}$, where $|\psi\rangle_A = \sum_{i=0}^{d-1} u_{i,0} |i\rangle_A$ and I_S is the identity acting on the system. Since U_A is applied to the all-zero state in the ancillary register, then the operation serves to extract the first column of the unitary operator U_A. The next step is to prepare a classical probability mixture on the system qubits, which can be done by applying CNOT gates between each ancilla and system qubit. This results in a state

$$\text{CNOT}_{AS}\left(|\psi\rangle_A \otimes |0\rangle_S^{\otimes n}\right) = \sum_{i=0}^{d-1} u_{i,0} |i\rangle_A \otimes |i\rangle_S. \qquad (11)$$

By then tracing out the ancillary qubits, we arrive at

$$\text{Tr}_A \left\{ \left(\sum_{i=0}^{d-1} u_{i,0} |i\rangle_A \otimes |i\rangle_S\right) \left(\sum_{j=0}^{d-1} u_{j,0}^* \langle j|_A \otimes \langle j|_S\right) \right\} = \sum_{i=0}^{d-1} |u_{i,0}|^2 |i\rangle\langle i|_S, \qquad (12)$$

ending up with a diagonal mixed state on the system, with probabilities given directly by the absolute square of the entries of the first column of U_A, that is, $p_i = |u_{i,0}|^2$. If the system qubits were traced out instead, we would end up with the same diagonal mixed state,

$$\text{Tr}_S \left\{ \left(\sum_{i=0}^{d-1} u_{i,0} |i\rangle_A \otimes |i\rangle_S\right) \left(\sum_{j=0}^{d-1} u_{j,0}^* \langle j|_A \otimes \langle j|_S\right) \right\} = \sum_{i=0}^{d-1} |u_{i,0}|^2 |i\rangle\langle i|_A. \qquad (13)$$

This implies that by measuring in the computational basis of the ancillary qubits, we can determine the probabilities p_i, which can then be post-processed to determine the von Neumann entropy $\mathcal{S}$ of the state ρ via Eq. (9) (since the entropy of A is the same as that of S). As a result, since U_A only serves to create a probability distribution from the entries of the first column, we can do away with a parametrized orthogonal (real unitary) operator, thus requiring less gates and parameters for the ancillary ansatz.

The unitary gate U_S then serves to transform the computational basis states of the system qubits to the eigenstates of the Gibbs state, once it is optimized by the VQA, such that

$$\rho = U_S \left(\sum_{i=0}^{d-1} |u_{i,0}|^2 |i\rangle\langle i|_S \right) U_S^\dagger = \sum_{i=0}^{d-1} p_i |\psi_i\rangle\langle \psi_i|, \quad (14)$$

where the expectation value $\mathrm{Tr}\mathcal{H}\rho$ of the Hamiltonian can be measured. Ideally, at the end of the optimization procedure, $p_i = \exp\left(-\beta E_i\right)/\mathcal{Z}(\beta, \mathcal{H})$ and $|\psi_i\rangle = |E_i\rangle$, so that we get

$$\rho(\beta, \mathcal{H}) = \sum_{i=0}^{d-1} \frac{e^{-\beta E_i}}{\mathcal{Z}(\beta, \mathcal{H})} |E_i\rangle\langle E_i|. \quad (15)$$

The VQA therefore avoids the entire difficulty of measuring the von Neumann entropy of a mixed state on a quantum computer, and instead transfers the task of post-processing measurement results to the classical computer, which is much more tractable. Finally, we define the objective function of our VQA to minimize the free energy (8), via our constructed PQC, to obtain the Gibbs state

$$\rho(\beta, \mathcal{H}) = \arg\min_{\boldsymbol{\theta},\boldsymbol{\varphi}} \mathcal{F}\left(\rho\left(\boldsymbol{\theta},\boldsymbol{\varphi}\right)\right) = \arg\min_{\boldsymbol{\theta},\boldsymbol{\varphi}} \left(\mathrm{Tr}\{\mathcal{H}\rho_S(\boldsymbol{\theta},\boldsymbol{\varphi})\} - \beta^{-1}\mathcal{S}\left(\rho_A(\boldsymbol{\theta})\right) \right). \quad (16)$$

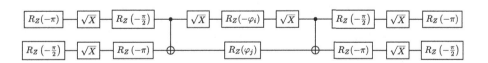

Fig. 2. Decomposed R_P gate in Eq. (18).

4 Preparing Gibbs States of the *XY* model

In this Section we assess the performance of the VQA for Gibbs state preparation of an XY model. The XY model [30] is defined as

$$\mathcal{H} = -\sum_{i=1}^{n} \left(\frac{1+\gamma}{2} \sigma_i^x \sigma_{i+1}^x + \frac{1-\gamma}{2} \sigma_i^y \sigma_{i+1}^y \right) - h \sum_{i=1}^{n} \sigma_i^z. \quad (17)$$

Here we only report one relevant property for implementing a problem-inspired ansatz for U_S. The Hamiltonian in Eq. (17) commutes with the parity operator $\mathcal{P} = \prod_{i=0}^{n-1} \sigma_i^z$. As a consequence, the eigenstates of $\mathcal{H}$ have definite parity, and so will the eigenstates of ρ_β.

To assess the performance of the VQA, we utilize the Uhlmann-Josza fidelity as a figure of merit [54], defined as $F(\rho, \sigma) = \left(\text{Tr}\{ \sqrt{\sqrt{\rho} \sigma \sqrt{\rho}} \} \right)^2$. This fidelity measure quantifies the "closeness" between the prepared state and the target Gibbs state, making it a commonly-employed metric for distinguishability. We use an alternating, entangling brick-wall PQC for the unitary U_A, composed of parametrized $R_Y(\theta_i)$ gates, and CNOTs as the entangling gates. This ansatz is hardware efficient and is sufficient to produce real amplitudes for preparing the probability distribution. Note that we require the use of entangling gates [16], as otherwise we will not be able to prepare any arbitrary probability distribution, including the Boltzmann distribution of the XY model.

For the unitary U_S, We employ a brick-wall structure solely using parity-preserving gates, denoted as

$$R_P(\varphi_i, \varphi_j) = \begin{pmatrix} \cos\left(\frac{\varphi_i+\varphi_j}{2}\right) & 0 & 0 & \sin\left(\frac{\varphi_i+\varphi_j}{2}\right) \\ 0 & \cos\left(\frac{\varphi_i-\varphi_j}{2}\right) & -\sin\left(\frac{\varphi_i-\varphi_j}{2}\right) & 0 \\ 0 & \sin\left(\frac{\varphi_i-\varphi_j}{2}\right) & \cos\left(\frac{\varphi_i-\varphi_j}{2}\right) & 0 \\ -\sin\left(\frac{\varphi_i+\varphi_j}{2}\right) & 0 & 0 & \cos\left(\frac{\varphi_i+\varphi_j}{2}\right) \end{pmatrix}. \quad (18)$$

The decomposed unitary is shown in Fig. 2, while Fig. 3 shows an example of a four-qubit PQC. Table 1 shows the scaling of the VQA assuming a closed ladder connectivity, for $n > 2$.

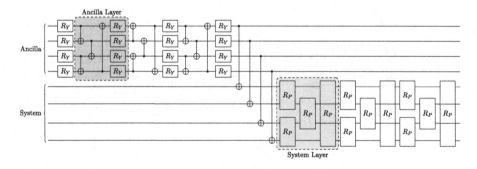

Fig. 3. Example of an eight-qubit PQC, consisting of three ancilla layers acting on a four-qubit register, and three system layers acting on another four-qubit register. Each R_Y gate is parametrized with one parameter θ_i, while each R_P gate has two parameters φ_i and φ_j. The R_P gate is defined in Eq. (18) with its decomposition shown in Fig. 2.

Variational Quantum Algorithms for Gibbs State Preparation 65

Table 1. Scaling of the VQA assuming a closed ladder connectivity, for $n > 2$, where l_A and l_S are the number of ancilla ansatz and system ansatz layers, respectively, and P is 2 when n is even and 3 when n is odd.

# of parameters	$n(l_A + 1) + 2nl_S$	$\mathcal{O}(n(l_A + l_S))$
# of CNOT gates	$nl_A + 2nl_S + n$	$\mathcal{O}(n(l_A + l_S))$
# of $\sqrt{X}$ gates	$2n(l_A + 1) + 6nl_S$	$\mathcal{O}(n(l_A + l_S))$
Circuit depth	$Pl_A + 2Pl_S + 1$	$\mathcal{O}(l_A + l_S)$

4.1 Statevector Results

Figure 4 shows the fidelity of the generated mixed state when compared with the exact Gibbs state of the XY model with $h = 0.5$, and $\gamma = 0.1, 0.5, 0.9$, across a range of temperatures for system sizes between two to seven qubits. The VQA was carried out using statevector simulations with the BFGS optimizer. We used $n - 1$ layers for both the ancilla ansatz and for the system ansatz. The number of layers was heuristically chosen to satisfy, at most, a polynomial scaling in quantum resources (but linear in depth), while achieving a fidelity higher than 99%. Furthermore, in order to alleviate the issue of getting stuck in local minima, ten random initial positions were chosen, each carrying out a local optimization procedure—which we call a 'run'—and finally taking the global minimum to be the minimum over all runs.

A total of ten runs per β were carried out to verify the reachability of the PQC, with Fig. 4 showcasing the maximal fidelity achieved for each β out of all runs. The results show that, indeed, our VQA, is able to reach a very high fidelity $F > 99\%$ for up to seven-qubit Gibbs states of the XY model. In the case of the extremal points, that is $\beta \to 0$ and $\beta \to \infty$, the fidelity reaches unity, for all investigated system sizes.

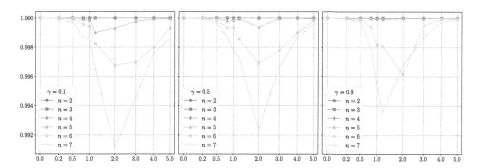

Fig. 4. Fidelity F, of the obtained state via statevector simulations with the exact Gibbs state, vs inverse temperature β, for two to seven qubits of the XY model with $\gamma = 0.1, 0.5, 0.9$, and $h = 0.5$. A total of ten runs are made for each point, with the optimal state taken to be the one that maximizes the fidelity.

5 Conclusion

In this manuscript, we provided a concise overview of various Gibbs state preparation and Gibbs-sampling algorithms. However, our primary focus was on investigating VQAs for Gibbs state preparation. To accomplish this, we conducted a benchmark study using one of the latest variational Gibbs state preparation algorithms on the XY model, considering a wide range of temperatures and γ coefficients. Following the approach proposed by Consiglio et al. [16], we leveraged the unique property of the Gibbs state as the state that minimizes the Helmholtz free energy, which serves as an appropriate objective function for the VQA. Through extensive statevector simulations, we achieved fidelities of $F > 99\%$ for system sizes up to seven qubits. Furthermore, we performed a qualitative analysis of the scalability of the VQA for implementation on a NISQ device. We found that the number of gates, iterations, and shots scale at most polynomially with the number of qubits in the XY model. The algorithm's scalability not only makes it suitable for near-term applications, but it holds significant potential for advancing quantum thermodynamic experiments on quantum computers, and ensuring the faithful preparation of Gibbs states for various computational tasks. The code for running the simulations can be found on GitHub [17].

Acknowledgments. MC would like to thank Jacopo Settino, Andrea Giordano, Carlo Mastroianni, Francesco Plastina, Salvatore Lorenzo, Sabrina Maniscalco, John Goold, and Tony J. G. Apollaro. MC acknowledges funding by TESS (Tertiary Education Scholarships Scheme), and project QVAQT (Quantum Variational Algorithms for Quantum Technologies) REP-2022-003 financed by the Malta Council for Science & Technology, for and on behalf of the Foundation for Science and Technology, through the FUSION: R&I Research Excellence Programme.

References

1. Ackley, D.H., Hinton, G.E., Sejnowski, T.J.: A learning algorithm for boltzmann machines. Cogn. Sci. **9**(1), 147–169 (1985). https://doi.org/10.1016/s0364-0213(85)80012-4
2. Aharonov, D., Arad, I., Vidick, T.: Guest column: the quantum PCP conjecture. SIGACT News **44**(2), 47–79 (2013). https://doi.org/10.1145/2491533.2491549
3. Amin, M.H., Andriyash, E., Rolfe, J., Kulchytskyy, B., Melko, R.: Quantum Boltzmann machine. Phys. Rev. X **8**, 021050 (2018). https://doi.org/10.1103/PhysRevX.8.021050
4. Anshu, A., Arunachalam, S., Kuwahara, T., Soleimanifar, M.: Sample-efficient learning of interacting quantum systems. Nat. Phys. **17**(8), 931–935 (2021). https://doi.org/10.1038/s41567-021-01232-0
5. Bardet, I., Capel, A., Gao, L., Lucia, A., Pérez-García, D., Rouzé, C.: Rapid thermalization of spin chain commuting hamiltonians. Phys. Rev. Lett. **130**, 060401 (2023). https://doi.org/10.1103/PhysRevLett.130.060401
6. Biamonte, J., Wittek, P., Pancotti, N., Rebentrost, P., Wiebe, N., Lloyd, S.: Quantum machine learning. Nature **549**(7671), 195–202 (2017). https://doi.org/10.1038/nature23474

7. Bilgin, E., Boixo, S.: Preparing thermal states of quantum systems by dimension reduction. Phys. Rev. Lett. **105**(17), 1–4 (2010). https://doi.org/10.1103/PhysRevLett.105.170405
8. Brandão, F.G.S.L., Kastoryano, M.J.: Finite correlation length implies efficient preparation of quantum thermal states. Commun. Math. Phys. **365**(1), 1–16 (2019). https://doi.org/10.1007/s00220-018-3150-8
9. Brandão, F.G.S.L., Svore, K.: Quantum speed-ups for semidefinite programming (2016). https://doi.org/10.48550/arxiv.1609.05537
10. Chen, C.F., Kastoryano, M.J., Ao, F.G.S.L.B., Gilyén, A.: Quantum thermal state preparation (2023). https://doi.org/10.48550/arXiv.2303.18224
11. Chiang, C.F., Wocjan, P.: quantum algorithm for preparing thermal Gibbs states - detailed analysis (2010). https://doi.org/10.48550/arXiv.1001.1130
12. Childs, A.M., Maslov, D., Nam, Y., Ross, N.J., Su, Y.: Toward the first quantum simulation with quantum speedup. Proc. Natl. Acad. Sci. **115**(38), 9456–9461 (2018). https://doi.org/10.1073/pnas.1801723115
13. Chowdhury, A.N., Low, G.H., Wiebe, N.: A variational quantum algorithm for preparing quantum Gibbs states (2020). https://doi.org/10.48550/arXiv.2002.00055
14. Chowdhury, A.N., Somma, R.D.: Quantum algorithms for gibbs sampling and hitting-time estimation. Quant. Inf. Comput. **17**(1-2), 41–64 (2017). https://doi.org/10.26421/qic17.1-2-3
15. Cohn, J., Yang, F., Najafi, K., Jones, B., Freericks, J.K.: Minimal effective Gibbs ansatz: a simple protocol for extracting an accurate thermal representation for quantum simulation. Phys. Rev. A **102**, 022622 (2020). https://doi.org/10.1103/PhysRevA.102.022622
16. Consiglio, M., et al.: Variational Gibbs state preparation on NISQ devices (2023). https://doi.org/10.48550/arXiv.2303.11276
17. Consiglio, M.: Variational Gibbs State Preparation (2023). https://github.com/mirkoconsiglio/VariationalGibbsStatePreparation
18. Coopmans, L., Kikuchi, Y., Benedetti, M.: Predicting gibbs-state expectation values with pure thermal shadows. PRX Quantum **4**, 010305 (2023). https://doi.org/10.1103/PRXQuantum.4.010305
19. Foldager, J., Pesah, A., Hansen, L.K.: Noise-assisted variational quantum thermalization. Sci. Rep. **12**(1), 3862 (2022). https://doi.org/10.1038/s41598-022-07296-z
20. França, D.S.: Perfect sampling for quantum Gibbs states. Quant. Inf. Comput. **18**(5-6), 361–388 (2018). https://doi.org/10.26421/qic18.5-6-1
21. Gacon, J., Zoufal, C., Carleo, G., Woerner, S.: Simultaneous perturbation stochastic approximation of the quantum fisher information. Quantum **5**, 567 (2021). https://doi.org/10.22331/q-2021-10-20-567
22. Getelina, J.C., Gomes, N., Iadecola, T., Orth, P.P., Yao, Y.X.: Adaptive variational quantum minimally entangled typical thermal states for finite temperature simulations (2023). https://doi.org/10.48550/arXiv.2301.02592
23. Guo, X.Y., et al.: Variational quantum simulation of thermal statistical states on a superconducting quantum processer. Chin. Phys. B **32**(1), 010307 (2023). https://doi.org/10.1088/1674-1056/aca7f3
24. Haug, T., Bharti, K.: Generalized quantum assisted simulator. Quant. Sci. Technol. **7**(4), 045019 (2022). https://doi.org/10.1088/2058-9565/ac83e7
25. Huijgen, O., Coopmans, L., Najafi, P., Benedetti, M., Kappen, H.J.: Training quantum Boltzmann machines with the β-variational quantum eigensolver (2023). https://doi.org/10.48550/arXiv.2304.08631

26. Kālis, M., Locāns, A., Šikovs, R., Naseri, H., Ambainis, A.: A hybrid quantum-classical approach for inference on restricted Boltzmann machines (2023). https://doi.org/10.48550/arXiv.2304.12418
27. Kastoryano, M.J., Ao, F.G.S.L.B.: Quantum Gibbs Samplers: the commuting case (2016). https://doi.org/10.48550/arXiv.1409.3435
28. Kieferová, M., Wiebe, N.: Tomography and generative training with quantum Boltzmann machines. Phys. Rev. A **96**, 062327 (2017). https://doi.org/10.1103/PhysRevA.96.062327
29. Lee, C.K., Zhang, S.X., Hsieh, C.Y., Zhang, S., Shi, L.: Variational quantum simulations of finite-temperature dynamical properties via thermofield dynamics (2022). https://doi.org/10.48550/arXiv.2206.05571
30. Lieb, E., Schultz, T., Mattis, D.: Two soluble models of an antiferromagnetic chain. Ann. Phys. **16**(3), 407–466 (1961). https://doi.org/10.1016/0003-4916(61)90115-4
31. Liu, J.G., Mao, L., Zhang, P., Wang, L.: Solving quantum statistical mechanics with variational autoregressive networks and quantum circuits. Mach. Learn. Sci. Technol. **2**(2), 025011 (2021). https://doi.org/10.1088/2632-2153/aba19d
32. Lu, S., Bañuls, M.C., Cirac, J.I.: Algorithms for quantum simulation at finite energies. PRX Quant. **2**, 020321 (2021). https://doi.org/10.1103/PRXQuantum.2.020321
33. Martyn, J., Swingle, B.: Product spectrum ansatz and the simplicity of thermal states. Phys. Rev. A **100**(3), 1–10 (2019). https://doi.org/10.1103/PhysRevA.100.032107
34. McArdle, S., Jones, T., Endo, S., Li, Y., Benjamin, S.C., Yuan, X.: Variational ansatz-based quantum simulation of imaginary time evolution. npj Quant. Inf. **5**(1), 75 (2019). https://doi.org/10.1038/s41534-019-0187-2
35. Metcalf, M., Moussa, J.E., de Jong, W.A., Sarovar, M.: Engineered thermalization and cooling of quantum many-body systems. Phys. Rev. Res. **2**, 023214 (2020). https://doi.org/10.1103/PhysRevResearch.2.023214
36. Motta, M., et al.: Determining eigenstates and thermal states on a quantum computer using quantum imaginary time evolution. Nat. Phys. **16**(2), 205–210 (2020). https://doi.org/10.1038/s41567-019-0704-4
37. Ozols, M., Roetteler, M., Roland, J.: Quantum rejection sampling. In: Proceedings of the 3rd Innovations in Theoretical Computer Science Conference, Itcs '12, pp. 290-308. Association for Computing Machinery, New York (2012). https://doi.org/10.1145/2090236.2090261
38. Poulin, D., Wocjan, P.: Sampling from the thermal quantum Gibbs state and evaluating partition functions with a quantum computer. Phys. Rev. Lett. **103**(22), 1–4 (2009). https://doi.org/10.1103/PhysRevLett.103.220502
39. Premaratne, S.P., Matsuura, A.Y.: Engineering a cost function for real-world implementation of a variational quantum algorithm. In: 2020 IEEE International Conference on Quantum Computing and Engineering (QCE), pp. 278–285 (2020). https://doi.org/10.1109/qce49297.2020.00042
40. Propp, J.G., Wilson, D.B.: Exact sampling with coupled Markov chains and applications to statistical mechanics. Rand. Struct. Algor. **9**(1–2), 223–252 (1996). https://doi.org/10.1002/(sici)1098-2418(199608/09)9:1/2<223::aid-rsa14>3.0.co;2-o
41. Rall, P., Wang, C., Wocjan, P.: Thermal State Preparation via Rounding Promises (2022). https://doi.org/10.48550/arXiv.2210.01670
42. Riera, A., Gogolin, C., Eisert, J.: Thermalization in nature and on a quantum computer. Phys. Rev. Lett. **108**(8) (2012). https://doi.org/10.1103/PhysRevLett.108.080402

43. Sagastizabal, R., et al.: Variational preparation of finite-temperature states on a quantum computer. npj Quant. Inf. **7**(1), 1–7 (2021). https://doi.org/10.1038/s41534-021-00468-1
44. Sewell, T.J., White, C.D., Swingle, B.: Thermal multi-scale entanglement renormalization ansatz for variational gibbs state preparation (2022). https://doi.org/10.48550/arXiv.2210.16419
45. Shabani, A., Neven, H.: Artificial quantum thermal bath: engineering temperature for a many-body quantum system. Phys. Rev. A **94**, 052301 (2016). https://doi.org/10.1103/PhysRevA.94.052301
46. Shor, P.: Algorithms for quantum computation: discrete logarithms and factoring. In: Proceedings 35th Annual Symposium on Foundations of Computer Science, pp. 124–134 (1994). https://doi.org/10.1109/sfcs.1994.365700
47. Shtanko, O., Movassagh, R.: Algorithms for Gibbs state preparation on noiseless and noisy random quantum circuits (2021). https://doi.org/10.48550/arXiv.2112.14688
48. Silva, T.L., Taddei, M.M., Carrazza, S., Aolita, L.: Fragmented imaginary-time evolution for early-stage quantum signal processors (2022). https://doi.org/10.48550/arXiv.2110.13180
49. Somma, R.D., Boixo, S., Barnum, H., Knill, E.: Quantum simulations of classical annealing processes. Phys. Rev. Lett. **101**(13), 1–4 (2008). https://doi.org/10.1103/PhysRevLett.101.130504
50. Sun, S.N., Motta, M., Tazhigulov, R.N., Tan, A.T., Chan, G.K.L., Minnich, A.J.: Quantum computation of finite-temperature static and dynamical properties of spin systems using quantum imaginary time evolution. PRX Quant. **2**, 010317 (2021). https://doi.org/10.1103/PRXQuantum.2.010317
51. Tan, K.C.: Fast quantum imaginary time evolution (2020). https://doi.org/10.48550/arXiv.2009.1223
52. Temme, K., Osborne, T.J., Vollbrecht, K.G., Poulin, D., Verstraete, F.: Quantum Metropolis sampling. Nature **471**(7336), 87–90 (2011). https://doi.org/10.1038/nature09770
53. Terhal, B.M., DiVincenzo, D.P.: Problem of equilibration and the computation of correlation functions on a quantum computer. Phys. Rev. A **61**, 022301 (2000). https://doi.org/10.1103/PhysRevA.61.022301
54. Uhlmann, A.: Transition probability (fidelity) and its relatives. Found. Phys. **41**(3), 288–298 (2011). https://doi.org/10.1007/s10701-009-9381-y
55. Verdon, G., Marks, J., Nanda, S., Leichenauer, S., Hidary, J.: Quantum Hamiltonian-based models and the variational quantum thermalizer algorithm (2019). https://doi.org/10.48550/arXiv.1910.02071
56. Wang, X., Feng, X., Hartung, T., Jansen, K., Stornati, P.: Critical behavior of Ising model by preparing thermal state on quantum computer (2023). https://doi.org/10.48550/arXiv.2302.14279
57. Wang, Y., Li, G., Wang, X.: Variational quantum gibbs state preparation with a truncated taylor series. Phys. Rev. Appl. **16**(5), 1 (2021). https://doi.org/10.1103/PhysRevApplied.16.054035
58. Warren, A., Zhu, L., Mayhall, N.J., Barnes, E., Economou, S.E.: Adaptive variational algorithms for quantum Gibbs state preparation (2022). https://doi.org/10.48550/arXiv.2203.12757
59. Watrous, J.: Quantum computational complexity (2008). https://doi.org/10.48550/arxiv.0804.3401
60. Wocjan, P., Temme, K.: Szegedy walk unitaries for quantum maps (2021). https://doi.org/10.48550/arXiv.2107.07365

61. Wu, J., Hsieh, T.H.: Variational thermal quantum simulation via thermofield double states. Phys. Rev. Lett. **123**(22), 1–7 (2019). https://doi.org/10.1103/PhysRevLett.123.220502
62. Yuan, X., Endo, S., Zhao, Q., Li, Y., Benjamin, S.C.: Theory of variational quantum simulation. Quantum **3**, 1–41 (2019). https://doi.org/10.22331/q-2019-10-07-191
63. Yung, M.H., Aspuru-Guzik, A.: A quantum-quantum Metropolis algorithm. Proc. Natl. Acad. Sci. U. S. A. **109**(3), 754–759 (2012). https://doi.org/10.1073/pnas.1111758109
64. Zhang, D., Bosse, J.L., Cubitt, T.: Dissipative quantum gibbs sampling (2023). https://doi.org/10.48550/arXiv.2304.04526
65. Zhu, D., et al.: Generation of thermofield double states and critical ground states with a quantum computer. Proc. Natl. Acad. Sci. U. S. A. **117**(41), 25402–25406 (2020). https://doi.org/10.1073/pnas.2006337117
66. Zoufal, C., Lucchi, A., Woerner, S.: Variational quantum Boltzmann machines. Quant. Mach. Intell. **3**(1), 1–15 (2021). https://doi.org/10.1007/s42484-020-00033-7

Towards a Parallel Code for Cellular Behavior in Vitro Prediction

Pasquale De Luca[1,2(✉)], Ardelio Galletti[2,3], and Livia Marcellino[2,3]

[1] Department of Mathematics, University Carlos III of Madrid, Leganes, Spain
pasquale.deluca@uniparthenope.it
[2] International PhD Programme / UNESCO Chair "Environment, Resources and Sustainable Development", Department of Science and Technology, Parthenope University of Naples, Centro Direzionale, Isola C4, (80143) Naples, Italy
{ardelio.galletti,livia.marcellino}@uniparthenope.it
[3] Science and Technologies Department, Parthenope University of Naples, Naples, Italy

Abstract. In recent years, there has been an increasing interest in developing in vitro models that predict the behavior of cells in living organisms. Mathematical models based on differential equations, and related numerical algorithms, have been provided to this aim. In this work, we present first experiences in designing parallel strategies for accelerating an algorithm for behavior prediction based on the Cellular Potts Model (CPM). In particular, we exploit the computational power of Graphic Process Units in CUDA environment to address main low-level kernels involved. Tests and experiments complete the paper.

Keywords: Cellular Potts Model · parallel strategies · GPGPU computing

1 Introduction

Studying cell behavior is crucial for developing treatments for various diseases, and this research spans numerous fields, such as molecular biology, genetics, biochemistry, and biophysics with applications in contexts like embryonic development, tissue regeneration, and cancer progression [8]. As a result, over the years, researchers have developed different mathematical models to characterize and simulate cellular evolution in order to enhance our understanding of biological processes. Commonly used models include agent-based models (ABMs), reaction-diffusion models (RDMs), and lattice-based models (LBMs).
  – ABMs simulate individual cells as discrete entities interacting with each other and their environment. A key advantage of ABMs is their ability to simulate individual cell behavior, providing insights into cellular process mechanisms [9]. – RDMs describe molecular diffusion and interactions within cells or tissues and are often used to study chemical signaling pathways and pattern formation in developing tissues as, for example, limb regeneration [10]. – LBMs represent

cells as connected lattice sites, with each site corresponding to a portion of the cell's membrane These models are helpful for studying cell behavior in simple environments, like cell cultures or artificial tissues, and incorporate physical parameters like cell-cell adhesion, cell-substrate adhesion, and cell deformability to simulate cellular processes such as cell migration and tissue morphogenesis [11]. Furthermore, LBMs can be computationally efficient because they require simulating, in general, only a small number of lattice sites representing individual cells interactions.

Among those models a well-known example is the Cellular Potts Model (CPM), which uses energy minimization principles depending on cell-cell interaction [12]. Several CPM extensions have been proposed to capture additional aspects of cellular behavior accurately, such as spatial heterogeneity, cell migration, and chemotaxis [13]. These extensions have enhanced the CPM's predictive capabilities and created new opportunities for exploring cellular behavior in complex biological systems. Researchers attempt to make some implementations of CPM in computer programs that are crucial to its application in research. One of the earliest implementations of the CPM was developed in C++ by Graner and Glazier in 1993 [12]. Their implementation was based on the Metropolis algorithm and included Monte Carlo simulations to calculate cell-cell interactions. However, their implementation was limited in terms of its scalability and speed, and it was not suitable for simulating large-scale biological processes. To overcome these limitations, several approaches have been proposed to optimize the implementation of the CPM. For example, Chen et al. developed a parallel implementation of the CPM using the Message Passing Interface (MPI) [14]. Their implementation utilized multiple processors to accelerate the simulation, and it was able to simulate larger systems than the original implementation by Graner and Glazier. This implementation suffers of the communication overhead between processors, which resulted in decreased performance for small-scale simulations. Hattne et al. developed a MATLAB implementation of the CPM for simulating protein crystal growth [15]. Their implementation included a graphical user interface that enabled users to interactively adjust simulation parameters and visualize the simulation results. However, the performance of their implementation was limited by MATLAB's interpretation overhead, and it was not suitable for large-scale simulations. Previous implementations, despite some ones are parallel, did not incorporate the finite element method (FEM) for improving accuracy in cell mechanics. In fact, this approach increases computational complexity of each related algorithm and parallel strategies and implementations become mandatory. However, to our knowledge, no parallel codes are present in literature to this aim. Therefore, in this paper, to fill this gap, we present a parallel GPU implementation of the CPM version which includes the cell mechanics description through FEM for evaluating the energy system.

This implementation takes advantage of the power of modern GPUs to efficiently simulate and analyze cellular behavior and interactions in a high-performance computing environment. To accomplish this, we employ parallel domain decomposition techniques, which allow for the efficient distribution of the simulation workload across multiple GPU cores. By partitioning the problem into smaller subdomains, we can harness the computational power of GPUs to achieve significant speedup and accelerate the execution of the CPM simulations. The parallel implementation of the CPM is developed using the CUDA [3] programming model, a widely adopted library for GPU computing. CUDA enables researchers to utilize the massively parallel architecture of GPUs, providing a flexible and efficient platform for accelerating scientific computations [4–7]. By using CUDA, we can optimize our implementation of the CPM, ensuring that it takes full advantage of the parallel processing capabilities of modern GPUs.

The rest of paper is organized as follows: in Sect. 2 a brief introduction of the Cellular Potts Model with related details and algorithm are given. The parallel algorithm and its details are addressed in Sect. 3. Finally, in Sect. 4 main results achieved by the parallel implementation are shown. Section 5 closes the paper with conclusions.

2 Cellular Potts Model: Main Features and Related Scheme

The Cellular Potts Model is a mathematical system that utilizes a spatial lattice-based structure to analyze the behavior of biological cell populations over time and space. This model proves useful when the details of interactions between cells are primarily dictated by individual cell size and shape, as well as the length of the contact area between adjacent cells.

Formally, the CPM is a time-discrete Markov chain in which, at each time step $\tau = 0, \ldots, N_\tau$, cells are represented as sets of pixels (or voxels in a 3D context) that form a simply-connected domain in the lattice.

In a two-dimensional square lattice $\Omega = \{1, 2, \ldots, N\}^2$, each point can be denoted by a pair of integer coordinates. By assuming that N_c different cells are in the medium, at each time step τ the complete lattice structure is represented as a index matrix $L = L^\tau$, where $L_{i,j}$ takes a integer value, the cell index, denoting the cell the pixel (i, j) belongs to: this value will be $L_{i,j} = 0$ for pixels (i, j) belonging to the medium, and $L_{i,j} = k$, for some $k \in \{1, 2, \ldots, N_c\}$, for pixels (i, j) representing the k-th cell. Hence in this model pixels belonging to same cell share the same cell index. The evolution of a CPM involves updating the configuration of cells one pixel at a time, by following a set of probabilistic rules: then, behaviour system changes over time (as τ increases) are represented through changes in L^τ values. In addition, at each pixel $P = (i, j)$ is associated a two-dimensional vector f_P which usually represents a number of biological factors, such as the direction of cell movement or the orientation of cell polarization. In mathematical terms, f_P can be seen as a vector that originates at the position (i, j) and points in the direction of the cell's movement or polarization. These dynamics can be thought of as membrane fluctuations, a process where the size of a cell decreases by one lattice site while a neighboring cell expands to occupy that site. The system transition rules, at each time step, leverage a modified version of the Metropolis algorithm [17] where the criterion involves a specific Hamiltonian function H, related to system energy. Since a probabilistic rule, based on H variation, influences the state changes, the CPM evolution is definitely a random dynamical system.

In the following, a specific CPM procedure, in which until to $M \geq 1$ pixel at time can be updated in their cell index, is described.

1. *Initialization*: this is a preliminary phase ($\tau = 0$) to set a simulated environment: each lattice site is assigned to medium or to a specific cell, that is L^0 is set.
2. *Modified metropolis algorithm*: for each time step, that is for $\tau = 1, \ldots, N_\tau$, the system tries to update one or more pixels at time. To this aim, for each $l = 1, \ldots, M$, where M represents the numbers of trials, the following procedures are repeated. More in particular, at each iteration l, the procedure:

    (a - PIXEL SELECTION) selects a random pixel P and one of its neighboring pixels, denoted as Q (Moore neighborhood). In a two-dimensional square lattice, P can be one of the eight pixels adjacent to Q.

    (b - COMPUTE ENERGY CHANGE) computes the change in the system's Hamiltonian, ΔH, if the state of pixel P were to change to the state of pixel Q. Formally, such a variation can be written as:

    $$\Delta H = H^{\text{NEW}} - H \qquad (1)$$

    where H represent the Hamiltonian for the current state of the system and H^{NEW} would be the new Hamiltonian value if update for P cell index were accepted.

    (c - METROPOLIS CRITERION) decides whether to accept or reject the change in state based on the following Metropolis criterion. If $\Delta H \leq 0$, the change is accepted; if $\Delta H > 0$, the change is accepted with a probability:

    $$p(\Delta H) = \exp\left(-\frac{\Delta H}{T}\right) \qquad (2)$$

    as well-known as Boltzmann probability, where T is a parameter analogous to temperature in physical systems.

(**d- UPDATE STATE**) if the change were accepted, updates the state of pixel P to the one of pixel Q. Otherwise, the state of pixel P remains the same.

The Hamiltonian H typically includes different terms, as:

$$H = H_{ad} + H_v + H_s. \qquad (3)$$

Here, H_{ad} refers to the whole cell adhesion defined as:

$$H_{ad} = J \cdot \sum_{(i,j) \in \Omega} \left(\sum_{(i',j') \in Nb\big((i,j)\big)} 1 - \delta(L_{i,j}, L_{i',j'}) \right), \qquad (4)$$

where $Nb\big((i,j)\big)$ is the set of neighbors of (i,j). and J is the adhesion energy between cells of different types. Last term $\delta(x,y)$ is the Kronecker delta, which equals 1 if $x = y$ and 0 otherwise, The second term H_v takes into account the cell volume, area in the 2D lattice, and it is defined as:

$$H_v = \lambda \sum_k \left(\frac{V_k - V_0}{V_0} \right)^2, \qquad (5)$$

where V_k represents the current volume of cell k, and V_0 is a target volume. The parameter λ is a value that controls the strength of the volume constraint. Last term H_s is the mechanical strain, generally defined as:

$$H_s = \mu \sum_{(i,j)} (\varepsilon(i,j) - \varepsilon_0(i,j))^2, \qquad (6)$$

where $\varepsilon(i,j)$ is the current strain at pixel (i,j), and $\varepsilon_0(i,j)$ is the target strain. The parameter μ is a value that controls the strength of the strain constraint. Accurately computing this term is crucial for capturing the influence of mechanical forces on cell behavior.

The strain ε in the system, is a measure of deformation or displacement from an equilibrium or 'rest' state. In a simple CPM, this could be calculated directly from the changes in the position or state of each cell. However, for more complex systems or systems where higher accuracy is desired, it may be beneficial to use a more sophisticated method like the Finite Element Method (FEM). It involves dividing the domain of interest (in this case, the cell or group of cells) into a set of smaller, simpler domains known as elements. The strain in each element is then computed, and these values are combined to compute the overall strain in the system. The displacements of the nodes, or points at the boundaries between elements, represent the deformation of the system.

The local stiffness matrix K_e, for each node or element e, represents the relationship between the displacements of the nodes and the forces acting on them. If we consider an uniform, linearly, elastic substrate, then K_e takes the form:

$$K_e = \int_{\Omega_e} B^T DB d\Omega_e, \qquad (7)$$

where Ω_e refers to the domain of each element e. The B matrix is the so-called strain-displacement matrix for a four-nodes quadrilateral element, while D is the material matrix for a 2D element under plane stress conditions (for more information see [18]). This matrix is a key component of the FEM, and the mechanical behavior of the system is represented by the solution of a generally large system of linear equations of the form:

$$Ku = f, \qquad (8)$$

where u is a vector of the displacements of all nodes, K is the global stiffness matrix [18]) which is assembled from the local stiffness matrices K_e and

$$f = (f_P)_{P \in \Omega} \qquad (9)$$

is a vector of the forces f_P acting on the all pixels P. More precisely, for each single pixel $P = (i,j)$, according to Lemmon and Romer's model [19], the force $f_{P,P'}$ exerted on this pixel by a neighboring pixel $P' = (i',j')$ is proportional to the difference in their states and is directed along the line joining the two cells, i.e. it is:

$$\|f_{P,P'}\|_2 = \alpha(L_{i,j} - L_{i',j'}) \cdot d_{P,P'}$$

where α is the tension per unit length, and $d_{P,P'} = \|(i,j) - (i',j')\|_2$ is the Euclidean distance between pixels P and P'. Hence, the total force f_P acting on pixel P is the vector sum of the forces exerted by all its neighboring pixels., i.e.:

$$f_P = \sum_{P' \in Nb(P)} f_{P,P'}. \qquad (10)$$

It is important to observe that K does not change over time τ making possible its assembling once only before CPM iterations. We also remark that the Metropolis criterion only concerns the Hamiltonian variation that can be decomposed as:

$$\Delta H = \Delta H_{ad} + \Delta H_v + \Delta H_s. \qquad (11)$$

It is no difficult to see that variations ΔH_{ad}, ΔH_v and ΔH_s no longer depend on the whole index matrix and cell distribution, but only on pixels involved in the change and on their neighbours. This issue is particularly useful because it makes computationally very convenient to evaluate ΔH.

Previous discussion allows us to propose the following Algorithm 1 which is the basis of the parallel implementation described in next section.

Algorithm 1 CPM Pseudo-algorithm

Input: N, N_τ, M, N_c

Output: $\{L^\tau\}_{\tau=1,\ldots,N_\tau}$
1: **STEP 0: environment setting at time** $\tau = 0$
2: Build: L^0
3: **STEP 1: FEM priming**
4: Compute: K_e % as in Eq. (7)
5: Assembly: K % once
6: **STEP 2: Cellular Potts Model**
7: **for** $\tau = 1, 2, \ldots, N_\tau$ **do** % Metropolis algorithm
8:   Compute: f_P % as in Eq. (10)
9:   Assembly: f % Force updating as in Eq. (9)
10:   Solve: $Ku = f$
11:   **for** l = 1,..., M **do** % Steps
12:     Select a random pixel on the grid as "the source" pixel P
13:     Select a random neighbor of P as the "target" pixel Q
14:     Compute: ΔH_{ad}
15:     Compute: ΔH_v
16:     Compute: ΔH_s
17:     Compute: ΔH % as in Eq. (11)
18:     try copy attempt (Boltzmann probability) % as in Eq. (2)
19:   **end for**
20: **end for**
---

3 Parallel Approach

In this section, the proposed parallel algorithm is discussed. Algorithm 1 was initially implemented in a sequential manner using a C language code, hence the parallel implementation arises from the parallelization of main steps of the CPU code. Details and some features about adopted parallel approach are addressed to perform a fast simulation.

Our implementation relies entirely on GPU execution, ensuring that all components of the algorithm, including cell initialization, lattice grid creation, and the Metropolis algorithm's execution, take full advantage of the GPU's processing power.

Notably, in order to grants us a fine-grained control over the algorithm's execution, we have chosen to manually implement all CUDA kernels, without relying on any specific libraries with the exception of the linear system (8) which is solved through the specific cuSOLVER library routine [20].

According to Algorithm 1, we parallelized the overall procedure, by using basic API routines of GPU environment. More in detail, the macro three steps work as follows:

- For the *Step 0*, related to the cells initialization and lattice grid creation, we have employed a thread-based parallelism approach. We assign groups of threads to handle distinct portions of the cells to be initialized. By associating threads with specific cell portions, we can efficiently distribute the workload across the GPU's many processing units. This approach also enables us to minimize data transfer between CPU and GPU, thereby reducing the overhead and latency associated with those transfers. Each thread is responsible for initializing a part of the cell, assigning it a unique identifier, and setting its initial state. The threads also create the lattice grid by associating each grid point with a specific cell. By using GPU threads to handle those tasks, we can significantly reduce the time it takes to initialize the simulation, especially for large-scale systems.
- About the *Step 1*, notice that, in order to build the matrix K arising from the FEM discretization for the strain, this operation has to be made once. We have developed CUDA kernels to assemble the global stiffness matrix, taking into account the contributions from each element in the mesh. This assembly process is performed in parallel, with each thread handling a specific portion of the elements. This specific parallel design allows us to obtain a very high degree of parallelism and then a remarkable speedup, in terms of execution time.
- *Step 2* is the core of the CUDA-based implementation, and it consists of a set of CUDA customized kernels to compute the required procedures for each step of the Metropolis algorithm. These kernels handle the following tasks: to update the forces vector as in (9), to solve the linear system (8) and to evaluate the energy change computation as in (11). More precisely:
  - the forces vector updating task is one of the pivotal aspects in the simulation and represents the active forces acting on each pixel and its neighboring pixels. Traditionally, the computation of this force vector has been carried out sequentially: firstly each pixel is individually analyzed, then its neighboring pixels identified, and finally the forces exerted on that specific pixel computed. While effective, this approach can be computationally expensive, especially when dealing with large-scale systems. Since these processes are independent among pixels, a suitable domain decomposition schema is modeled to parallelize computations and for saving up execution times. More specifically, our CUDA kernel creates a matrix where each row corresponds to a unique pixel. The forces are then computed in parallel across all pixels, thus eliminating the need to sequentially traverse each pixel and calculate the force. This approach harnesses the high throughput of GPU architectures to perform numerous computations simultaneously, exploiting data parallelism to a large extent. After the force values for each pixel have been computed asynchronously, they are stored in the corresponding positions of the force vector f. The asynchronous nature of this phase further increases the performance by overlapping computations and memory operations, which is a key advantage of CUDA environment;

- to solve the linear system (8), exploiting that K is strictly diagonally dominant, we have chosen as solver the Jacobi method that is well-suited for the characteristics of the problem at hand. In particular, we employed the `gesvdj` arising from API routines of cuSOLVER library and we have optimized it to achieve high performance on the GPU;
- computing the energy change is a crucial step in the Metropolis algorithm. Specifically, we have developed CUDA kernels that allow for the simultaneous computation of the $\Delta H_{ad}, \Delta H_v,$ and ΔH_s terms, exploiting a asynchronous paradigm of parallelism. The first two terms, ΔH_{ad} and ΔH_v, are straightforward in their formulation and this fact enables their fast computation. For this reason, both tasks are demanded to a single thread. On the other hand, the computation of the ΔH_s term requires a retrieving operation about the previously computed displacement u and force vectors f. This task can be more computation-intensive in a single-processor machine. Here, by using the computational power of GPU ed a parallel strategy adopting a *dynamic parallelization*, we significantly accelerate the whole process. The parallel approach consists only of two threads activation. Each thread computes asynchronously the update of ΔH_s by retrieving the displacement information from u (only one about the "source" pixel) and computing the requested operations for the strain features. At the end, one thread updates the information about ΔH_s. Despite the computational complexity involved, the proposed CUDA kernels efficiently handle the overall computation of the ΔH_s term, by leveraging parallelism and the high throughput capabilities of GPU architecture.

4 Results

In this section, some preliminary experiences to validate and assess the performance of the implemented parallel software are presented. Since the trial number required for real health-care application in general becomes significantly large, and consequently the computational complexity of the Metropolis algorithm dramatically increases, our primary objective is to demonstrate the efficiency, in terms of execution time gain, of the proposed parallel code, that is how much the GPU-parallel algorithm can reach a strong time performance improvement. To do this, in the first experimental tests we compare execution times between the CPU code and the parallel one. The computations were carried out on the high-performance machine MARCONI-100 offered by CINECA equipped with:

- dual 16-core IBM POWER9 AC922 CPUs running at 3.1 GHz;
- 4 NVIDIA Volta V100 GPUs with Nvlink 2.0 and 16 GB of memory each;
- 256 GB of RAM.

The input parameters for the problem under consideration have been configured based on a relevant literature study, thus ensuring that tests are grounded in a real-world scenario and that related results are directly comparable with [2].

We have configured the parameters to represent a range of scenarios that reflect different applications where our method might be used. Here, we report only a representative testing scheme wherein we systematically vary two critical parameters: the number of trials M, which in this case is set equal to the cells number, and the Metropolis time steps number N_τ. These parameters have been chosen as they are known to have a significant influence on the performance of both parallel and sequential algorithms.

To present a comprehensive view of our findings, Table 1 collects the execution times of both implementations under different combinations of these parameters. In all tests presented in the Table, we set:

- $N_c = 400$ different cells ad time step $\tau = 0$;
- $\Omega = \{1, 2, \ldots, 600\}^2$, that is $N = 600$;
- N_τ varies on $\{3000, 6000, 9000\}$
- M varies on $\{150, 300, 450, 750, 900\}$

The results clearly demonstrate the efficacy of the parallel approach, highlighting the potential for significant computational time reductions, particularly as the complexity of the system increases. Our parallel implementation has shown a distinct speed-up with respect to the sequential version, as it is evident from the empirical data. Let us observe that the computational cost of the problem increases significantly as both M and N_τ increase. This growth is primarily due to the workload needed to solve N_τ times the large linear system $Ku = f$, where both K and f have sizes depending on N which is large too for real simulations. These satisfying results can be explained by observing that the parallel code has the ability to distribute the computation across multiple processing units, by exhibiting a clear advantage. Thus, as the system size grows, our parallel implementation consistently outperforms the sequential approach, offering a compelling case for its adoption in larger, more complex simulations.

The Fig. 1 just illustrates how the Cellular Potts Model system evolves as the time parameter τ increases, without taking into account any consideration about time execution. Specifically, the three figures depict distinct stages of the system's progression as τ varies and for values $N = 600$, $M = 300$ and $N_c = 400$. All cells are represented in red, while medium is in white. The subfigure (a), with $\tau = 0$, shows the system in its original state, with the cellular model yet to undergo any changes. This works as a reference point from which the subsequent transformations of the system can be clearly observed.

As we progress through the illustrations, we see the effects of simulation over time on the cellular system. These simulations are carried out up until $\tau = 3000$, offering us a visual journey of the system's evolution over this span. The value of 3000 is designated as the parameter N_τ, which has been predetermined as the test value for this series.

In the ensuing figures ($\tau = 1500, 3000$), the impact of increasing time on the Cellular Potts Model is evident. The transformation of the system from its original state to a more complex arrangement reflects the ongoing simulation until $\tau = N_\tau = 3000$. Each figure, therefore, represents a snapshot of the system

Table 1. Execution times (in seconds) comparison: CPU vs. GPU. Here, for all simulations it is $N = 600$, $N_c = 400$.

		$N_\tau = 3000$	$N_\tau = 6000$	$N_\tau = 9000$
M=150	GPU	771.13	1243.55	2149.81
	CPU	1089.88	2591.12	4119.16
M=300	GPU	1002.84	1992.53	2899.16
	CPU	2300.12	4811.92	7910.15
M=450	GPU	1754.97	3486.89	5073.53
	CPU	4301.22	8998.39	14791.85
M=750	GPU	3071.97	6101.45	8877.75
	CPU	8043.28	16826.98	27660.75
M=900	GPU	454.06	9457.85	14221.84
	CPU	14799.67	31802.99	52294.97

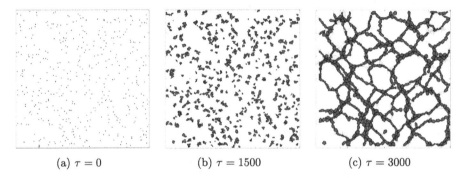

(a) $\tau = 0$ (b) $\tau = 1500$ (c) $\tau = 3000$

Fig. 1. Cellular Potts Model - Metropolis time evolution. Both the x and y axes represent the dimensions of the symmetrical lattice grid.

at a certain point in time, providing an overview of some of the changes the Cellular Potts Model undergoes as time progresses.

5 Conclusions

In this paper a novel parallel algorithm for the CPM model and its GPU implementation have been introduced. By analysing the main tasks of Metropolis approach to perform solution for the CPM model, our idea is to use GPU-based parallel computing in order to accelerate the way to approach solution at a complex cellular simulation problem. In fact, cellular behavior simulation can become a problem with high computational complexity as the number of simulations increases. Here, to overcome this, we have developed a parallel approach using the CUDA platform for large-scale and real-time cellular behavior simulation.

This approach significantly differs from the sequential one. Thanks to the GPU environment, specific strategy as domain decomposition and task decomposition (also combined with each other) ensuring faster computation times. Preliminary tests allow us to observe an evident gain of performances, in terms of execution time, using our GPU-parallel software, with respect to the sequential version and encourage next experiments.

Acknowledgment. This paper has been supported by project *"Intelligenza Artificiale e Calcolo ad Alte Prestazioni per applicazioni avanzate e big data"*, while computational resources are offered by project *"Advanced parallel numerical methods for environmental problems"* (APNE2023).

References

1. Graner, F., Glazier, J.A.: Simulation of biological cell sorting using a two-dimensional extended Potts model. Phys. Rev. Lett. **69**(13) (1992)
2. van Oers, R.F., Rens, E.G., LaValley, D.J., Reinhart-King, C.A., Merks, R.M.: Mechanical cell-matrix feedback explains pairwise and collective endothelial cell behavior in vitro. PLoS Comput. Biol. **10**(8), e1003774 (2014)
3. David, K.: NVIDIA CUDA software and GPU parallel computing architecture. In :Proceedings of the 6th international symposium on Memory management (ISMM '07). Association for Computing Machinery, New York, NY, USA, pp. 103–104 (2007)
4. Fiscale, S., et al.: A GPU algorithm for outliers detection in TESS light curves. In: International Conference on Computational Science, pp. 420–432. Cham: Springer International Publishing (2021)
5. De Luca, P., Galletti, A., Marcellino, L.: Parallel solvers comparison for an inverse problem in fractional calculus. In: 2020 Proceeding of 9th International Conference on Theory and Practice in Modern Computing (TPMC 2020), pp. 197–204 (2020)
6. Cuomo, S., Galletti, A., Marcellino, L.: A GPU algorithm in a distributed computing system for 3D MRI denoising. In: 2015 10th International Conference on P2P, Parallel, Grid, Cloud and Internet Computing (3PGCIC), pp. 557–562. IEEE (2015)
7. Cuomo, S., De Michele, P., Galletti, A., Marcellino, L.A.: GPU parallel implementation of the Local Principal Component Analysis overcomplete method for DW image denoising. Proc.-IEEE Symp. Comput. Commun. art. no. 7543709. **10**, 26–31 (2016)
8. Anderson, A.R., Quaranta, V.: Integrative mathematical oncology. Nat. Rev. Cancer **8**(3), 227–234 (2008)
9. Li, Y., Yang, X., Wu, J., Sun, H., Guo, X., Zhou, L.: Discrete-event simulations for metro train operation under emergencies: A multi-agent based model with parallel computing. Phys. A **573**, 125964 (2021)
10. Ahmed, N., Rafiq, M., Adel, W., Rezazadeh, H., Khan, I., Nisar, K.S.: Structure preserving numerical analysis of HIV and CD4+ T-cells reaction diffusion model in two space dimensions. Chaos, Solitons Fract. **139**, 110307 (2020)
11. Giverso, C., Ciarletta, P.: Tumour angiogenesis as a chemo-mechanical surface instability. Sci. Rep. **6**(1), 1–11 (2016)
12. Glazier, J.A., Graner, F.: Simulation of the differential adhesion driven rearrangement of biological cells. Phys. Rev. E **47**(3), 2128 (1993)

13. Tosin, A.N., Ambrosi, D.A., Preziosi, L.U.: Mechanics and chemotaxis in the morphogenesis of vascular networks. Bull. Math. Biol. **68**, 1819–36 (2006)
14. Chen, N., Glazier, J.A., Izaguirre, J.A., Alber, M.S.: A parallel implementation of the Cellular Potts Model for simulation of cell-based morphogenesis. Comput. Phys. Commun. **176**(11–12), 670–681 (2007)
15. Hattne, J., et al.: Analysis of global and site-specific radiation damage in cryo-EM. Structure **26**(5), 759–766 (2018)
16. Albert, P.J., Schwarz, U.S.: Dynamics of cell shape and forces on micropatterned substrates predicted by a cellular Potts model. Biophys. J. **106**(11), 2340–2352 (2014)
17. Metropolis, N., Rosenbluth, A.W., Rosenbluth, M.N., Teller, A.H., Teller, E.: Equation of state calculations by fast computing machines. J. Chem. Phys. **21**(6), 1087–1092 (1953)
18. Zienkiewicz, O.C., Taylor, R.L.: The finite element method for solid and structural mechanics. Elsevier (2005)
19. Lemmon, C.A., Romer, L.H.: A predictive model of cell traction forces based on cell geometry. Biophys. J . **99**(9), L78–L80 (2010)
20. https://docs.nvidia.com/cuda/cusolver/index.html

Combinators as Observable Presheaves: A Characterization in the *Grossone* Framework

Rocco Gangle[1], Fernando Tohmé[2], and Gianluca Caterina[1(✉)]

[1] Endicott College, Beverly, MA 01915, USA
{rgangle,gcaterin}@endicott.edu
[2] Universidad Nacional del Sur and Conicet, Bahía, Blanca, Argentina

Abstract. Combinators, as defined originally by Moses Schönfinkel, give rise to a Turing-complete model of computation. This paper presents a diagrammatic representation of combinators as presheaves defined over a category of *generic figures*. We adopt Sergeyev's *grossone* numeral system, which, together with our categorical representation of combinators, ensures a sharper characterization of non-halting combinators. As a result of our analysis, we show how certain "infinite" combinators can be recast in the grossone formalism using the notion of *observability*, which captures the general concept of tractable properties of sequences with length less than or equal to grossone.

Keywords: Combinatory Logic · Generic Figures · Category Theory · Grossone

1 Introduction

Among the Turing complete models of computation, the one based on the logic of *combinators* has recently drawn renewed attention from computer scientists, logicians, and other researchers (see [37]). Albeit being one of the first formal systems developed for the expression of computation (it was introduced in 1920 by Moses Schönfinkel in [35]), it is far less well known than related approaches, like λ-calculus. Combinators may be understood not only as elementary structures for generating arbitrarily complex logical and computational operations but also, as new approaches like that of [22] have shown, as theoretical frameworks for modeling representational and reasoning processes in general.

In this more general framework, the practical utility of using combinators to model and understand diverse types of cognitive processes is unfortunately inhibited by an unwieldy and difficult-to-read standard syntax. More diagrammatic representations of combinators (as in [37]) show promise in this regard, but their operational dynamics can be difficult to track in a rigorous and precise mathematical way.

This contribution aims to provide a more natural representation of combinatory terms in which reductions and other operations can be applied in an elegant and homogeneous way. For that, we have to define a category $\mathcal{CB}$ such that each combinatory term is a contravariant functor $F : \mathcal{CB}^{op} \to \mathbf{Set}$ and the morphisms are natural transformations between those functors.[1] It is a standard result in category theory that any category of presheaves constitutes an *elementary topos* and thus provides a rich environment for modeling practically all ordinary mathematical structures and operations.

As we recast combinators with their correspondent presheaves, combinators that are "infinite" – in a sense that will be specified later in the paper – are associated with presheaves whose base category has an infinite cardinality (in a Cantorian sense). The logical interpretation of such combinators is indeterminate (for a loose comparison, think of a divergent sequence). It would be desirable to make properties of these indeterminate combinators that are obscured by standard representations of infinity more tractable. This issue can be addressed by applying the notion of *grossone*, first defined by Yaroslav Sergeyev [31]. In Sergeyev's approach, the set of natural numbers is $\mathbb{N} = \{1, 2, 3, \ldots, ① - 2, ① - 1, ①\}$.

The basic idea is that *grossone* allows for a "finer" control on infinite denumerable sequences than the classic Cantorian approach (for details, see Sergeyev's seminal works [26,28,31,33]). Within the *grossone* framework it becomes possible to deal computationally with infinite quantities, in a way that is both new (in the sense that previously intractable problems become implementable) and natural.

A grossone category of presheaves over $\mathcal{CB}$ constitutes a sound and treatable version of combinatory logic, ensuring the interpretability of the combinators corresponding to the presheaves.

2 Combinators: an Introduction

A combinator can be thought of as a higher-order function, that is, a function whose arguments are functions. Here the notion of function as having a domain and a codomain is replaced by formal strings of parenthesized symbols which, in essence, constitute the primal form of what a function is supposed to calculate. These strings are called *combinatory terms*, which can be recursively defined. Each combinatory term is either:

– A primitive combinator.

---

[1] A mathematical *category* consists of a class of *objects* together with *morphisms* or arrows between objects subject to axioms of *identity* (every object is equipped with an identity morphism that composes inertly), *composition* (head-to-tail morphisms compose to a unique morphism) and *associativity* (paths of morphisms compose uniquely). For a comprehensive introduction and details filling out this rough characterization, see [19].

- The application of a combinator to a combinatory term.

Primitive combinators may be represented by a stock of variables (traditionally, capital letters $A, B, C \ldots$). The application of one term to another is represented by juxtaposing them from left to right. That is, AB represents the application of A to B.

From a syntactical point of view, then, according to the definition above, combinatory terms are simply finite strings of letters of a denumerable alphabet along with a given parenthesis structure:

$$A(AB), (AB)A, A(BC), (AB)(C(DD))$$

are examples of simple combinatory terms. For readability, we employ the standard convention of implicit parenthesizing from the left; thus $AB(CD)E$ is shorthand for $((AB)(CD))E$.

Of particular interest are those primitive combinators whose actions on other terms may be expressed by general rules that may be notated with equations. The equations specify in an axiom-like manner how finite sequences of successive terms are rearranged and/or deleted by the application of these combinators. Such combinators are called *proper*.[2] Such structures, from a logical standpoint, can be described using a new type of variable and equational rewriting rules. We use lowercase x, y, z, to denote variables standing for arbitrary combinatory terms and bold font capitalized letters to represent the proper combinators themselves. For instance, we specify the action of proper combinators **K** and **S** with the following rules:

- $\mathbf{K}xy = x$
- $\mathbf{S}xyz = xz(yz)$

where x, y and z stand for arbitrary combinatory terms. Again, we use the convention that we parenthesize starting from the left. For instance, $\mathbf{K}xy$ should be read as $(\mathbf{K}x)y$, i.e. that **K** is first applied on x to yield a new function $\mathbf{K}x$ which is then applied on y to result in x. Similarly, **S** is applied such that x, y and z are successive arguments, that is, $(((\mathbf{S}x)y)z)$, yielding a function xz which is applied on the term yz.

K and **S** can be used to build new combinators: for instance, consider the identity combinator **I**:

$$\mathbf{I}x = x$$

which can be obtained as **SKK**:

$$\mathbf{SKK}x = \mathbf{K}x(\mathbf{K}x) = x = \mathbf{I}x.$$

What is truly remarkable is that **K** and **S** are *functionally complete*, that is, any combinator whatsoever can be built via nested applications of **K** and

---

[2] For details, see any introductory text on combinatory logic, for instance [2]

S. Moreover, it can be shown that the terms obtained in the **K** - **S** system are equivalent to those of λ-calculus. It constitutes, therefore, a Turing complete system, so that **K** and **S** suffice to represent all Turing computable functions.

The abstract generality, elegance, and simplicity of the combinator-based model of computation come with a price. Since every single entity in the system boils down to a structured (that is, parenthesized) string of **S** and **K** symbols, calculations are typically very long and, more importantly, the actual structure of the nested operations is not clearly evident in the usual linear notation.

In the next section, we show how to address this issue by developing a new mathematical formalization of the syntax of combinators in terms of presheaves.

3 The *Generic Figures* Approach

One way of solving the interpretative shortcomings of the standard syntax of combinators is by employing the "generic figures" approach introduced by Reyes, Reyes, and Zolfaghari in [23] based on categories of presheaves. The main advantage of this approach is that it supports a visually intuitive and easily interpreted diagrammatic syntax for combinators on the one hand and yet remains in itself a precise and mathematically rigorous formalization on the other.

To keep the paper self-contained, we give now a quick description of the main categorical notions to be used in the rest of the paper. For further details see [19].

A *category* consists of a class of *objects* together with *morphisms* or arrows between objects. Given two objects a and b a morphism f between them will be denoted either $f : a \to b$ or $a \xrightarrow{f} b$. A category is subject to axioms of *identity* (every object a is equipped with an identity morphism, $a \xrightarrow{1_a} a$), *composition* (two morphisms, $a \xrightarrow{f} b$ and $b \xrightarrow{g} c$ compose to a unique morphism $a \xrightarrow{g \circ f} c$, where ∘ indicates the operation of composition) and *associativity* (paths of morphisms compose uniquely, i.e. given three arrows $f : a \to b, g : b \to c$ and $h : c \to d$, $h \circ (g \circ f) = (h \circ g) \circ f$, given the same morphism from a to d).

A functor F is a map between two categories (say **A** and **B**), sending objects to objects and morphisms to morphisms. If for any morphism $a \to a'$ in **A**, $F(a \to a')$ is mapped to a morphism $F(a) \to F(a')$ in **B**, F is said *covariant*. If instead $F(a \to a')$ maps to $F(a) \leftarrow F(a')$, F is *contravariant*. A contravariant functor can be seen as a covariant functor $F : \mathbf{A}^{Op} \to \mathbf{B}$, where $\mathbf{A}^{Op}$ is obtained from **A** by reversing the direction of all its morphisms.

Given two functors $F, G : \mathbf{C} \to \mathbf{D}$, a *natural transformation* $\tau : F \to G$ is such that:

- For each object X in category **C**, there exists a morphism in **D**, $\tau_X : F(X) \to G(X)$.
- Given a morphism in **C**, $f : X \to Y$ the following diagram commutes:

$$\begin{array}{ccc} F(X) & \xrightarrow{\tau_X} & G(X) \\ F(f) \downarrow & & \downarrow G(f) \\ F(Y) & \xrightarrow{\tau_Y} & G(Y) \end{array}$$

meaning that $G(f) \circ \tau_X = \tau_Y \circ F(f)$.

If a contravariant functor F has as codomain the category of sets, i.e. $F: \mathbf{A}^{Op} \to \mathbf{Set}$, F is called a *presheaf*. Given a fixed category $\mathbf{A}$, the category in which the objects are all the presheaves over $\mathbf{A}$ while the morphisms are the natural transformations between presheaves is called a *topos*.[3]

The key mathematical construction underlying the generic figures approach derives from the fact that each object of $\mathcal{GF}$, as a result of the Yoneda lemma, is associated with its *representable set*, and therefore each structure can be thought of as a category of its figures. This new category is constructed as a gluing of several copies of the representable sets.

An example of how the generic figures approach works is that of directed graphs. Each such graph can be seen as a contravariant functor $G : \mathcal{G}^{op} \to \mathbf{Set}$, where $\mathcal{G}$ consists of exactly two objects V and A and two non-identity arrows $s, t : V \to A$. This category of generic figures is depicted here, with the identity arrows on V and A not shown:

$$V \underset{t}{\overset{s}{\rightrightarrows}} A$$

Given $G : \mathcal{G}^{Op} \to \mathbf{Set}$, $G(V)$ is the set of graph vertices and $G(A)$ is the set of graph arrows, $G(s)$ and $G(t)$ are then two functions $G(A) \to G(V)$ assigning a source-vertex and a target-vertex, respectively, to each arrow in the directed graph.

In this presheaf category, the representable set h_V is the directed graph pictured here:

•

while the representable set h_A for this category is:

• ⟶ •

The structure of any directed graph G may be articulated exhaustively in terms of morphisms $h_V \to G$ and $h_A \to G$ together with incidence relations among these.

4 The Generic Figures of Combinators

In our case, the generic figures category $\mathcal{CB}$ can be depicted as follows (with the identity morphisms omitted):

---

[3] More precisely, this category is an *elementary topos*, a category with finite *limits* and *colimits*, *exponentials* and a *subobject classifier*.

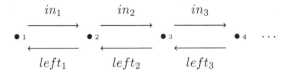

$$left_k \circ in_k = Id_k$$

More precisely, the *objects* of $\mathcal{CB}$ are the natural numbers $\mathbb{N}$ and in addition to *identity morphisms* Id_n for all $n \in \mathbb{N}$, there are *morphisms* $in_n : n \to n+1$ (capturing the order $\leq$ of $\mathbb{N}$) and $left_n : n+1 \to n$, for all $n, n+1 \in \mathbb{N}$, such that $left_n \circ in_n = Id_n$. Besides these, the only other morphisms in $\mathcal{CB}$ are the necessarily-defined compositional morphisms, which by a slight abuse of notation we designate $in_{n,m} : n \to m$ and $left_{n,m} : m \to n$ for all $m > n$.

Given a functor $F : \mathcal{CB}^{op} \to \mathbf{Set}$, for each $n \in \mathbb{N}$, $F(n)$ is a set. The elements of $F(n)$ may be interpreted as nested boxes at depth n or, equivalently, as nodes at level n of a *forest*. Given any morphism in_n, $F(in_n) : F(n+1) \to F(n)$ yields, for any given element $x \in F(n+1)$, its *parent* in the forest (or the box represented by $b \in F(n)$ in which the box x is nested) [15].

In turn, for each $left_n$ morphism, $F(left_n) : F(n) \to F(n+1)$ is a function such that to $x \in F(n)$ is assigned a single element $y \in F(n+1)$. This element y is called the *leftmost* child of x (or the leftmost box inside box x).

Finally, we call an element s of $F(n)$ a *singleton* if for all $x \in F(n+1)$ such that $F(in_n)(x) = s$, x is leftmost. In other words, an element s is a singleton if it has only a single (necessarily leftmost) child.

Combinatory terms can be represented by a particular presheaf $F : \mathcal{CB}^{op} \to \mathbf{Set}$ by defining functions from the singletons of F into a chosen alphabet of primitives, for instance, the alphabet $\{\mathbf{S}, \mathbf{K}\}$.

The system of functions comprised of $F(left_n)$ and $F(in_n)$ may then be understood to characterize the relevant nestings of combinatory operations in an abstract way, with standard combinators always characterized by binary trees. These in turn may be translated into various diagrammatic representation schemes.

In this framework, the straightforward use of morphisms in $\mathbf{Set}^{\mathcal{CB}^{op}}$, that is, natural transformations $\nu : F \to F'$ for presheaves F and F' (in particular, natural transformations from *representable sets*, or *h-sets* h_n for each object n in $\mathcal{CB}$), allows for the characterization of all essential structural features of combinators and their operational dynamics.

5 Finite and Infinite Combinators

There are some amendments to this framework to address some issues in the representation of combinators. One arises when we try to represent the presheaf corresponding to the earlier example of the combinatory term **SKK**. It is natural to represent it diagrammatically as either the nested box structure on the right or, equivalently, the tree structure on the left (see Fig. 1).

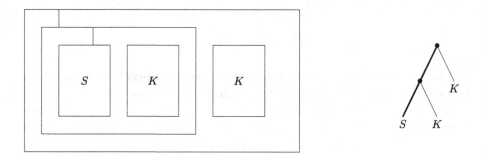

Fig. 1. Diagrammatic representations of **SKK**

We interpret the "box version" by noticing that, within each box (except the innermost ones), there are two boxes, exactly one of which is marked with a line connecting it to its enveloping box. Those marked boxes are understood as acting "from the left" on the adjacent unmarked box. That resolves the ambiguity of having to read the diagrams from the left to the right. The "binary tree version" similarly has a highlighted (thickened) branch, to indicate the action of its "left" node on its "right" one. Notice that in the binary tree representation, only the leaves of the trees are labeled, whereas the nodes can be thought of as a system of rewriting determined by the structure of the tree itself.

If these were representations of a presheaf $F : \mathcal{CB}^{op} \to$ **Set** we would have that $F(n) = \emptyset$ for $n > 3$. But then, each $F(in_n)$ for $n \geq 3$ would not be functions, since they all would have an empty domain.

This issue can be addressed by considering a category **Set**$_*$, which is the category of *pointed sets*, in which a distinguished set $*$ is such that a total function f with domain $*$ is equivalent to a partial function f', i.e. one that may not be defined for certain elements of its domain.

The **SKK** example can then be seen as representing a functor $F' : \mathcal{CB}^{op} \to$ **Set**$_*$ such that $F'(n) = *$ for any $n > 3$. Furthermore, each $F'(in_n)$ is a partial function (with domain $*$ for $n \geq 3$). Conveniently, the graphical representation of F' trims the corresponding tree or the nested boxes at the levels at which $F'(n) = *$.

This could be understood as meaning that the presheaf representation (i.e. where the functors have codomain **Set**). But the *Yoneda extension lemma*

(Proposition 2.7.1 in [16]) can be applied here.[4] Then, we have the following result:

Proposition 1. *There exists a functor $\mathcal{F} : \mathbf{Set}^{\mathcal{CB}^{op}} \to \mathbf{Set}_*$ such that given a functor*
$$F' : \mathcal{CB}^{op} \to \mathbf{Set}_*$$
there exists a presheaf $F : \mathcal{CB}^{op} \to \mathbf{Set}$ such that
$$\mathcal{F} \circ F \simeq F'$$
and viceversa.

Proof: *Since any presheaf F can be obtained by gluing the representable sets for $\mathcal{CB}$, the gluing of the corresponding $\mathbf{Set}_*$-representable objects obtained according the Yoneda extension lemma yields F' and viceversa.*

This means that the diagrammatic representation given above for **SKK**, obtained as a functor $F' : \mathcal{CB}^{op} \to \mathbf{Set}_*$ can be equivalently obtained as a presheaf $F : \mathcal{CB}^{op} \to \mathbf{Set}$. On the other hand, the equivalence predicated by Proposition 1 allows to incorporate the use of partial functions, exhibiting the essential computational nature of combinators.

The other problem is perhaps more fundamental. As a computational tool, a combinator should yield a definite output. But this works only if the corresponding tree or nested boxes representation has finite depth. The case in which it is infinite, equivalent to the case of a Turing machine running forever, needs to be addressed if the additional property of *observability* (see [32]) is to be predicated of combinators. Roughly speaking, what is "observable" depends on the counting system that we agree upon, so that we can adjust our framework depending on the degree of accuracy requested by the framework that we want to represent.

To deal with this issue of determining or "observing" certain properties of such infinitely nested combinators, we turn to Sergeyev's *grossone* approach.

6 The *Grossone* Framework

At a very basic level, the main idea behind Yaroslav Sergeyev's *grossone* numeral system involves determining infinite sequences. More precisely, Sergeyev builds up a fully-fledged theory that allows to attach numerical value (expressed in a suitable numeral system) to sequences that, otherwise, would simply be indeterminate in the standard numeral systems. Since lengths are specifiable in a numeral system, sequences carry a determinate length when a given numeral system specifies it. For instance, in base ten, every finite sequence carries a length

---

[4] This result indicates that to a *representable set* on a small category it corresponds a *representable $\mathcal{A}$-object* and viceversa, for any category $\mathcal{A}$ with small inductive limits. $\mathbf{Set}_*$ satisfies this condition (inheriting it from $\mathbf{Set}$).

but every infinite sequence is indeterminate. The grossone-based numeral system can be introduced as a way of increasing the store of determinate sequences, thus rendering certain of their otherwise indeterminate properties "observable".

As a sequence in base ten, $\mathbb{N}$ is indeterminate. In contrast, according to Sergeyev's treatment of sequences, it is postulated that such a sequence is determined in an explicitly given number system.

In slightly more detail, the sequence of natural numbers is represented in the grossone as follows:

$$\mathbb{N} = \{1, 2, 3, \ldots, ① - 2, ① - 1, ①\}.$$

The conditions governing arithmetic in the new numeral system are:

1. *Infinity*: any finite number n is less than the grossone. That is $n < ①$ for every finite natural number n.
2. *Identity*: there are several relations linking ① to the identity elements 0 and 1, namely

$$0 \cdot ① = ① \cdot 0 = 0 \; ; \; ① - ① = 0 \; ; \; \frac{①}{①} = 1 \; ; \; ①^0 = 1^① = 1 \; ; \; 0^① = 0$$

3. *Divisibility*: for any finite number n, if $1 \leq k \leq n$, define $\mathbb{N}_{n,k} = \{k, k+n, k+2n, k+3n, \ldots\}$. The class $\{\mathbb{N}_{n,k}\}_{1 \leq k \leq n}$ satisfies that $\cup_{k=1}^n \mathbb{N}_{n,k} = \mathbb{N}$ and each $\mathbb{N}_{n,k}$ has a number of elements indicated by the numeral $\frac{①}{n}$.

An additional axiom ([38], [17]), characterizes *induction* in this framework:

- If a statement $F(n)$ is valid for all $n > N$, for a large finite N, then $F(①)$ is valid.

Sergeyev proves (see [28]) the following result:

Theorem: *The number of elements in an infinite sequence is less or equal to ①.*

This result has an interesting consequence. Given a sequence $\{a_n\}$, it is not enough to give a formula for each a_n. We must also determine the first and last elements in the sequence. Thus, given two sequences

$$\{a_n\} = \{5, 10, 15, \ldots, 5(① - 1), 5①\}$$

$$\{b_n\} = \{5, 10, 15, \ldots, 5(\frac{2①}{5} - 1), 5\frac{2①}{5}\}$$

even if $a_n = b_n = 5n$ they are different because they have a different number of elements, in each case indexed by the last element in the sequence divided by five. So, in the grossone framework, the first sequence has ① elements whereas the second has $\frac{2①}{5}$ elements.

Notice that, while ① represents *infinity*, it does not imply that there do not exist infinite numbers smaller or larger than ①. For instance, the set of even or odd natural numbers has $\frac{①}{2}$ elements (as an immediate consequence of the *Divisibility* property) while $\mathbb{Z}$, the class of integer numbers, has $2①+1$ elements [33].

Thus, the introduction of ① may require the augmentation of $\mathbb{N}$ to a larger set $\hat{\mathbb{N}}$, where

$$\hat{\mathbb{N}} = \{1, 2\ldots, ①-1, ①, ①+1, ①+2, \ldots, 2\cdot①, \ldots, ①^2-1, ①^2, ①^2+1, \ldots\}$$

For the purposes of this paper, we will not need to make use of $\hat{\mathbb{N}}$, since we assume that the objects of $\mathcal{CB}$ correspond to the elements of $\mathbb{N}$.

Evidence of the efficacy of this approach is highlighted by its successful application to many fields of applied mathematics, including optimization (see [4,5,10,34]), fractals and cellular automata (see [3,7–9]) as well as infinite decision-making processes, game theory, and probability (see [11,24,25]), whereas logical approaches to *grossone* have been investigated in [18,20,21].

7 Combinators in the *Grossone* Framework

In the *grossone* framework every functor $F : \mathcal{CB}^{op} \to \mathbf{Set}$ (as equivalent to a functor $F' : \mathcal{CB}^{op} \to \mathbf{Set}_*$) would yield a structure (tree or nested boxes) of a depth that can be represented in base ①. To see this, consider a presheaf F that can be translated, into a tree structure T_F. Each node a in this tree can be assigned an index:

$$\iota : \mathrm{Nodes}(T_F) \to \mathbb{N}$$

where $\mathrm{Nodes}(T_F)$ is the set of nodes of T_F where $\mathbb{N}$ is defined in the *grossone* framework, i.e. $\mathbb{N} = \{1, 2, \ldots, ①-2, ①-1, ①\}$, in base ①. $\iota(a)$ is the *depth* of the pre-terminal node a. Thus, the root node r is such that $\iota(r) = 1$

We have to redefine now the concept of *terminal* node for T_F. We say that a node t is terminal if it has no child t' with $\iota(t') > \iota(t)$.

We know that in the *grossone* setting the following is true:

Proposition 2. *Every node a in a tree T_F is either terminal, or it has a sequence of descendants[5] $\{a_1, \ldots, a_K\}$ ($k \in \mathbb{N}$) such that a_K is a terminal node, under the condition that $\{1, \ldots, K\}$ is a sequence in $\mathbb{N}$ and K is in base ①.*

Proof: *Consider by contradiction that there exists a node a in T_F, which is not terminal and every one of its descendants has at least one child. But then,*

---

[5] A node t' is a *descendant* of node t, if it belongs to the transitive closure of the relation $child(x, y)$ ("x is a child of y").

each sequence of descendants $\{a_1, a_2, \ldots\}$ is such that $\iota(a_{k+1}) > \iota(a_k)$. Now by enumeration, there will exist a node in the sequence $a_{①}$ such that $\iota(a_{①}) = ①$. Now by hypothesis $a_{①}$ has at least one child, say $a_{①+1}$. But this would mean that $\iota(a_{①+1}) > ①$, contradicting Theorem 3.1 in [32] (Theorem 1 in [33]), which states that the number of elements in a sequence indexed over the natural numbers is less or equal than ①.

An immediate corollary of Proposition 2 is the following:

Corollary 1. *A node* **a** *in T_F such that $i(\mathbf{a}) = ①$ cannot have a child.*

Proof: *Assume that there exists a node* **a**′ *that is a child of node* **a**. *Then* $\iota(\mathbf{a}) = ① = \iota(\mathbf{a}')$. *On the other hand, notice that the depth function* depth $: T_F \to \mathbb{N}$ *satisfies the following property:*

$P(n)$: if any node a of finite depth depth(a), has a child a′, depth(a′) = depth(a) + 1 and thus, depth(a′) ≠ depth(a).

*According to the induction axiom for grossone, depth(**a**′) ≠ depth(**a**′), but then $\iota(\mathbf{a}) \neq \iota(\mathbf{a}')$. Contradiction.*

Recalling that the **K − S** system is Turing complete, we can examine how this extends to the ① setting. A Turing machine can be informally described as a *head* that can read and write symbols on a (potentially infinitely long) tape, moving the tape left and right, one cell of the tape at a time. At each step, the head is in an internal state. Then, it reads the symbol at the current cell and applies a *transition function* that, given the state and the symbol it is reading, specifies (i) a new symbol to write on the tape, (ii) whether to move the tape to a contiguous cell or not, (iii) the next state of the head. In the general case of a non-deterministic Turing machine, its behavior can be represented as a tree in which the root corresponds to the initial configuration while each path represents a sequence of configurations generated in a possible run of the machine. If the Turing machine can work for an infinite time and it can produce outputs of an infinite length it is called *imaginary*. If its output has a length less or equal than ①, it is called *observable*. We highlight the fact if a sequence has a length less than ①, it does not imply that such length is finite, as $① - 1, ① - 2, \ldots$ are still infinite numbers less than ①.

Theorem 5.1 in [32] states the conditions that ensure the observability of outputs of imaginary Turing machines. These are:

(i) No more than ① computational steps of an imaginary Turing machine can be observed in a sequence.
(ii) Observability is ensured by the fact that the depth k of the computational tree is $k \leq ①$.

Then, it is immediate from this and Proposition 2:

Corollary 2. *Any combinator $F \in \mathbf{Set}^{\mathcal{CB}^{op}}$ is observable.*

Example 1. *The grossone framework allows to analyze combinatorial terms considered* undecidable *in the classical setting [1]. For instance, consider [36]:*

$$AAA \quad \text{where} \quad A = \mathbf{SSS}$$

Defining $Y_0 = \mathbf{S}A(\mathbf{S}AA)$ and $Y_{k+1} = AY_k$, it can be shown that the reduction process (i.e. the computation carried out with an initial input AAA) generates a sequence of combinatorial terms that can be represented as trees:

$$AAA \to \mathbf{S}A(\mathbf{S}A)A \to AA(\mathbf{S}AA) \to \mathbf{S}A(\mathbf{S}A)(\mathbf{S}AA) \to A(\mathbf{S}AA)Y_0 \to$$
$$\to \mathbf{S}(\mathbf{S}AA)(\mathbf{S}AA)(\mathbf{S}(\mathbf{S}AA))Y_0 \to \mathbf{S}AAY_0* \to Y_1Y_1* \to \ldots Y_kY_k* \to \ldots$$

Where $*$ represents a leaf in the corresponding tree. Thus it generates a sequence of trees of increasing depth. The undecidability of AAA is associated to the impossibility of defining a "last" tree.

In the grossone framework, Corollary 2 indicates that there exists a tree

$$Y_{①}Y_{①}*$$

such that all paths end at nodes of depth at most $①$.

8 Conclusions

The technical presheaf machinery introduced above provides a completely precise mathematical framework that represents the intuitive (and diagrammatically picturable) notion of a syntactical combinatory term as a complex arrangement of nested operations.

Proposition 1 ensures the equivalence of the presheaf-defined combinators and Turing machines. And Proposition 2 ensures the observability of their computation. Thus, the additional representational resources of the category of presheaves (and thus a *topos* [23]) $\mathbf{Set}^{\mathcal{CB}^{op}}$ provide for the possibility of representing *choices of argument from among given alternatives* at each level of a combinatory term.

In this respect, the presheaf framework is not only sufficient to represent all standard combinatory terms; it also allows for a more general type of combinatory expression, which we call a *multi-combinator*. The dynamics of these multi-combinators support a more elegant and visually attractive representation of the computation of the reduced form of a term and are especially useful for capturing the notion of *hypotheses* as operative in computational settings, as discussed in [22]. In coordination with the fine tracking of properties of infinite sequences made possible via the grossone numeral system, a variety of new approaches to infinite (non-halting) computational processes may become available for further research.

References

1. Barendregt, H.: The Lambda Calculus, its Syntax and Semantics. Elsevier, Amsterdam (1984)
2. Bimbó, K.: Combinatory Logic: Pure. Applied and Typed, Routledge/Chapman and Hall, New York (2011)
3. Caldarola, F.: The Sierpinski curve viewed by numerical computations with infinities and infinitesimals. Appl. Math. Comput. **318**, 321–328 (2018)
4. Cococcioni, M., Pappalardo, M., Sergeyev, Y.D.: Lexicographic multi-objective linear programming using grossone methodology: theory and algorithms. Appl. Math. Comput. **318**, 298–311 (2018)
5. Cococcioni, M., Cudazzo, A., Pappalardo, M., Sergeyev, Y.D.: Solving the lexicographic multi-objective mixed-integer linear programming problem using branch-and-bound and grossone methodology. Commun. Nonlinear Sci. Numer. Simul. **84**, 1–20 (2020)
6. Coecke, B., Kissinger, A.: Picturing Quantum Processes: A First Course in Quantum Theory and Diagrammatic Reasoning. Cambridge University Press, Cambridge, UK (2017)
7. D'Alotto, L.: Cellular automata using infinite computations. Appl. Math. Comput. **218**, 8077–8082 (2012)
8. D'Alotto, L.: A classification of two-dimensional cellular automata using infinite computations. Indian J. Math. **55**, 143–158 (2013)
9. D'Alotto, L.: A classification of one-dimensional cellular automata using infinite computations. Appl. Math. Comput. **255**, 15–24 (2015)
10. De Cosmis, S., De Leone, R.: The use of grossone in mathematical programming and operations research. Appl. Math. Comput. **218**, 8029–8038 (2012)
11. Fiaschi, L., Cococcioni, M.: Numerical asymptotic results in game theory using sergeyev's infinity computing. Int. J. Unconvent. Comput. **14**, 1–25 (2018)
12. Gangle, R., Caterina, G., Tohmé, F.: A generic figures reconstruction of peirce's existential graphs (alpha). Erkenntnis **87**(2), 623–656 (2022)
13. Goldblatt, R.: Topoi: The Categorical Analysis of Logic. Elsevier, Amsterdam (1984)
14. Iudin, D., Sergeyev, Y.D., Hayakawa, M.: Interpretation of percolation in terms of infinity computations. Appl. Math. Comput. **218**, 8099–8111 (2012)
15. Kasangian, S., Vigna, S.: The topos of labelled trees: a categorical semantics for SCCS. Fund. Inform. **32**(1), 27–45 (1997)
16. Kashiwara, M., Schapira, P.: Categories and Sheaves. Grundlehren der mathematischen Wissenschaften **332** (2006)
17. Kauffman, L.: Infinite computations and the generic finite. Appl. Math. Comput. **255**, 25–35 (2015)
18. Lolli, G.: Infinitesimals and infinities in the history of mathematics: a brief survey. Appl. Math. Comput. **218**, 7979–7988 (2012)
19. Mac Lane, S.: Categories for the Working Mathematician. Springer-Verlag, Berlin (1998)
20. Margenstern, M.: Using grossone to count the number of elements of infinite sets and the connection with bijections. p-Adic Numbers, Ultrametric Anal. Appl. **33**, 196–204 (2011)

21. Montagna, F., Simi, G., Sorbi, A.: Taking the Pirahã seriously. Commun. Nonlinear Sci. Numer. Simul. **21**(1–3), 52–69 (2015)
22. Piantadosi, S.T.: The computational origin of representation. Mind. Mach. **31**(1), 1–58 (2021)
23. Reyes, M., Reyes, G., Zolfaghari, H.: Generic Figures and their Glueings. Polimetrica, Milan (Italy) (2004)
24. Rizza, D.: A study of mathematical determination through bertrand's paradox. Philos. Math. **26**, 375–395 (2018)
25. Rizza, D.: Numerical methods for infinite decision-making processes. Int. J. Unconvent. Comput. **14**, 139–158 (2019)
26. Sergeyev, Y.: Arithmetic of Infinity. Edizioni Orizzonti Meridionali (2003)
27. Sergeyev, Y.: Blinking fractals and their quantitative analysis using infinite and infinitesimal numbers. Chaos, Solitons Fract. **33**, 50–75 (2007)
28. Sergeyev, Y.: A new applied approach for executing computations with infinite and infinitesimal quantities. Informatica **19**, 567–596 (2008)
29. Sergeyev, Y.: Computer system for storing infinite, infinitesimal, and finite quantities and executing arithmetical operations with them. USA patent **7**(860), 914 (2010)
30. Sergeyev, Y.: Solving ordinary differential equations by working with infinitesimals numerically on the infinity computer. Appl. Math. Comput. **219**(22), 10668–10681 (2013)
31. Sergeyev, Y.: Numerical infinities and infinitesimals: methodology, applications, and repercussions on two hilbert problems. EMS Surv. Math. Sci. **4**, 219–320 (2017)
32. Sergeyev, Y., Garro, A.: Observability of turing machines: a refinement of the theory of computation. Informatica **21**, 425–454 (2010)
33. Sergeyev, Y., Garro, A.: Single tape and multi-tape turing machines through the lens of the grossone methodology. J. Supercomput. **65**, 645–663 (2013)
34. Sergeyev, Y., Kvasov, D.E., Mukhametzhanov, M.S.: On strong homogeneity of a class of global optimization algorithms working with infinite and infinitesimal scales. Commun. Nonlinear Sci. Numer. Simul. **59**, 319–330 (2018)
35. Schoenfinkel, M.: Über die Bausteine der mathematischen Logik. *Mathematische Annalen* **92**: 305-316. English version (1967): On the Building Blocks of Mathematical Logic, pp. 355-366 in Van Heijenoort, J. (ed.) From Frege to Gdel, a Source Book in Mathematical Logic, 1878-1931, Harvard University Press, Cambridge, MA (1924)
36. Waldmann, J.: The Combinator **S**. Inf. Comput. **159**, 1–21 (2000)
37. Wolfram, S.: Combinators: a Centennial View. Wolfram Media Inc., Champaign, IL (2021)
38. Zhigljavsky, A.: Computing Sums of Conditionally Convergent and Divergent Series using the Concept of Grossone. Appl. Math. Comput. **218**, 8064–8076 (2012)

Applying Variational Quantum Classifier on Acceptability Judgements: A QNLP Experiment

Raffaele Guarasci[✉][iD], Giuseppe Buonaiuto, Giuseppe De Pietro, and Massimo Esposito

National Research Council of Italy (CNR), Institute for High Performance Computing and Networking (ICAR), 80131 Naples, Italy
{raffaele.guarasci,giuseppe.buonaiuto,giuseppe.de.pietro, massimo.esposito}@icar.cnr.it

Abstract. The newborn Quantum Natural Language Processing (QNLP) field has experienced tremendous growth in recent years. The possibility of applying quantum mechanics to critical aspects of language processing has dramatically impacted many tasks, ranging from theoretical approaches to algorithms implemented on real quantum hardware. From a methodological point of view, the possibility offered by applying quantum mechanics to NLP problems is well suited to classification tasks. This work aims to test the potential computational advantages of a hybrid algorithm, namely the Variational Quantum Classifier (VQC), to perform classification on a classical Linguistics task: acceptability judgments. VQC is a quantum machine learning algorithm able to infer the relations between input features and the associated belonging class using a parametrized quantum circuit and an encoding layer that embeds classical data into quantum states. An acceptability judgment is defined as the ability to determine whether a sentence is considered as natural and well-formed by a native speaker. The approach has been tested on sentences extracted from ItaCoLa, a corpus that collects Italian sentences labeled with their acceptability judgment. The evaluation phase has considered quantitative metrics and qualitative analysis to investigate further the algorithm's behavior on specific linguistic phenomena included in the corpus.

Keywords: Quantum Machine Learning · Quantum Natural Language Processing · Variational Quantum Classifier

1 Introduction

Recently, there were significant advancements in all Natural Language Processing (NLP) tasks, ranging from machine translation [41], text classification [34], coreference resolution [10,12,19,28] or multi-language syntactic analysis [5,13,14]. This is primarily due to the explosion of neural language models based on deep learning architecture, especially Transformers-based models such as BERT [7]. However, this improvement comes with an increasing complexity of models, which requires a considerable amount of data and computation resources to be efficiently trained [3,9]. Besides, there are open issues related to what these models learn about language, how they encode this information [15], and how much of it is interpretable [17]. Quantum machine learning (QML) is gaining attention as an alternative approach that exploits powerful aspects borrowed from quantum mechanics to overcome the computational limitations of current approaches. A strand derived from QML is the sub-field of Quantum Natural Language Processing (QNLP) [6], which aims to solve natural language-related tasks using quantum properties or applying algorithms derived from quantum theory or testing new approaches using real quantum hardware.

However, quantum-based approaches currently suffer the limitations of available hardware, and there are open issues about scalability because the quantum circuit can be considered a linear or at most sub-linear model in the feature space [31]. There is an ongoing research effort concerning the encoding of classical data into quantum computers, which in principle allows to extract information and make computations more efficiently than classical computer, but with a strong backside when considering large dataset or data represented as large vectors in feature space. Nowadays hardware has a limited amount of qubits and, when the number of qubits is sufficient for embedding data, the lack of fault tolerance, i.e., the quantum noise associated with the computation, hinders the performances and destroys the potential advantages of quantum computing. To overcome these limitations, novel hybrid approaches combining classic pre-trained models and quantum techniques have been proposed [20,21]. This type of approach offers the advantage of implementing specific layers of models on a quantum device, while classical models for non-linear operations handle intermediate results. A successful example of such hybrid approaches is the so-called classical-quantum transfer learning [24]. It encodes input features in a multi-qubit state, then a quantum circuit transforms and measures such features. In this pipeline, output probabilities are projected to the task label space, and losses are backpropagated to update parameters.

Starting from the approach proposed in [21], this work proposes a hybrid transfer learning model for QNLP applied to a binary classification task. It uses the pre-trained Sentence-BERT model and fine-tunes it to perform the classification. This approach has been tested on current noisy intermediate-scale quantum (NISQ) machines [35]. It exploits the advantage and robustness of classic pre-trained language features already well-known in the literature [29] and integrates quantum encodings to reduce noisy devices. As already demonstrated in [21], avoiding tensor product of vectors is useful to allow a scalable QNLP model and improve representativeness facing natural language sentences of various length and complexity. The Variational Quantum Classifier (VQC) [4] is the algorithm chosen to perform a binary classification on acceptability judgments, a task which has gained much popularity in recent years in the field of NLP [22,39]. This particular task has been chosen because QNLP has proven particularly effective in binary classification [26,33]. For the purpose of this work, which focuses on the Italian language, the dataset chosen has been ItaCoLa [2,36]. It is the largest existing resource for this task in Italian, collecting sentences labeled with their judgments by expert linguists. The evaluation phase has considered quantitative metrics.

The paper is organized as follows: in Sect. 2, the research works available in the literature related to what is presented are described, while in Sect. 3 the dataset and applied methodologies are presented, then in Sect. 4, the results obtained are exposed, and relevant aspects are discussed, and finally, overall conclusions are drawn in Sect. 5.

2 Related Work

2.1 Quantum Machine Learning

Numerous examples have received increasing interest in recent years concerning the adaptation of classical machine learning algorithms through the use of properties and techniques borrowed from quantum mechanics.

One notable approach is Quantum Support Vector Machines (QSVMs), which aim to enhance the performance of traditional Support Vector Machines (SVMs) by utilizing quantum algorithms. QSVMs have shown promising results in various text classification tasks, including sentiment analysis, topic classification, and document classification. Another area of research is quantum-inspired algorithms for text classification. Quantum-inspired algorithms, such as Quantum-Inspired Genetic Algorithm (QGA) and Quantum-Inspired Particle Swarm Optimization (QPSO), draw inspiration from quantum mechanics and apply quantum-like principles to improve traditional optimization techniques. These algorithms have been explored in the context of feature selection and parameter optimization for text classification, demonstrating their potential to enhance classification accuracy and efficiency. Moreover, quantum embeddings have gained attention as a means to represent and analyze textual data. Quantum embeddings leverage the concepts of quantum superposition and entanglement to capture semantic relationships between words or documents. These embeddings aim to capture more nuanced and context-dependent information compared to traditional word embeddings. By utilizing quantum representations, classification models can benefit from enhanced semantic understanding, improving performance in various NLP tasks.

Furthermore, quantum machine learning algorithms, such as Quantum Neural Networks (QNNs) and Quantum Boltzmann Machines (QBMs), have been explored in text classification. These quantum-inspired models leverage the unique computational capabilities of quantum systems to perform complex computations efficiently. Although still in its early stages, quantum machine learning holds the potential to address the computational challenges associated with large-scale language data and to provide more powerful models for classification tasks.

2.2 Quantum Natural Language Processing

The growing interest in exploring the potential of QML in language-related tasks has led to the birth of the QNLP. As a sub-field of QML, QNLP offers a new paradigm focused on processing and analyzing language data by leveraging the principles of quantum mechanics.

QNLP approaches proposed so far can be tested on real datasets using classical hardware (quantum-inspired approaches) or on currently available quantum machines (quantum-computer approaches). Quantum-inspired approaches integrate the advancements of quantum mechanics into an existing model. They are usually tested on benchmark datasets surpassing the current state-of-the-art. By contrast, Quantum-computer approaches are tested on NISQ hardware. Due to the limitations of current quantum hardware, these approaches are tested on simple NLP tasks and only on small-to-medium scale datasets. Several studies have investigated the application of quantum computing in classification tasks within NLP; a detailed review is described in [11].

An alternative way proposed to push through the limitations inherent in the scalability of such experiments is represented by hybrid approaches. A hybrid classical-quantum scheme using a quantum self-attention neural network (QSANN) has been recently proposed [20]. Even if it introduces the possibility of non-linearity, significantly improving over other QNLP models [23], this approach is limited by the continuous switching between quantum and classical hardware at each self-attention layer needed to run the network. [21] proposes a more viable approach to solve the low non-linearity issue for QNLP models using the classical-quantum transfer learning paradigm [24]. Using the classical-quantum transfer mechanism and pre-trained quantum encodings seems the most promising approach to develop scalable QNLP models, paving the way for the possibility of being implemented on real quantum hardware. Given this high potential and the increasing appeal of this approach, it was chosen for use in this work.

2.3 Acceptability Judgements Task

Automatically assessing acceptability tasks has always been a popular task in Linguistics. However, first resources built have been limited to small dataset theoretical purposes or within psycholinguistic experiments [18,25]. Since the release of the CoLa corpus [39], the first large-scale corpus of English acceptability, containing more than 10k sentences, the task has garnered increasing interest from researchers in the field of NLP.

The CoLA corpus has been presented with a number of experiments to assess the performance of neural networks on a novel binary acceptability task. Furthermore, it has been included in the GLUE dataset [38], a very popular multi-task benchmark for English natural language understanding, and an acceptability challenge has been launched on Kaggle,[1]. For such reasons, the number of studies dealing with binary acceptability has remarkably increased, although proposed approaches have often used different metrics for evaluation, so comparisons cannot always be made.

Starting with the same methodology introduced in COLA, similar resources have been released in recent years in various typologically different languages,

---

[1] https://www.kaggle.com/c/cola-in-domain-open-evaluation/.

ranging from Italian [36] - which is the subject of this work - Norwegian [16], Swedish [37] and Russian [27], Japanese [32], and Chinese [40].

3 Materials and Methods

3.1 ItaCola Dataset

The dataset used for this work is ItaCoLA, the Italian Corpus of Linguistic Acceptability [36]. This corpus was designed to represent a broad range of linguistic phenomena while distinguishing between sentences that are considered acceptable and those that are not. The process used to create the corpus was modeled as closely as possible on the methodology employed for the English CoLA dataset [39]. ItaCoLA consists of approximately 9,700 sentences drawn from various sources that cover numerous linguistic phenomena.

Table 1. Example sentences from the ItaCoLA dataset. 1 = acceptable, 0 = not acceptable. The symbol * is conventionally used in Linguistics to mark unacceptable sentences.

Label	Sentence
0	*Edoardo è tornato nella sua l'anno scorso città. (*Edoardo returned to his last year city)
1	Ho voglia di salutare Maria (I want to greet Maria)

The acceptability annotation for these sentences is based on Boolean judgments formulated by experts who authored the different data sources. The sentences are sourced from various linguistic publications spanning four decades, transcribed manually and released in digital format. An example of how the data are structured in the corpus is shown in Table 1. ItaCola is divided into training, validation, and test split, including respectively 7,801, 946, and 975 examples. In [36] a baseline using LSTM with FastText embeddings and an Italian version of BERT has been released.

3.2 Methodology

This section outlines the methodology for incorporating Italian BERT embeddings in Quantum Natural Language Processing (QNLP) pipeline for acceptability judgements classification task. The proposed pipeline is articulated in following steps:

- **Data Preprocessing**: Sentences extracted from ItaCola needs to be preprocessed and prepared for embedding generation.

- **Embedding Generation**: For the purpose of this work, pretrained Sentence-BERT [30] embeddings have been used. The embedding model is trained on a large corpus of text data to learn the semantic relationships between words, using siamese and triplet network to represent a semantically meaningful embedding for the a whole sentence. Specifically, Sentence-BERT is trained for the classification of pairs of sentences, where the scope of the learning procedure is to optimize a similarity score between sentences. This training process involves learning the contextual information surrounding each word in the corpus. Feature vectors extracted via Sentence-BERT have 768 real-valued entries. These values need to be encoded in a quantum state for training a quantum classifier.

- **Quantum Embedding Construction**: Quantum embeddings aim to leverage the principles of quantum mechanics to represent language data in a quantum state. These embeddings capture more nuanced and context-dependent information compared to classical embeddings. One approach to constructing quantum embeddings is to utilize quantum superposition and entanglement. The embeddings are generated by encoding the classical embeddings into quantum states, where the amplitudes represent the weights or probabilities associated with each word or document. Quantum circuits or operators are employed to perform transformations on the quantum state, enabling the manipulation and analysis of the quantum embeddings.

- **Quantum Classification Model**: Once the quantum embeddings are constructed, they can be used as inputs to quantum classification models for various NLP tasks. In this work, the variational quantum classifier is constructed with Quantum-inspired neural networks (QNNs). These are generally composed of a series of single and two-qubit gates with free parameters that are trained in the learning phase. The structure of these circuits needs to be sufficiently complex to accommodate the possible solutions of the task, and yet simple enough to prevent detrimental effects from quantum noise. There is currently an ongoing research effort aiming to find the optimal strategy for the construction of a variational quantum circuits [8]: one of the key element to take into account is a certain level of entanglament, i.e. quantum correlation, between qubits so that information is shared among each element of the computation and complex solution are potentially explored during training. Here, a basic entanglement ansatz is used, where each qubit is going to be forced in a quantum correlated state with another qubit, pairwisely.

- **Model Training and Evaluation**: The quantum classification model, along with the quantum embeddings, is trained using a labeled dataset. This dataset consists of instances with their corresponding class labels. Model training involves optimizing the parameters of the quantum classification model to minimize a predefined loss function, such as hinge loss for SVMs or cross-entropy loss for neural networks. The trained model is evaluated using appropriate evaluation metrics, such as accuracy, precision, recall, and F1 score, to assess its performance on unseen data.

- **Evaluation**: To assess the effectiveness of quantum embeddings in QNLP tasks, it is beneficial to compare their performance against classical embedding-based models. Classical models, such as traditional machine learning algorithms or neural networks, can be trained and evaluated using the same dataset and evaluation metrics. The performance of the quantum embedding-based model can be compared to these classical models to determine the impact and benefits of quantum embeddings in QNLP tasks. It is worth noting that the specific implementation details of the methodology may vary depending on the chosen embedding technique, quantum embedding construction approach, and the specific quantum classification model employed. The above steps provide a general framework for integrating embeddings in Quantum Natural Language Processing.

In this work three classification pipelines are used, so that the performances out of each strategy can be compared. A schematic of each model can be seen in Fig. 1 and a detailed explanation is provided in followings.

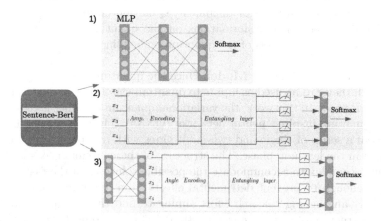

Fig. 1. Schematic of the different training strategies. In **1)** embeddings from Sentence-BERT are the inputs of a Multi-Layer perceptron, i.e. in a complete classical approach, trained for binary classification (**SentenceC**). In **2)** emebeddings vectors are encoded in the amplitude of a quantum state, then the information is passed to a parametrized quantum circuit. The outcome of the measurements is then used as input of a small Multi-Layer perceptron for binary classification (**SentenceQF**). In **3)** emebeddings vectors are first passed through a small Multi-Layer perceptron for dimensionality reduction and then encoded in the rotation angle of single-qubit quantum gates. Then the information is passed to a parametrized quantum circuit. The outcome of the measurements is finally used as input of a small Multi-Layer perceptron for binary classification (**SentenceQS**).

SentenceC. First a fully classical pipeline is implemented (which is in the following is named for convenience **SentenceC**). Here, the preprocessed dataset is converted into a collection of vector using Sentence-BERT and then ingested into a standard multi layer perceptron for the classification. In this sense, *SentenceC* is a classical deep learning algorithm, as it does not contain any quantum layer during training.

SentenceQF. The second and the third pipelines are hybrid quantum-classical classifier, where the quantum layers are implemented using the *Pennylane* library [1]. Specifically, **SentenceQF** (which stands for Sentence Quantum Full) takes the vector obtained via Sentence-BERT and encodes them in a quantum circuit making use of the *Amplitude encoding* strategy: here, numerical data are encoded into the amplitude of the wavefunction (cit), i.e. in the coefficients of a linear combination of basis vectors in a complex-valued space. Hence, a set of normalized N-dimensional data points are encoded in a the coefficient of the combination of $n = log_2(N)$ qubits. This strategy allows to encode large dimensional feature vector in a relatively small number of qubit.

SentenceQS. The last pipeline implemented, named **SentenceQS** (Sentence Quantum Smooth), relies on the *Angle encoding* protocol. This strategy allows to encode a single number into an angular value of a rotation gate applied to a single qubit, for instance, given a single value, x_i^j of the jth feature vector:

$$x_i^j \to R(x_i^j)_Z|0\rangle = e^{-x_i^j \sigma_Z/2}, \tag{1}$$

where σ_Z is the Pauli Z operator. The rotations are periodic up to a global phase, hence, when using this protocol, data are conveniently normalized between $[0, \pi]$. The down side of this encoding lies on the fact that, in principle, the number of qubit needed is equivalent to the dimension of the feature vector $n = N$, thus preventing its use on currently available devices. However, it is possible to smooth this requirement by performing a dimensionality reduction on the feature vectors extracted with Sentence-BERT: this dimensionality reduction can be achieved either via standard PCA or via a fully connected neural network placed at the input of the encoding quantum circuit. Both quantum encoding strategies provide a mean to insert classical data into quantum states, which are then used for quantum computation.

Information encoded in the quantum state is processed by a parametrized quantum circuit, generally called variational ansatz, whose parameters are iteratively updated via a classical optimizer via gradient descent. Designing an optimal ansatz for quantum machine learning tasks is still an open problem: however, it is generally fundamental to find the right balance between an expressive citcuit, i.e. containing an important number of parameters and entangling gate, so that the space of solutions can be properly explored, while minimizing the circuit depth. The last requirement is rather fundamental for deploying quantum algorithms on currently available devices, as an excessive circuit depth can increase the overall noise and hence hinder the computation. In this work, the same quantum variational ansatz is used for both *SentenceQF* and *SentenceQS*, which is a **BasicEntangledLayer** provided by Pennylane. This ansatz consist in layers of rotation gates with a trainable parameter, applied on each qubit involved, followed by a set of CNOT gates applied on pairs of consecutive qubits. Such chain of CNOT generates entanglement among the qubits involved, thus allowing to explore, while training, large sectors of the energy landscape defined by the loss function. The result of the computation is obtained performing measurements on all or a sub-set of the qubits, after the application of the parametrized circuit. The measurements outcome is thus used to evaluate a problem specific loss function, and then passed to a classical optimizer, which updates the weights both of the classical and of the quantum layers of the computation.

4 Results and Discussion

For convenience, the binary data corresponding to the accepted phrase (1) and the rejected phrase (0) are one-hot encoded into a two dimensional tensor. The embeddings from the pre-trained Sentence-BERT are properly standardized to facilitate the convergence during training. Given the strong class unbalance (approximately three times more 1 than 0), the loss function is appropriately weighted, in order to prevent bias in the classification phase. Three different experiments have been performed for the Acceptability Judgements on the Itacola Dataset. For each of them, the main properties, hyperparameters and training strategies are listed in the following:

- **SentenceC**: for the full classical protocol, which serves as a benchmark for comparing the quantum strategies, the embedding extracted from Sentence-BERT are fed into a Multi-Layer perceptron, composed of three layers, each of them with an *ReLu* activation function, with input-output dimension respectively of 768 − 384, 384 − 192, 192 − 96. Last, a classification layer is applied, with a *Softmax* activation function, and dimension of the layer 96 − 2. The whole training is performed with an Adam optimizer, with learning rate 0.01, batchsize of 16 and a weighted categorical cross entropy as objective function.
- **SentenceQF**: here the embeddings extracted from Sentence-BERT are encoded in a quantum circuit using an Amplitude encoding strategy. To encode all the 768 dimensions of the BERT feature vectors, the minimum number of qubit required is $n = \lceil log_2(768) \rceil = 10$, where the rest of the 1024 amplitudes are padded all with 0.01. After the encoding phase, the state obtained is used as an input for a two layers of parametrized *BasicEntangledLayer*. The outcome from the measurements are then passed to a Multi-Layer perceptron with a *Softmax* activation function, and input-output dimension of 10−2, for the classification. The whole training is performed with an Adam optimizer, with learning rate 0.01, batchsize of 16 and a weighted categorical cross entropy as objective function.

- **SentenceQS**: here the embeddings extracted from Sentence-BERT are encoded first in a Multi-layer perceptron for dimensionality reduction. The network is made of three layers, each of them with an *ReLu* activation function, with input-output dimension respectively of 768−384, 384−192, 192−10. Then a quantum circuit for Angle encoding is used, with $n = 10$ qubits, that encodes the classical data into the Z-rotation gates angles. After the encoding phase, the state obtained is used as an input for a two layers of parametrized *BasicEntangledLayer*. The outcome from the measurements are then passed to a Multi-Layer perceptron with a *Softmax* activation function, and input-output dimension of $10 − 2$, for the classification. The whole training is performed with an Adam optimizer, with learning rate 0.01, batchsize of 16 and a weighted categorical cross entropy as objective function.

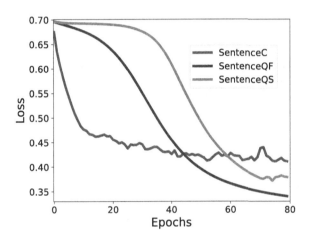

Fig. 2. Experimental results of the training for each pipeline. Training on the ItaCola Dataset for the acceptability judgements. The *SentenceC* stands for the full classical pipeline, where the embeddings from Sentence-BERT are inputs of a multi-layer perceptron designed for binary classification. *SentenceQF* refers to the quantum pipeline with amplitude embedding, where each component of the feature vector is encoded in the amplitude of a quantum state made of $n = 10$ qubit. *SentenceQS* refers to the smooth-quantum pipeline, i.e. an Angle encoding approach. A small multi layer perceptron is used before the quantum encoding to reduce the dimensionality of the feature vector from 768 to 10, so that the new data can be accommodated in the single qubits rotational gates, with $n = 10$ qubits overall.

Results of the various training are showed in Fig. 2: all the aforementioned strategies are trained for 80 Epochs. It is evident from the training phase that, while the *SentenceC* strategy reaches convergence faster compared to the quantum approaches, the steady value of the loss function is higher, indicating that the learning capacity of the model is generally smaller than the other considered. Concerning the two quantum strategies in use, the loss function in Fig. 2 indicates that, when the total information extracted from Sentence-BERT is used in the quantum computation (as in *SentenceQF*), the training is not only faster compared to the *SentenceQS* approach, but the loss reaches a global minimum, stating an overall stronger learning ability. Conversely, when considering *SentenceQS*, the dimensionality reduction operated by the Multi-Layer percepetron and the subsequent implementation of a reduced encoding in the quantum circuit, hinder the computation, decreasing the leaning capabilities of the model. It is worth to notice that in this case the loss requires more epochs to reach convergence: this is due to the fact that, adding a whole classical layer to the quantum circuit, generates a more complex energy landscape, making the training, i.e. the finding of the optimal minimum, expensive and complicated. In this sense, it is here pointed out, that is generally advisable, when using quantum computing for transfer learning, to make use of the most information as possible, coming from the pre-trained models. Hence, it is necessary to use in this regards a proper quantum encoding strategy that prevents information loss during the computation.

Table 2. Classification results on the ItaCoLA test set using LSTM and BERT and the three training strategies: SentenceC, SentenceQF, SentenceQS.

Model	Accuracy
LSTM	0.794
ITA-BERT	0.904
SentenceC	0.67
SentenceQF	0.81
SentenceQS	0.71

The performances of each training are captured by the accuracy of the classification on the test set. The accuracy might be biased by the presence of class unbalance: however, both the weighting in the loss function and the balanced sampling on the test set are able to prevent biased classification. The results are listed in Table 2: while *ITA-BERT* is still the most accurate model for the Acceptability Judgement task, both the quantum classifier in use, with transfer learning from Sentence-BERT, show improved performances compared to the classical pipeline *SentenceC*. Furthermore, *SentenceQF* is able to outperform quite significantly the LSTM. This is a promising result for two particular reasons: first it shows the potential of quantum computing in general terms for machine learning tasks. In fact, compared to an LSTM, the *SentenceQF* model has fewer parameters but it still able to learn more efficiently. Second it shows that hybrid quantum-classical methods can play already an important role for NLP tasks. The ability of quantum circuit to represent language data in complex high dimensional vector spaces can fuel the most sophisticated NLP deep learning model with higher expressive capabilities, hence potentially impacting the performances of task beyond the binary classification hereby investigated.

5 Conclusion

In this work, a novel approach for transfer learning with quantum computing has been investigated, focused on the Acceptability Judgment task on Italian language using an annotated dataset. Specifically, using a whole sentence embedding strategy, provided by Sentence-BERT, a quantum classifier has been trained, with two different encoding strategies. Results demonstrate the potential of this approach: while the metrics are not yet comparable with the most advanced classical models, such as ITA-BERT, they have the advantage of being good enough even with a relatively simple ansatz. Furthermore, they show an advantage on the specific task considered both over a simple multilayer perceptron, and when compared with more complicated benchmarks, such as the LSTM.

Future works on the same pipelines hereby proposed, will focus on two elements: first, overcoming the sentence based embedding, provided by sentence-BERT, and focusing on single word embedding that in principle entail more information about the semantics and the properties of each element of the phrase. Second, on the quantum level, increasing the complexity of the ansatz in order to rise up the overall expressivity of the model. These two elements combined might be able to show quantum advantage on NLP transfer learning tasks. From a strictly linguistic point of view, the prediction and classification of acceptability judgments in NLP play a crucial role in improving the quality and fluency of various NLP applications, ranging from language models to language generation. Overall, By advancing the understanding and modeling of acceptability judgments, researchers aim to enhance the effectiveness and naturalness of human-computer interactions, making NLP systems more linguistically informed and aligned with human expectations.

Although still in its infancy and with numerous hardware limitations, quantum computing in NLP classification tasks offers a novel perspective on addressing NLP challenge. Quantum-inspired algorithms, quantum embeddings, and quantum machine learning models present exciting avenues for improving classification accuracy, semantic understanding, and computational efficiency. Further research and development in this interdisciplinary field have the potential to unlock new possibilities for quantum-assisted natural language processing.

Acknowledgments. We acknowledge financial support from the project PNR MUR project PE0000013-FAIR.

References

1. Bergholm, V., et al.: PennyLane: automatic differentiation of hybrid quantum-classical computations. arXiv e-prints arXiv:1811.04968 (2018). https://doi.org/10.48550/arXiv.1811.04968
2. Bonetti, F., Leonardelli, E., Trotta, D., Guarasci, R., Tonelli, S.: Work hard, play hard: collecting acceptability annotations through a 3d game. In: Proceedings of the Thirteenth Language Resources and Evaluation Conference, pp. 1740–1750. European Language Resources Association (2022)

3. Brown, T., et al.: Language models are few-shot learners. In: Larochelle, H., Ranzato, M., Hadsell, R., Balcan, M., Lin, H. (eds.) Advances in Neural Information Processing Systems, vol. 33, pp. 1877–1901. Curran Associates, Inc. (2020). https://proceedings.neurips.cc/paper/2020/file/1457c0d6bfcb4967418bfb8ac142f64a-Paper.pdf
4. Chen, S.Y.C., Huang, C.M., Hsing, C.W., Kao, Y.J.: Hybrid quantum-classical classifier based on tensor network and variational quantum circuit. arXiv preprint arXiv:2011.14651 (2020)
5. Chi, E.A., Hewitt, J., Manning, C.D.: Finding universal grammatical relations in multilingual bert. arXiv preprint arXiv:2005.04511 (2020)
6. Coecke, B., Sadrzadeh, M., Clark, S.: Mathematical foundations for a compositional distributional model of meaning. arXiv preprint arXiv:1003.4394 (2010)
7. Devlin, J., Chang, M.W., Lee, K., Toutanova, K.: BERT: pre-training of deep bidirectional transformers for language understanding. In: Proceedings of the 2019 Conference of the North American Chapter of the Association for Computational Linguistics: Human Language Technologies, Volume 1 (Long and Short Papers), pp. 4171–4186. Association for Computational Linguistics, Minneapolis, Minnesota (2019). https://doi.org/10.18653/v1/N19-1423
8. Du, Y., Huang, T., You, S., Hsieh, M.H., Tao, D.: Quantum circuit architecture search for variational quantum algorithms. NPJ Quantum Inf. **8**(1), 62 (2022). https://doi.org/10.1038/s41534-022-00570-y
9. Floridi, L., Chiriatti, M.: GPT-3: its nature, scope, limits, and consequences. Mind. Mach. **30**(4), 681–694 (2020)
10. Gargiulo, F., et al.: An electra-based model for neural coreference resolution. IEEE Access **10**, 75144–75157 (2022). https://doi.org/10.1109/ACCESS.2022.3189956. https://www.scopus.com/inward/record.uri?eid=2-s2.0-85134194779&doi=10.1109%2fACCESS.2022.3189956&partnerID=40&md5=57417b400106b1a21ff6e21924ed0a61
11. Guarasci, R., De Pietro, G., Esposito, M.: Quantum natural language processing: challenges and opportunities. Appl. Sci. **12**(11), 5651 (2022)
12. Guarasci, R., Minutolo, A., Damiano, E., De Pietro, G., Fujita, H., Esposito, M.: Electra for neural coreference resolution in Italian. IEEE Access **9**, 115643–115654 (2021)
13. Guarasci, R., Silvestri, S., De Pietro, G., Fujita, H., Esposito, M.: Assessing bert's ability to learn Italian syntax: a study on null-subject and agreement phenomena. J. Ambient Intell. Humanized Comput. 1–15 (2021)
14. Guarasci, R., Silvestri, S., De Pietro, G., Fujita, H., Esposito, M.: Bert syntactic transfer: a computational experiment on Italian, French and English languages. Comput. Speech Lang. **71**, 101261 (2022)
15. Jawahar, G., Sagot, B., Seddah, D.: What does BERT learn about the structure of language? In: Proceedings of the 57th Annual Meeting of the Association for Computational Linguistics, pp. 3651–3657. ACL, Florence, Italy (2019). https://doi.org/10.18653/v1/P19-1356. https://www.aclweb.org/anthology/P19-1356
16. Jentoft, M., Samuel, D.: Nocola: the norwegian corpus of linguistic acceptability. In: Proceedings of the 24th Nordic Conference on Computational Linguistics (NoDaLiDa), pp. 610–617 (2023)
17. Jiang, Z., Xu, F.F., Araki, J., Neubig, G.: How can we know what language models know? Trans. Assoc. Comput. Linguist. **8**, 423–438 (2020)
18. Lau, J.H., Clark, A., Lappin, S.: Measuring gradience in speakers' grammaticality judgements. In: Proceedings of the Annual Meeting of the Cognitive Science Society, vol. 36 (2014)

19. Lee, K., He, L., Lewis, M., Zettlemoyer, L.: End-to-end neural coreference resolution. In: Proceedings of the 2017 Conference on Empirical Methods in Natural Language Processing, pp. 188–197. Association for Computational Linguistics, Copenhagen, Denmark (2017). https://doi.org/10.18653/v1/D17-1018. https://aclanthology.org/D17-1018
20. Li, G., Zhao, X., Wang, X.: Quantum self-attention neural networks for text classification. arXiv preprint arXiv:2205.05625 (2022)
21. Li, Q., Wang, B., Zhu, Y., Lioma, C., Liu, Q.: Adapting pre-trained language models for quantum natural language processing. arXiv preprint arXiv:2302.13812 (2023)
22. Linzen, T.: What can linguistics and deep learning contribute to each other? Response to pater. Language **95**(1) (2019)
23. Lloyd, S., Schuld, M., Ijaz, A., Izaac, J., Killoran, N.: Quantum embeddings for machine learning. arXiv preprint arXiv:2001.03622 (2020)
24. Mari, A., Bromley, T.R., Izaac, J., Schuld, M., Killoran, N.: Transfer learning in hybrid classical-quantum neural networks. Quantum **4**, 340 (2020)
25. Marvin, R., Linzen, T.: Targeted syntactic evaluation of language models. In: Proceedings of the Society for Computation in Linguistics (SCiL), pp. 373–374 (2019)
26. Meichanetzidis, K., Toumi, A., de Felice, G., Coecke, B.: Grammar-aware question-answering on quantum computers. arXiv preprint arXiv:2012.03756 (2020)
27. Mikhailov, V., Shamardina, T., Ryabinin, M., Pestova, A., Smurov, I., Artemova, E.: Rucola: Russian corpus of linguistic acceptability. arXiv preprint arXiv:2210.12814 (2022)
28. Minutolo, A., Guarasci, R., Damiano, E., De Pietro, G., Fujita, H., Esposito, M.: A multi-level methodology for the automated translation of a coreference resolution dataset: an application to the Italian language. Neural Comput. Appl. **34**(24), 22493–22518 (2022). https://doi.org/10.1007/s00521-022-07641-3
29. Qiu, X., Sun, T., Xu, Y., Shao, Y., Dai, N., Huang, X.: Pre-trained models for natural language processing: a survey. SCIENCE CHINA Technol. Sci. **63**(10), 1872–1897 (2020)
30. Reimers, N., Gurevych, I.: Sentence-bert: sentence embeddings using siamese bert-networks. In: Proceedings of the 2019 Conference on Empirical Methods in Natural Language Processing. Association for Computational Linguistics (2019). https://arxiv.org/abs/1908.10084
31. Schuld, M., Petruccione, F., Schuld, M., Petruccione, F.: Quantum models as kernel methods. In: Machine Learning with Quantum Computers, pp. 217–245 (2021)
32. Someya, T., Oseki, Y.: Jblimp: Japanese benchmark of linguistic minimal pairs. In: Findings of the Association for Computational Linguistics: EACL 2023, pp. 1536–1549 (2023)
33. Sordoni, A., Nie, J.Y., Bengio, Y.: Modeling term dependencies with quantum language models for IR. In: Proceedings of the 36th International ACM SIGIR Conference on Research and Development in Information Retrieval, pp. 653–662 (2013)
34. Sun, C., Qiu, X., Xu, Y., Huang, X.: How to fine-tune BERT for text classification? In: Sun, M., Huang, X., Ji, H., Liu, Z., Liu, Y. (eds.) CCL 2019. LNCS (LNAI), vol. 11856, pp. 194–206. Springer, Cham (2019). https://doi.org/10.1007/978-3-030-32381-3_16
35. Torlai, G., Melko, R.G.: Machine-learning quantum states in the NISQ era. Ann. Rev. Condensed Matter Phys. **11**, 325–344 (2020)

36. Trotta, D., Guarasci, R., Leonardelli, E., Tonelli, S.: Monolingual and cross-lingual acceptability judgments with the Italian CoLA corpus. In: Findings of the Association for Computational Linguistics: EMNLP 2021, pp. 2929–2940. Association for Computational Linguistics, Punta Cana, Dominican Republic (2021). https://aclanthology.org/2021.findings-emnlp.250
37. Volodina, E., Mohammed, Y.A., Klezl, J.: DaLAJ – a dataset for linguistic acceptability judgments for Swedish. In: Proceedings of the 10th Workshop on NLP for Computer Assisted Language Learning, pp. 28–37. LiU Electronic Press, Online (2021). https://aclanthology.org/2021.nlp4call-1.3
38. Wang, A., Singh, A., Michael, J., Hill, F., Levy, O., Bowman, S.: GLUE: a multitask benchmark and analysis platform for natural language understanding (2018). https://doi.org/10.18653/v1/W18-5446. https://aclanthology.org/W18-5446
39. Warstadt, A., Singh, A., Bowman, S.R.: Neural network acceptability judgments. Trans. Assoc. Comput. Linguist. **7**, 625–641 (2019). https://doi.org/10.1162/tacl_a_00290
40. Xiang, B., Yang, C., Li, Y., Warstadt, A., Kann, K.: CLiMP: a benchmark for Chinese language model evaluation. In: Proceedings of the 16th Conference of the European Chapter of the Association for Computational Linguistics: Main Volume, pp. 2784–2790. Association for Computational Linguistics, Online (2021). https://doi.org/10.18653/v1/2021.eacl-main.242. https://aclanthology.org/2021.eacl-main.242
41. Zhu, J., et al.: Incorporating bert into neural machine translation. In: International Conference on Learning Representations (2019)

Unimaginable Numbers: a Case Study as a Starting Point for an Educational Experimentation

Francesco Ingarozza[1]([✉])[iD], Gianfranco d'Atri[2], and Rosanna Iembo[2][iD]

[1] Liceo Scientifico Filolao, 88900 Crotone, Italy
francesco.ingarozza@filolao.edu.it
[2] Università della Calabria (University of Calabria), 87036 Arcavacata di Rende, Italy
datri@mat.unical.it, rosannaiembo@libero.it

Abstract. How big is a large number? How far can our imagination go to think or to imagine a large number? This work is part of the study of didactic approaches aimed at the knowledge of unimaginable numbers in secondary school. The topic is very interesting because this numbers very large, but finite numbers can be transitioned in the passage of the concrete transition from the concept of finite to that of infinity. This work is placed in continuity with the previous case studies carried out on the computational arithmetic of infinity. Specifically, a further case study is shown here: the planning phase contains identification regarding research questions and additional aspects including choice of case study method and discovering strengths or limitations. The case study design linked together from the beginning until the end, including everything from hypothesis and question design to data analysis and conclusion. In addition, it is showed the unit of analysis concerning the case study, the method by which these numbers together with their characteristics will be exhibiting order to give an idea of their estimated size, even with regard to the physical universe. Finally, the effects and results in terms of educational implications will be evaluated through the analysis of the results that emerged from the administration of a questionnaire proposed to the students of the classes involved.

Keywords: Unimaginable numbers · Knuth up-arrow notation · grossone

1 Introduction

The regions of Southern Italy known as Magna Graecia have made a great contribution from a scientific-philosophical point of view to the study of infinity. Evidently already at the time the most illustrious thinkers were used to asking themselves questions such as the following:

- How big is a large number and ?
- How far can our imagination go to think or imagine a large number?

© The Author(s), under exclusive license to Springer Nature Switzerland AG 2025
Y. D. Sergeyev et al. (Eds.): NUMTA 2023, LNCS 14478, pp. 113–126, 2025.
https://doi.org/10.1007/978-3-031-81247-7_9

The answers to these questions allow us to introduce the topic of this work: the so called *unimaginable numbers*. From a historical point of view, the idea of manipulating extremely large numbers, what today we could somehow define as the forerunners or ancestors of unimaginable numbers, has deep roots. We can say that the first of those numbers that today we would call unimaginable was a quantity proposed by Archimedes of Syracuse. This number is well known as Ψ. It could also be found under the name of *Arenarius* or *Send Reckoner*. We have to note that the right name is Ψ while different names are due to a misunderstanding between the name of this number and the name of the book in which the same number is described. We well call this number Arenarius'Ψ (or simply Ψ) from now on. Regarding to this number, we find $\Psi = 10^{8 \times 10^{16}}$ (from the homonyms work, as we have already said, that is simply a number represented by *1* followed by *80 billions millions zeros*, a number used to identify the grain contained in the entire universe). To date, the best known notation for their representation is the one called "Knuth's up-arrow notation". More than a known notation in an absolute sense, it is a relatively known notation since this sector of mathematics is still little traveled in the academic field and almost unexplored in the rigorously scholastic field. This work is aimed to present an unknown topic as unimaginable numbers, to secondary school students and evaluate the didactic results by analyzing results emerged fro a test carried out by the same students. We want to remember that we don't submit a *zero knowledge* test. Students carried out the test after a presentation of unimaginable numbers and their properties. Such an approach has already been proposed for other topics. In particular, possible forms of experimentation related to the computational arithmetic of infinity have been carried out (see [3,18]). The aim of this work is therefore similar. The object of the study is no longer a form of computational arithmetic of infinity (①) but a form of arithmetic that stops an instant before: the arithmetic of unimaginable numbers. For these reasons we divide this work into three principal sections. Each of them is a fundamental part of the case of study and needs of a detailed description. In Sect. 2 we present unimaginable numbers and their properties. we have chosen to make a detailed presentation, starting from a meticulous description of the historical part and of both the purely algebraic and the analytical parts (including the hyperoperations, the different representations); besides we represents some properties as the results of exercises proposed to the students. We re proposed these example even on this paper, for discursive completeness, without leaving any step that could be useful to repeat a similar work in future. In Sect. 3 we do a little presentation of the properties of computational arithmetic of infinity (①), and then, finally, in Sect. 4 we recognize the results of the test, after a presentation of the set of questions which constitute the same test.

2 Unimaginable Numbers

In this section we describe unimaginable numbers and their properties in the same way they have been presented to the students of the sample before they carried out the test. The unimaginable numbers are finite but extremely large quantities. In this section, the unimaginable numbers will be presented.

2.1 Historical and Scientific Background

Since the period of the Hellenic colonization of southern Italy, some schools that arose in the colonies dealt with representing extremely large numbers. Subsequently, it seems that all traces referring to this topic have been lost in official and sector literature. In fact we return to talk about unimaginable numbers only from 1900 thanks to the work of some of Hilbert's disciples (W. Ackermann) and then G. Sudan, R. Robinson, R. Peter and finally by R. Goodstein around 1940. Then again a blank period until the works of D. E. Knuth (1976). In this section, some ways of representing unimaginable numbers will be described. For some of these modes of representation some arithmetic properties are also foreseen as amply demonstrated (see [11,12,21,22]); only a mention will be given to these latter properties given that the objective of the work is to proceed with a case of study preparatory to a real didactic experimentation aimed at disseminating the topic in the school environment. The basic idea is to express a number, large or small, as the result of an operation suitably coded in such a way as to be defined *recursively*. We introduce the concept of *hyperoperation*, already originally proposed by Ackermann. It is developed starting from the Ackermann function. In a such way, any operation can be expressed recursively starting from known results. About Ackermann function, it is as follows:

$$f_A : \mathbb{N}^2 \to \mathbb{N}, \qquad (m;n) \mapsto f_A(m;n). \tag{1}$$

It is defined recursively as follows:

$$\begin{cases} f_A(0;n) &= 1; \\ f_A(m+1;0) &= f_A(m;1); \\ f_A(m+1;n+1) &= f_A(m; f_A(m+1;n)). \end{cases} \tag{2}$$

Starting from Ackermann function we can define hyperoperations as follows:

$$H_n(a;b) = \begin{cases} b+1 & \text{if } n=0; \\ a & \text{if } n=1 \text{ and } b=0; \\ 0 & \text{if } n=2 \text{ and } b=0; \\ 1 & \text{if } n \geq 3 \text{ and } b=0; \\ H_{n-1}(a; H_n(a; b-1)) & \text{otherwise.} \end{cases} \tag{3}$$

We have to remember that $H_n(a;b) = a[n]b$. By previous definition for hyperoperation we can reach important results (whose demonstration has been gift to students as an exercise!):

$$H_1(a;1) = a+1. \tag{4}$$

$$H_n(a;1) = a \qquad \forall n \geq 2. \tag{5}$$

Knuth came to propose a simple method for expressing sufficiently or arbitrarily large quantities as the result of well-known and well-defined operations. From Ackermann notation to Knuth's up-arrow notation:

$$\uparrow^n = \underbrace{\uparrow\uparrow \ldots \ldots \uparrow}_{n \text{ copies}} = H_{n+2} \tag{6}$$

since

$$n_{Ackerman} = n_{Knuth} + 2. \tag{7}$$

According to (6) and (7) we will intend

$$a \uparrow^n b = a \underbrace{\uparrow\uparrow \ldots \ldots \uparrow}_{n \text{ copies}} b = H_{n+2}(a; b). \tag{8}$$

A such representation is known in the reference literature as *Knuth's up-arrow notation*:

$$a \uparrow^n b = \begin{cases} a \times b & \text{if } n = 0; \\ 1 & \text{if } n \geq 1 \text{ and } b = 0; \\ a \uparrow^{n-1} (a \uparrow^n (b-1)) & \text{if } n \geq 1 \text{ and } b \geq 1. \end{cases} \tag{9}$$

Such a definition provides that operations are defined recursively. We conclude that the general term of (9) is

$$a \uparrow^n b = a \uparrow^{n-1} (a \uparrow^n (b-1)). \tag{10}$$

Stating from (6) and (7) students will simply proof (as an exercise!) *Knuth's version* for (4) and (5). Another very important exercise is to prove that:

$$a \uparrow^n b = \underbrace{a \uparrow^{n-1} (a \uparrow^{n-1} (\ldots (a \uparrow^{n-1} a) \ldots))}_{b-1 \text{ copies of } a\uparrow^{n-1} \text{ and one of } a}. \tag{11}$$

Note that the innermost operation (inside $(b-2)^{th}$ brackets) is $(a \uparrow^{n-1} a)$.

Observation

This mode of representation, known as Knuth's vertical arrows representation is also known as *krata*. The term krata is the plural form of *kratos*, an ancient word of Greek origin which means *power* [7,12]. We have to highlight that kratos is a function *and it is only for simplicity that we use it as Knuth's vertical arrows notation*. Therefore, as far as the representation in the form of power is concerned, the hyperoperation can therefore be expressed in the form $B \uparrow^d T$ for which:

- B is the *base*
- d is the *depth* i.e. which has to be understood as a "power"
- T is the *tag* and indicates the number of copies of the operation defined not by d but starting by d

In detail, the base represents the numerical value on which we perform the operations, the depth is somehow which is associated with the operations themselves and finally the tag gives a measure of the iterations. We must pay attention to this last aspect as the tag does not really constitute the measure of the iterations of the operations, that is the number of repetitions of the operations themselves if not in a particular representation. In fact, there are at least three equivalent representations in krata or Knuth's up-arrow notation. These three different equivalent representations can be generalized by (11). For discursive completeness, some practical results will now be presented which are useful for giving a measure of the capacity of compact representation or, if you like, of the computational power of the notation that makes use of Knuth's arrows.

Product ($n = 0$)

$$a \uparrow^0 b = a \times b$$

or *sum* between a and $b - 1$ *copies* of a or, definitively, *sum* of b *copies* of a.

Exponentiation ($n = 1$)

$$a \uparrow^1 b = \underbrace{a \times a \times ... a}_{b \text{ copies}} = a^b$$

or *product* between a and $b - 1$ *copies* of a or, definitively, *product* of b *copies* of a.

Tetraction ($n = 2$)

$$a \uparrow^2 b = H_4(a; b) = \underbrace{a^{a^{\cdot^{\cdot^{\cdot a}}}}}_{b \text{ copies}}$$

or *exponentiation (power)* between a and $b - 1$ *copies* of a or, definitively, *power (recursive!)* of b *copies* of a.

Pentation ($n = 3$)

$$a \uparrow^3 b = H_5(a; b) = \underbrace{a \uparrow^2 (a \uparrow ...(a \uparrow^2 a))}_{b \text{ copies}}$$

or *tetraction* between a and $b-1$ *copies* of a or, definitively, *tetraction (recursive!)* of b *copies* of a.

Exaction ($n = 4$)

$$a \uparrow^4 b = H_6(a; b) = \underbrace{a \uparrow^3 (a \uparrow ...(a \uparrow^3 a))}_{b \text{ copies}}$$

or *pentaction* between a and $b - 1$ *copies* of a or, definitively, *pentaction (recursive!)* of b *copies* of a. Of course the process can be iterated over and over again.

For example

$$3 \uparrow^3 5 = 3 \uparrow^2 (3 \uparrow^2 (3 \uparrow^2 (3 \uparrow^2 3))).$$

Since $3 \uparrow^2 3 = 3^{27} = 7625597484987$, we can conclude that

$$3 \uparrow^3 5 = 3 \uparrow^2 (3 \uparrow^2 (3 \uparrow^2 7625597484987))$$

2.2 Some Famous Very Big Numbers

In this section we will present some large and more or less known numbers. From a purely historical point of view it can be said that the progenitor of the family of unimaginable numbers is Archimedes' Arenarius'Ψ.

Arenarius Ψ
It is equal to $\Psi = 10^{8 \times 10^{16}}$. In Knuth's notation: $\Psi = ((10 \uparrow 8) \uparrow^2 2) \uparrow (10 \uparrow 8)$. This number is so big that it is represented, in decimal notation, with 80 *million billion* digits (first digit is 1 and the others are 0). A word page, written in *Calibri* font and with size 11 has about 4000 characters (exactly 3956). We need

$$\frac{8 \times 10^{16}}{4 \times 10^3} = 2 \times 10^{13}$$

sheets, for a height of $2 \times 10^{13} \times 8 \times 10^{-5} = 1.6 \times 10^9 m$ (more than 4 times the distance between Hearth and Moon!).

Googol and googolplex.
The googol number is a quantity introduced by the American mathematician Edward Kasner in 1938 to give an estimate of the unimaginable magnitude of infinity in comparison with large but finite quantities. *Googol* is equal to 10^{100} and it represents an upper bound for the size of the physical universe (since the number of elementary particles of the physical universe does not go beyond 10^{90}). This means that a googol is about 10 *billion* times the number of particles of physical universe. Starting from googol we define googolplex as follows: $10^{googol} = 10^{10^{100}}$.

Mega and megiston.
The megiston (according to Steinhaus-Moser representation) is a very large unimaginable number. It is smaller than Graham's number. The Steinhaus-Moser representation for mega is .

Graham's number (G).
Graham's number, the greatest unimaginable numbers which has been used for a mathematical demonstration. Among the various unimaginable numbers, it is the one that presents itself as a solution to a problem which is Graham's problem (hence the name Graham's number!). In Knuth's up-arrow notation, the Graham number is defined by the following recursive representation:

$$G = \left. \begin{matrix} \underbrace{3 \uparrow \dots \uparrow 3} \\ \vdots \\ \underbrace{3 \uparrow^4 3} \end{matrix} \right\} 64 \text{ levels}, \qquad (12)$$

where the number of arrows appearing at each level from the second onwards is given by the number expressed in the next lower level. In other words we have $G = g_{64}$ where g_n is recursively defined by

$$\begin{cases} g_1 = 3 \uparrow^4 3 \\ g_n = 3 \uparrow^{g_{n-1}} 3 & \text{if } n \geq 2 \end{cases}.$$

In order to give an idea of the size of the Graham number, we must remember that $g_1 = 3 \uparrow^4 3$ represents an *exaction* that is $g_1 = 3 \uparrow^4 3 = 3 \uparrow^3 (3 \uparrow^3 3)$. In next section we will speak about other properties of unimaginable numbers.

2.3 When a Number Could Be Consider *Unimaginable*?

In this section, we will present some arithmetic properties of unimaginable numbers and a *threshold* for unimaginable numbers.

– When an obviously large number can be considered unimaginable?

This question is important as it allows us to have an effective estimate, beyond the numerical data itself but related to a more profound interpretation of it, of the kratic representation. Therefore, if we set the *googol* as the minimum limit, as the *threshold*, we find that there are only 58 [7,12] numbers that have a *non-trivial* kratic representation (i.e. in the form of non-trivial powers according to Knuth's notation) and which at the same time are lower than the set limit. Conversely, if the threshold rises further and sets the limit 10^{10000} then it has been demonstrated [12] that there are only 2893 numbers below the threshold 10^{10000} that have a non-trivial kratic representation. Of these, 2888 are of the type $a \uparrow^2 2$ while the remaining 5 do not have a representation of this type. Still for unimaginable numbers, given the function $k(BDT) = B \uparrow^D T$, i.e. the function that associates a kratic representation to each triad of the type *(Base,Depth,Tag)*, various arithmetic properties exist and have been tested, in particular with respect to a periodicity with respect to modular arithmetic. For example, if we suppose $B, D, T \geq 2$, the sequences:

- $\{B \uparrow^D n\}_n$
- $\{B \uparrow^n T\}_n$
- $\{B \uparrow^n n\}_n$

they become constants modulo M (for a fixed positive integer M, see [8]). Also, the sequences $\{n \uparrow^D T\}_n$; $\{n \uparrow^D n\}_n$; $\{n \uparrow^n T\}_n$ and $\{n \uparrow^n n\}_n$ are all periodic modulo M. Finally, there is an algorithm to be able to calculate $\{B \uparrow^D T\}$ modulo M (see [8]). Finally, there are alternative representations with respect to the kratic representation. Among these we can mention the *box notation*, the *superscript and subscript notation* (see [10,11]), the *extended operations* [12], *Nambiar notations* [13], or *Cutler's bar notation* also called *Cutler's circular notation* (see [14,15]). At the end we find *Conway's chained arrows*: which could be used similarly to Knuth's up-arrow notation or when a number is to big that even Knuth's notation could be inappropriate!

3 The *grossone*-Based Numerical System

About 20 years ago, Y. Sergeyev introduced a new computational system based on the so-called *grossone*, whose symbol is ①. This new system is able to perform computations not only using the ordinary (finite) real number, but also by using infinite and infinitesimal quantities. Sergeyev's system is also very easy to use as the familiar system of natural or real numbers. Roughly speaking, the grossone-based system is made up on two fundamental units: the familiar unit 1 to obtain finite numbers (integers, rationals and reals) and a new unit ①, called *grossone*, which is used to write infinite and infinitesimal numbers. The reader can find many details on the new system in introductory surveys as [29,31] and also in the book [27] written in a popular way. Since the new system was proposed about 20 years ago, it has immediately found a large number of applications in many fields of both mathematics and experimental sciences. For example see [1,5,15, 16,20,31], for connection with Fibonacci numbers and applications [19,24], for applications to ordinary differential equations, optimization, cellular automata and game theory, [4,8,9,13,14,26,28,30] for applications to fractals, space filling curves and summations, [10,23,31] for some discussions on logic foundations, paradoxes and their solutions, etc. Recently the grossone-based system has been also tested for educational purposes in high schools both in Italy and abroad: see for example [2,3,17,25] and [18] where the same school as this paper was involved). In the next section and in the conclusions we will discuss and compare the results obtained 4 years later in the same school, Liceo Scientifico Statale Filolao (KR, Italy), on similar tests regarding very basic computations involving the *grossone* system.

4 The *Case of Study*

In this section we present the core of the case of study i.e. the results of tests in order to measure the effectiveness of a didactic approach (discussed in Sect. 2) related to the knowledge of unimaginable numbers. After discussing of the topics of the case of study we presents the sample of the students, the test and its results. The basic idea is to present the unimaginable numbers to high school students. The didactic proposal starts from two elements: the history of unimaginable numbers and the need for a compact representation that overcomes the computational constraint of decimal representation but also of exponential representation. The students are then introduced to the concept of hyperoperations, i.e. operations defined recursively with references to operations already known and sufficiently used (for example, zeroing instead of the successor, *unaction* instead of addition and so on). For completeness, the known unimaginable numbers will also be presented with examples capable of giving an idea of their magnitude. Finally, in order to evaluate the didactic impact, the sample will be administered a test and then the results will be presented and discussed. We have to remember that for the aim of this work we used the model proposed in [6] which has been adapted and for the needs of a high school.

4.1 The Sample

The sample object of the case study discussed in this work is made up of the pupils of 2 classes of the Filolao scientific high school of Crotone. Pupils are at fourth class. In details we are submitted the test to the students of two classes for a total of 48 pupils.

4.2 The Test

The test to be administered to the pupils who make up the sample of the case study is a test of 10 questions with multiple choice. The questions tend to evaluate the following indicators:

- the acquisition of skills in the representation of unimaginable numbers;
- the knowledge of some famous unimaginable numbers and therefore the knowledge of the aspects that could arouse more curiosity or that could act as catalysts towards the knowledge of these particular aspects of mathematics;
- sensitivity towards the measurement of unimaginable numbers arithmetic skills in the operations of transition from one form of representation to another.

The test is administered anonymously in the form of multiple choice questions. Both the questions and the associated answers will be distributed randomly in order to avoid cheating phenomena. The set was elaborated in the same way as a set of questions proposed for the investigation of the arithmetic of infinity (*grossone*). Since in [18] it has been demonstrated that it is useful to administer the questionnaire after the cycle of lessons, we decided to carry out the test only after the presentation of the arguments. Now we present a sample of questions and multiple choice answers referred both to *grossone* test and to *unimaginable numbers* test.

Grossone test (an example of)

1. Let consider the expression $3① + 4①$. It's equal to (select among multiple choice):
    A $7①$.
    B $4①$.
    C $①$.
    D without sense.
    E neither of the previous answers.
5. Let consider the expression $A = ①, B = ① + ①$ e $C = ① \times ①$. Select the right order relationship:
    A $A < B < C$.
    B $A < C < B$.
    C $B < C < A$.
    D $B < A < C$.
    E $A = B = C$.

9 The number of elements of $\mathbb{Z}$ set is (select among multiple choice):
   A ①.
   B ① − 1.
   C ① + 1.
   D ∞.
   E neither of the previous answers.

The questions number 1, 5, 9 are the original ones in the students' test.
  Unimaginable numbers test (an example of)

1 *megiston* is
   A The biggest knonw unimaginable number as solution of a known problem.
   B The number whose value is $10^{8 \times 10^{16}}$.
   C ⑩.
   D ②.
   E ①.

3 What does the following notation represent

$$\begin{cases} g_1 = 3 \uparrow^4 3 \\ g_{64} = 3 \uparrow^{63} 3 \end{cases} \tag{13}$$

   A Arenarius' Ψ.
   B googlplex.
   C mega.
   D megiston.
   E Graham's number.

4 The result of $H_1(a;b)$ is:
   A 0.
   B a.
   C $a \times b$.
   D $a + b$.
   E $b + 1$.

5 The result of $H_3(a;b)$ is
   A a.
   B a^b.
   C $a \times b$.
   D $a + b$.
   E $b + 1$.

4.3 The Results

In this section we presents test results. After a short presentation of unimaginable numbers and their properties pupils were invited to carry out the test on that topics and a second test referred to *grossone*. Both the tests have been presented in previous sections. We did not submit a *zero knowledge* test. In such a way we could have a direct measurement of the effectiveness of the presentation of the topic (since unimaginable numbers represent a topic which is not covered in high schools). In Figs. 1 and 2 we highlight the results of single students regarding to both the tests. Since the size of the sample is small, 48 units or students (for this reason we present a case of study) we can presents single results for each students. This method give us an instantaneous idea of the didactic impact on a single students. In fact, the scores achieved by individual pupils have led to averages specific to the spheres of excellence. We can therefore conclude the following: a detailed theoretical introduction accompanied by some significant examples and some elements of the history of mathematics, used to capture the students' attention, provided the starting point for reaching acceptable levels of knowledge. This result, certainly positive, must certainly be contextualized to the single case study but it can certainly be taken as a model for new and more in-depth experiences in the didactic field. We remember that every test related to this work is made up by ten questions (while for [18] we have 46 questions); for every question, every students scores 1 point for each correct answer and zero points for each non given or incorrect answer. We remember *grossone* test the expected value is 8.67/10 points while for *unimaginable numbers* test it is 7.92/10 points. For [18] the expected value is 42.59/46 points. We have to note that for *grossone* test we find a mean value which reaches 87% of maximum value (normalized on a set of 10 questions) while in [18] it reaches 92% but it is related to a greater set of questions (46 questions instead of 10). For *unimaginable numbers* test we have an expected value of 79% (strictly closed to *grossone* test results). Besides, in the better case we have to remember that the same test has been carried out twice (before speaking about *grossone* and after!) and on a greater sample (3 classes instead of 2). These are very important results since if we consider, over and over again, that nobody of the students in the sample have any knowledge about unimaginable numbers before this didactic experimentation.

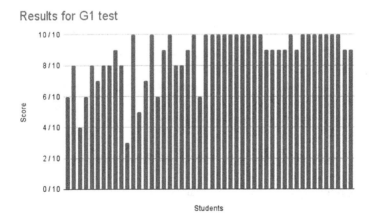

Fig. 1. Statistics and results for *grossone* test

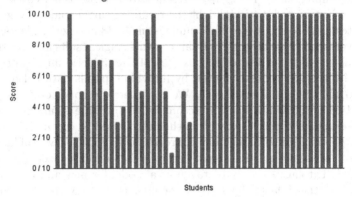

Fig. 2. Statistics and results for *unimaginable numbers* test

5 Conclusions

In this work we presents the results of a case of study carried out on students of Italian high school. We expose and some elements of theory related to unimaginable numbers (first of all the idea of hype operations and notations used to define them). We have to remember a very important aspects: unimaginable numbers are not studied or known by students of Italian high school. Nevertheless the test administered had a very positive outcome. The results (the *positive results* as shown in previous section!) of this test and the approach of this case of study, similarly to the previous [18] should be used in order to do other case of study for unexplored topics in secondary school or to be a starting points for didactic experiment involving a great number of students. We have to remember that this work is only a case of study and a starting point for future educational aims. For this reason we can use it for further and future step:

- *zero knowledge* and *not zero knowledge* tests only related to unimaginable numbers
- *zero knowledge* and *not zero knowledge* tests related to both unimaginable numbers and *grossone*
- comparison between previous tests results

We remember that this is only a starting point from which we can lead different didactic experimentations by using well defined proceedings.

Acknowlegments. This research is partially supported by BlockchainLab s.r.l., Milan (IT).

References

1. Amodio, P., Iavernaro, F., Mazzia, F., Mukhametzhanov, M.S., Sergeyev, Y.D.: A generalized Taylor method of order three for the solution of initial value problems in standard and infinity floating-point arithmetic. Math. Comput. Simul. **141**, 24–39 (2017). https://doi.org/10.1016/j.matcom.2016.03.007
2. Antoniotti, L., Astorino, A., Caldarola, F.: Unimaginable numbers and infinity computing at school: an experimentation in northern Italy. In: Sergeyev, Y.D., Kvasov, D.E., Astorino, A. (eds.) NUMTA 2023. LNCS, vol. 14478, pp. xx–yy. Springer, Cham (2024). https://doi.org/10.1007/978-3-031-81247-7_17
3. Antoniotti, L., Caldarola, F., d'Atri, G., Pellegrini, M.: New approaches to basic calculus: an experimentation via numerical computation. In: Sergeyev, Y.D., Kvasov, D.E. (eds.) NUMTA 2019. LNCS, vol. 11973, pp. 329–342. Springer, Cham (2020). https://doi.org/10.1007/978-3-030-39081-5_29
4. Antoniotti, L., Caldarola, F., Maiolo, M.: Infinite numerical computing applied to Peano's, Hilbert's, and Moore's curves. Mediterr. J. Math. **17**, 99 (2020). https://doi.org/10.1007/s00009-020-01531-5
5. Astorino, A., Fuduli, A.: Spherical separation with infinitely far center. Soft. Comput. **24**(23), 17751–17759 (2020)
6. Bertacchini, F., Bilotta, E., Caldarola, F., Pantano, P.: The role of computer simulations in learning analytic mechanics towards chaos theory: a course experimentation. Int. J. Math. Educ. Sci. Technol. **50**, 100–120 (2019). https://doi.org/10.1080/0020739X.2018.1478134
7. Blakley, G.R., Borosh, I.: Knuth's iterated powers. Adv. Math. **34**, 109–136 (1979). https://doi.org/10.1016/0001-8708(79)90052-5
8. Caldarola, F.: The exact measures of the Sierpiński d-dimensional tetrahedron in connection with a Diophantine nonlinear system. Commun. Nonlinear Sci. Numer. Simul. **63**, 228–238 (2018). https://doi.org/10.1016/j.cnsns.2018.02.026
9. Caldarola, F.: The Sierpiński curve viewed by numerical computations with infinities and infinitesimals. Appl. Math. Comput. **318**, 321–328 (2018). https://doi.org/10.1016/j.amc.2017.06.024
10. Caldarola, F., Cortese, D., d'Atri, G., Maiolo, M.: Paradoxes of the infinite and ontological dilemmas between ancient philosophy and modern mathematical solutions. In: Sergeyev, Y.D., Kvasov, D.E. (eds.) NUMTA 2019. LNCS, vol. 11973, pp. 358–372. Springer, Cham (2020). https://doi.org/10.1007/978-3-030-39081-5_31
11. Caldarola, F., d'Atri, G., Maiolo, M.: What are the "unimaginable numbers"? In: Caldarola, F., d'Atri, G., Maiolo, M., Pirillo, G. (eds.) Proceedings of the International Conference on "From Pitagora to Schützenberger", pp. 17–29. Pellegrini Editore, Cosenza (IT) (2020)
12. Caldarola, F., d'Atri, G., Mercuri, P., Talamanca, V.: On the arithmetic of Knuth's powers and some computational results about their density. In: Sergeyev, Y.D., Kvasov, D.E. (eds.) NUMTA 2019. LNCS, vol. 11973, pp. 381–388. Springer, Cham (2020). https://doi.org/10.1007/978-3-030-39081-5_33
13. Caldarola, F., Maiolo, M.: On the topological convergence of multi-rule sequences of sets and fractal patterns. Soft. Comput. **24**, 17737–17749 (2020). https://doi.org/10.1007/s00500-020-05358-w
14. Caldarola, F., Maiolo, M., Solferino, V.: A new approach to the Z-transform through infinite computation. Commun. Nonlinear Sci. Numer. Simul. **82**, 105019 (2020). https://doi.org/10.1016/j.cnsns.2019.105019

15. Cococcioni, M., Cudazzo, A., Pappalardo, M., Sergeyev, Y.D.: Solving the lexicographic multi-objective mixed-integer linear programming problem using branch-and-bound and grossone methodology. Commun. Nonlinear Sci. Numer. Simul. **84**, 105177 (2020). https://doi.org/10.1016/j.cnsns.2020.105177
16. D'Alotto, L.: Infinite games on finite graphs using grossone. Soft. Comput. **24**, 17509–17515 (2020)
17. Iannone, P., Rizza, D., Thoma, A.: Investigating secondary school students' epistemologies through a class activity concerning infinity. In: Bergqvist, E., Österholm, M., Granberg, C., Sumpter, L. (eds.) Proceedings of the 42nd Conference of the International Group for the Psychology of Mathematics Education, vol. 3, pp. 131–138. PME, Umeå, Sweden (2018)
18. Ingarozza, F., Adamo, M.T., Martino, M., Piscitelli, A.: A grossone-based numerical model for computations with infinity: a case study in an Italian high school. In: Sergeyev, Y.D., Kvasov, D.E. (eds.) NUMTA 2019. LNCS, vol. 11973, pp. 451–462. Springer, Cham (2020). https://doi.org/10.1007/978-3-030-39081-5_39
19. Ingarozza, F., Piscitelli, A.: A new class of octagons. In: Sergeyev, Y.D., Kvasov, D.E. (eds.) Proceedings of the 4th International Conference on "Numerical Computations: Theory and Algorithms". Lecture Notes in Computer Science. Springer, Cham (2024). https://doi.org/10.1007/978-3-031-81247-7_10
20. Iudin, D., Sergeyev, Y.D., Hayakawa, M.: Infinity computations in cellular automaton forest-fire model. Commun. Nonlinear Sci. Numer. Simul. **20**, 861–870 (2015)
21. Leonardis, A., d'Atri, G., Caldarola, F.: Beyond Knuth's notation for unimaginable numbers within computational number theory. Int. Electron. J. Algebra **31**, 55–73 (2022). https://doi.org/10.24330/ieja.1058413
22. Leonardis, A., d'Atri, G., Zanardo, E.: Goodstein's generalized theorem: from rooted tree representations to the Hydra game. J. Appl. Math. Inform. **40**, 833–896 (2022). https://doi.org/10.14317/jami.2022.883
23. Lolli, G.: Metamathematical investigations on the theory of grossone. Appl. Math. Comput. **255**, 3–14 (2015)
24. Margenstern, M.: Fibonacci words, hyperbolic tilings and grossone. Commun. Nonlinear Sci. Numer. Simul. **21**, 3–11 (2015)
25. Mazzia, F.: A computational point of view on teaching derivatives. Inform. Educ. **37**, 79–86 (2022)
26. Pepelyshev, A., Zhigljavsky, A.: Discrete uniform and binomial distributions with infinite support. Soft. Comput. **24**, 17517–17524 (2020)
27. Sergeyev, Y.D.: Arithmetic of Infinity. Edizioni Orizzonti Meridionali, Cosenza (2003, 2nd ed 2013)
28. Sergeyev, Y.D.: Blinking fractals and their quantitative analysis using infinite and infinitesimal numbers. Chaos Solitons Fractals **33**(1), 50–75 (2007)
29. Sergeyev, Y.D.: Lagrange lecture: methodology of numerical computations with infinities and infinitesimals. Rendiconti del Seminario Matematico dell'Università e del Politecnico di Torino **68**, 95–113 (2010)
30. Sergeyev, Y.D.: Using blinking fractals for mathematical modelling of processes of growth in biological systems. Informatica **22**, 559–576 (2011)
31. Sergeyev, Y.D.: Numerical infinities and infinitesimals: methodology, applications, and repercussions on two Hilbert problems. EMS Surv. Math. Sci. **4**, 219–320 (2017)

Some Notes on a Continuous Class of Octagons

Francesco Ingarozza[✉] and Aldo Piscitelli

Liceo Scientifico Filolao, 88900 Crotone, Italy
{francesco.ingarozza,aldo.piscitelli}@filolao.edu.it

Abstract. This paper deals with a continuous class of octagons, built from the so-called "sequence of Carboncettus octagons". We first explain the building techniques, then we study the main properties and make comparisons with the original sequence (which gives a discrete family of octagons) and some related ones. Finally, in the last part of this work, we also study the behavior of this new continuous family of octagons by adopting the lens of the infinity computing.

Keywords: Octagons · Fibonacci numbers · Golden section · Carboncettus octagons · Infinity computing

1 Introduction

This paper contains some notes on a new continuous family of octagons that arises from the so-called sequence of *Carboncettus octagons* (see [10,11]). But while the sequence of Carboncettus octagons is a discrete family, in this paper we will study a family that originates from it, and whose elements depend on a parameter that varies in a continuous interval, i.e. an interval of real numbers.

We call our continuous family and its element *Carboncettus-like octagons*, often abbreviated in the following as *CL octagons*. For the historical origins of the name *Carboncettus* we refer the reader to [29].

In this work we also make some comparisons between the new octagons family, the original sequence $\{C_n\}_n$ of Carboncettus octagons, and the sequence of the "normalized" Carboncettus octagons $\{C_n^N\}_n$ (see [11]). In this perspective the use of infinity computing assumes a very important role and gives very interesting advantages, as we show in Sect. 4.

In order to construct our sequences, for some calculations and for the figures necessary to our discussion, we have used the software *GeoGebra* (see [18,19]). We point out that, in this context, both the use of infinity computing and the use of a software like GeoGebra can have very interesting educational implications. For example, they could be very useful in the study of the transition phase from discrete to continuum mathematics (see [17,20,23,24,34–37]).

2 The Original Octagons Sequences $\{C_n\}_n$ and $\{C_n^N\}_n$

In this section we recall the construction of the original sequence $\{C_n\}_n$ of Carboncettus octagons (see [10,11]). The building procedure is as follows:

- We take two concentric circles with their centers at the origin of the axes and we draw four tangent lines to the inner circumference, each one parallel to a coordinate axis.
- The points where the tangents intersect the outer circle are the vertices of an octagon.

The main characteristic of the sequence $\{C_n\}_n$ is that each octagon is obtained by using two consecutive Fibonacci numbers, but both with odd or even indexes. In other words, the n-th Carboncettus octagon C_n is originated by starting from an inner circumference of radius equal to the n-th Fibonacci number φ_n, and an outer circumference of radius φ_{n+2}. We remark that at least starting from C_4, all the octagons of the sequence $\{C_n\}_n$ are almost regular, i.e. completely indistinguishable from a regular octagon (see [10,11,29]).

Regarding to the sequence $\{C_n^N\}_n$, the normalized radii of the circumferences are φ_n/φ_n (for the internal radius) and φ_{n+2}/φ_n (for the external one). For the external normalized radius we get:

$$\lim_{n\to\infty} \frac{\varphi_{n+2}}{\varphi_n} = \lim_{n\to\infty} \frac{\varphi_{n+2}}{\varphi_{n+1}} \cdot \frac{\varphi_{n+1}}{\varphi_n} = \varphi^2,$$

where

$$\varphi = \frac{1+\sqrt{5}}{2} \qquad (1)$$

is the golden ratio and, consequently,

$$\varphi^2 = \frac{3+\sqrt{5}}{2}. \qquad (2)$$

Hence we can conclude that the sequence $\{C_n^N\}_n$ converges to a "limit octagon" inscribed inside a circumference with radius φ^2.

3 The New Continuous Family of Octagons

In this section we present a new class of octagons, whose constructive model is derived from the one used for $\{C_n\}_n$ and $\{C_n^N\}_n$. Then we will discuss some properties and characteristics of the elements of the new family. In Fig. 1 the building model for the CL octagons is represented. In Fig. 2 a detail of the CL octagon when $r = 0.1$.

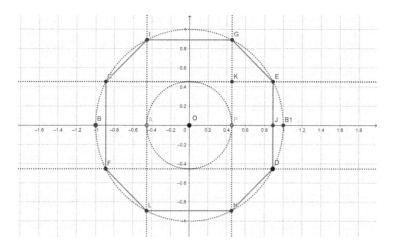

Fig. 1. The output of GeoGebra for the CL octagons construction.

3.1 General Characteristics of the CL Octagons

We can summarize the building procedure for the new class of octagons in the following steps:

- We identify two concentric circumferences.
- We draw the two horizontal tangents and the two vertical tangents to the inner circumferences.
- Each tangent line intersects the external circumference at two points.
- We join these points and find an octagon.

The resulting octagon is called a *CL octagon* to remember the used procedure to build it. For this new class of octagons the outer circle is kept fixed with unit radius, while the radius of the inner circle varies *continuously* within the interval $[0, 1]$.

At an operational level, with reference to Fig. 1, we can say:

- It arises by construction

$$|OB| = |OB_1| = 1, \qquad |OP| = r, \qquad r \in [0, 1].$$

- Always by construction it arises

$$|AP| = 2r.$$

It can be observed that the construction conditions of the CL octagons are absolutely compatible with geometric constraints of general validity. For example, the chord $|OB|$ will always be not greater than $|OB_1|$, where the last is the radius of the outer circumference.

Since

$$|OB| = |OB_1| = 1,$$

a consequence of Pythagoras theorem is

$$|OJ| = \sqrt{1-r^2}. \tag{3}$$

Then we find

$$|PJ| = |OJ| - |OP| = \sqrt{1-r^2} - r. \tag{4}$$

From the properties of parallelograms we get

$$|GK| = |KE| = |PJ| \tag{5}$$

and then

$$|GE| = \sqrt{2}|PJ| = \sqrt{2}(\sqrt{1-r^2} - r). \tag{6}$$

So it is possible to identify three functions, defined below, which *measure* in some way the ratios between the sides of the octagon CL thus obtained:

$$\theta(r) = \frac{|ED|}{|GE|} = \frac{2r}{\sqrt{2}(\sqrt{1-r^2} - r)} = \frac{\sqrt{2}r(\sqrt{1-r^2} + r)}{1 - 2r^2}, \tag{7}$$

$$\rho(r) = \frac{|GE|}{|ED|} = \theta(r)^{-1} = \frac{(\sqrt{1-r^2} - r)}{\sqrt{2}r}, \tag{8}$$

$$\delta(r) = |GE| - |ED| = \sqrt{2}(\sqrt{1-r^2} - r) - 2r. \tag{9}$$

In this way it is possible to study the evolution of these CL octagons when the radius $r = |OP|$ of the inner circumference varies. In the next section we will studies the properties and the behavior of these functions.

3.2 The Functions $\theta(r)$, $\rho(r)$ and $\delta(r)$

Some important information can be deduced from the study of the functions $\theta(r)$, $\rho(r)$ and $\delta(r)$ defined above. First of all it should be noted that the functions will be studied in relation to their "geometric meaning". For $\theta(r)$, (see (7)) we have:

$$\begin{cases} 1 - r^2 \geq 0 \\ 1 - 2r^2 \neq 0. \end{cases} \tag{10}$$

The function $\rho(r)$ instead represents the inverse ratio, in comparison with $\theta(r)$. Referring to $\rho(r)$ (see (8)), we have:

$$\begin{cases} 1 - r^2 \geq 0 \\ r \neq 0. \end{cases} \tag{11}$$

Regarding to $\delta(r)$ (see (9)) we get:

$$1 - r^2 \geq 0. \tag{12}$$

Hence we can rewrite the functions $\theta(r)$, $\rho(r)$ and $\delta(r)$ as follows

$$\theta: D_\theta(r) \to \mathbb{R}, \qquad r \mapsto \frac{|ED|}{|GE|},$$

$$\rho: D_\rho(r) \to \mathbb{R}, \qquad r \mapsto \frac{|GE|}{|DE|},$$

$$\delta: D_{\delta(r)} \to \mathbb{R}, \qquad r \mapsto (|GE| - |DE|),$$

where, by solving (10), (11) and (12), we get

$$D_{\theta(r)} = [0,1] \setminus \left\{\frac{\sqrt{2}}{2}\right\}, \tag{13}$$

$$D_{\rho(r)} = (0,1], \tag{14}$$

$$D_{\delta(r)} = [0,1]. \tag{15}$$

We now make some considerations for the functions $\theta(r)$ and $\rho(r)$. Note that the function $\rho(r)$ does not exists when $r = 0$, and the same for the function $\theta(r)$ when $r = \sqrt{2}/2$. Geometrically this occurs with the disappearance of one of the sides $|DE|$ or $|GE|$, respectively. When the internal circumference has radius $r = 0$, it becomes a single point and the CL octagon becomes a square. Note that when $r = \sqrt{2}/2$ the CL octagon becomes a square as well.

By convenience, we denote an element of the new continuous family of octagons by O_r, where the subscript $r \in [0,1]$ represents the radius of the internal circumference. In Fig. 2, the graphs of the functions $\theta(r)$, $\delta(r)$ and $\rho(r)$ are represented by different colors.

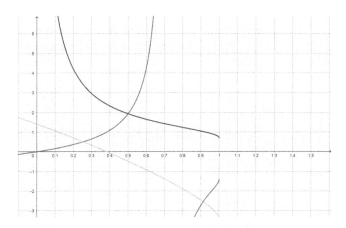

Fig. 2. The graph of the function $\theta(r)$ (red), $\rho(r)$ (blue) and $\delta(r)$ (light blue). (Color figure online)

Obviously all the elements of the sequence $\{C_n^N\}_n$ is contained in the new continuous family $\mathcal{O} := \{O_r : r \in [0,1]\}$. In other words, this means that every normalized Carboncettus octagon C_n^N, for all natural numbers n, can be retrieved in the new family $\{O_r : r \in [0,1]\}$ for a suitable value of the parameter/radius r. In fact, to get the n-th normalized Carboncettus octagon C_n^N we need the value $r = \varphi_n/\varphi_{n+2}$ for the continuous parameter r. A particular case occurs when $\delta(r) = 0$ (see (9)), which yields

$$r = \frac{\sqrt{2-\sqrt{2}}}{2}. \tag{16}$$

A further important fact is that we can find the "limit normalized octagon" C_∞^N in our family $\mathcal{O} = \{O_r : r \in [0,1]\}$, see Eq. (19) below (Table 1).

Table 1. In the second column $r_{int}(C_n)$, which is equal to φ_n, represents the value of the internal circumference used to construct the n-th Carboncettus octagon C_n. Similarly, $r_{ext}(C_n)$ is the radius of the external one. Them, the meaning of $r_{int}(C_n^N)$ and $r_{ext}(C_n^N)$ are now obvious.

n	$r_{int}(C_n)$	$r_{ext}(C_n)$	$r_{int}(C_n^N)$	$r_{ext}(C_n^N)$	φ_n/φ_{n+2}
1	1	2	1	2	1/2
2	1	3	1	3	1/3
3	2	5	1	5/2	2/5
4	3	8	1	8/3	3/8
5	5	13	1	13/5	5/13
6	...	...	...	...	...
$n \to \infty$	...	...	...	φ^2	$1/\varphi^2$

In order to understand better the evolution of the octagon O_r when the parameter r varies, we divide the unitary segment $[0,1]$ into the following 5 different subsets:

$$[0,1] = \left[0, \frac{1}{3}\right) \cup \left[\frac{1}{3}, \frac{1}{2}\right] \cup \left(\frac{1}{2}, \frac{\sqrt{2}}{2}\right) \cup \left\{\frac{\sqrt{2}}{2}\right\} \cup \left(\frac{\sqrt{2}}{2}, 1\right]. \tag{17}$$

Note that dome special octagons can now easily recovered in the suitable subset appearing in the decomposition (17). For example, all the octagons of the sequence $\{C_n^N\}_n$ lie in the interval

$$\left[\frac{1}{3}, \frac{1}{2}\right], \tag{18}$$

and this is a quite remarkable fact. In particular, note that the minimum of the interval (18) gives C_2^N and the maximum C_1^N, i.e. the first two elements of the sequence $\{C_n^N\}_n$. Within the interval (18) we also find the limit octagon C_∞^N obtained in correspondence of the value

$$r = \frac{1}{\varphi^2}, \tag{19}$$

(see (2)). Note also the singleton $\{\sqrt{2}/2\}$ in the decomposition (17): we in fact already know that the value $r = \sqrt{2}/2$ corresponding to the case where O_r degenerate to a square. The same phenomenon and the same square (with side $\sqrt{2}/2$) occurs for $r = 0$ too, but the square has its vertices on the coordinate axes in this case. Figure 3 shows the CL octagon $O_{1/10}$ which is close to coinciding with the square O_0.

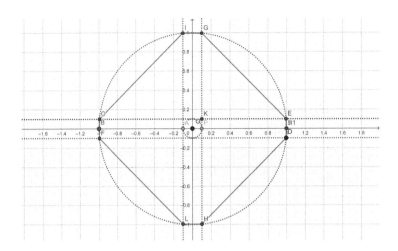

Fig. 3. The CL octagon $O_{1/10}$ obtained by choosing $r = 1/10$.

Recall that in (16) we found the value of r correspondent to the regular octagon, i.e.
$$r = \frac{\sqrt{2-\sqrt{2}}}{2} \approx 0.38268343.$$
Note that even such a value of r belongs to the "principal interval" $[1/3, 1/2]$. Finally we remark another important fact: the last of the five subsets shown in the decomposition (17), i.e. the interval $(\sqrt{2}/2, 1]$ corresponds to the most strange octagons, where they lose their convexity (see Fig. 4). Looking at Fig. 4, this phenomenon is due to the fact that a line segment like DE will become secant for the inner circumference when $r > \sqrt{2}/2$. In fact, in the range of values $(\sqrt{2}/2, 1]$ for r we find a family of *self intersecting octagons*.

To resume ideas, in Table 2 we list some of the most representatives octagons O_r belonging to our family $\mathcal{O}$.

4 The Sequence $\{C_n\}_n$ $\{C_n^N\}_n$ Viewed Through the Lens of Infinity Computing

In this section we want to give some hints for the study of the original Carboncettus sequence $\{C_n\}_n$ when n grows, by using a newly introduced methodology called *infinity computing* or *grossone*-based numerical system.

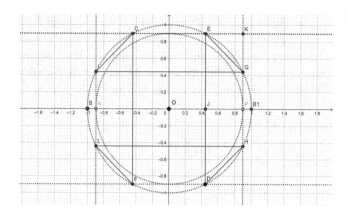

Fig. 4. The CL octagon $O_{9/10}$ obtained by choosing $r = 9/10$. It is a self-intersecting polygon.

Table 2. Values of the radius r of the internal circumference associated with O_r (first column) and the relative value of n as an element of the sequence $\{C_n\}_n$, when such n exists (second column).

Value of r	Element of $\{C_n^N\}_n$	Resulting figure
0	NA	Square
1/3	C_2^N	Octagon
3/8	C_4^N	Octagon
...	...	Octagon
$\varphi^2 = \frac{3+\sqrt{5}}{2}$	C_∞^N	Limit octagon
$\sqrt{\frac{2-\sqrt{2}}{2}}$	NA	Regular octagon
...	...	Octagon
5/13	C_5^N	Octagon
2/5	C_3^N	Octagon
1/2	C_1^N	Octagon
1	NA	Square

In the early 2000s Y. Sergeyev introduced a new numerical system able to perform computations with infinite and infinitesimal number, in a very easy and handle way, as we ordinarily do with natural and real numbers. Roughly speaking, such a new system is constructed on two fundamental units: the ordinary 1 for finite numbers and a new unit ①, called *grossone*, for infinite (and infinitesimal) quantities. We refer the reader to [31,33] for introductory surveys and also to [30] for a book written in a popular way. The grossone-based system has today a number of applications in pure and applied mathematics, as well as in experimental sciences. For example see [1,5,15,16,33] for applications to differential equations, game theory and optimization, [4,7,8,13,14,28,32] to fractals, space filling curves and summations, [9,26,33] for some discussions on foundations and paradoxes, etc. In the last few years Sergeyev's system has also been used in didactic experiments in high schools (see for example [2,3,21,22]) and a very interesting connection with Fibonacci numbers has been explored in [27]. Regarding to [2,22], innovative educational experimentation joining *unimaginable numbers* and *grossone* are described in [12,25]. As said in the previous sections, in [11] there are some hints to apply infinite computing and grossone to the sequence $\{C_n\}_{n\in\mathbb{N}}$. Following them, we first point out to the reader that Fibonacci sequence $\{\varphi_n\}_{n\in\mathbb{N}}$ has a least element $\varphi_①$ in the grossone system, and it can be written for instance as

$$\varphi_① = \frac{\left(\frac{1+\sqrt{5}}{2}\right)^① - \left(\frac{1-\sqrt{5}}{2}\right)^①}{\sqrt{5}}. \tag{20}$$

The previous formula is derived from the well known Binet formula for φ_n (see [6]). The very relevant thing is that, through the new system, we can precisely compute various measures of infinitely large octagons. For example the last element $C_①$ of the $\{C_n\}$ octagons sequence has the following diameter

$$\mathrm{diam}(C_①) = 2\varphi_{①+2} = \frac{(1+\sqrt{5})^{①+2} - (1-\sqrt{5})^{①+2}}{2^{①+1}\sqrt{5}}, \tag{21}$$

which, for instance, is easily comparable with the one of $C_{①-2}$:

$$\mathrm{diam}(C_{①-2}) = 2\varphi_① = \frac{(1+\sqrt{5})^① - (1-\sqrt{5})^①}{2^{①-1}\sqrt{5}}. \tag{22}$$

The reader can note that, without the new system, we can just say that

$$\lim_{n\to\infty} \mathrm{diam}(C_n) = +\infty. \tag{23}$$

Many more computations and evaluations are allowed by the new grossone-based system and with different levels of precision. For example, if we denote by $\approx_i$ the equality up to infinitesimals, we have from (21) and (22)

$$\mathrm{diam}(C_①) \approx_i \frac{(1+\sqrt{5})^{①+2}}{2^{①+1}\sqrt{5}}, \qquad \mathrm{diam}(C_{①-2}) \approx_i \frac{(1+\sqrt{5})^①}{2^{①-1}\sqrt{5}}.$$

And we can compute quite easily, up to infinitesimals, the perimeter, the area, and other measures of $C_①, C_{①-2}$, etc. Anyway, a deeper discussion of such things and more demanding calculations are beyond the scope of this paper: we plan to do this and to give full examples in a future work.

5 Conclusions

In this work we introduced a new family of octagons $\mathcal{O}$ whose elements vary continuously on dependence of a real parameter $r \in [0,1]$. We showed that the previous known Carboncettus normalized octagons C_n^N introduced in [11], are all recoverable $\mathcal{O}$ for suitable values of r belonging to the subinterval $[1/3, 1/2] \subset [0, 1]$. We called an element of the greater family $\mathcal{O}$ a Carboncettus-like octagon (CL octagon for short). We have studied the main properties of the new family $\mathcal{O}$, also with the help of three suitably defined functions $\theta(r)$, $\delta(r)$ and $\rho(r)$, but much work remains to do and we aim to do so in the near future.

In Sect. 4 we finally used the grossone based system to perform a first study of the original sequence $\{C_n\}_n$ inside the greater family $\mathcal{O}$ when n goes to assume infinite values. In this direction there are many aspects that can be investigated in the future as well.

Even possible employs, of the material here exposed, in mathematical education can be examined in the future. And, lastly, from a theoretical and purely mathematical point of view, possible connections with Blaschke's theorem and subsequent results could be studied.

Acknowlegments. This research is partially supported by BlockchainLab s.r.l., Milan (IT).

Special thanks to the anonymous reviewers for their suggestions and patience during the long revision phase of this work.

References

1. Amodio, P., Iavernaro, F., Mazzia, F., Mukhametzhanov, M.S., Sergeyev, Y.D.: A generalized Taylor method of order three for the solution of initial value problems in standard and infinity floating-point arithmetic. Math. Comput. Simul. **141**, 24–39 (2017). https://doi.org/10.1016/j.matcom.2016.03.007
2. Antoniotti, L., Astorino, A., Caldarola, F.: Unimaginable numbers and infinity computing at school: an experimentation in Northern Italy. In: Sergeyev, Y.D., Kvasov, D.E., Astorino, A. (eds.) NUMTA 2023. LNCS, vol. 14478, pp. xx–yy. Springer, Cham (2024). https://doi.org/10.1007/978-3-031-81247-7_17
3. Antoniotti, L., Caldarola, F., d'Atri, G., Pellegrini, M.: New approaches to basic calculus: an experimentation via numerical computation. In: Sergeyev, Y.D., Kvasov, D.E. (eds.) NUMTA 2019. LNCS, vol. 11973, pp. 329–342. Springer, Cham (2020). https://doi.org/10.1007/978-3-030-39081-5_29
4. Antoniotti, L., Caldarola, F., Maiolo, M.: Infinite numerical computing applied to Peano's, Hilbert's, and Moore's curves. Mediterr. J. Math. **17**, 99 (2020). https://doi.org/10.1007/s00009-020-01531-5
5. Astorino, A., Fuduli, A.: Spherical separation with infinitely far center. Soft. Comput. **24**, 17751–17759 (2020). https://doi.org/10.1007/s00500-020-05352-2
6. Ball, K.M.: Strange Curves, Counting Rabbits, and Other Mathematical Explorations. Princeton University Press, Princeton (2003)
7. Caldarola, F.: The exact measures of the Sierpiński d-dimensional tetrahedron in connection with a Diophantine nonlinear system. Commun. Nonlinear Sci. Numer. Simul. **63**, 228–238 (2018). https://doi.org/10.1016/j.cnsns.2018.02.026
8. Caldarola, F.: The Sierpiński curve viewed by numerical computations with infinities and infinitesimals. Appl. Math. Comput. **318**, 321–328 (2018). https://doi.org/10.1016/j.amc.2017.06.024

9. Caldarola, F., Cortese, D., d'Atri, G., Maiolo, M.: Paradoxes of the infinite and ontological dilemmas between ancient philosophy and modern mathematical solutions. In: Sergeyev, Y.D., Kvasov, D.E. (eds.) NUMTA 2019. LNCS, vol. 11973, pp. 358–372. Springer, Cham (2020). https://doi.org/10.1007/978-3-030-39081-5_31
10. Caldarola, F., d'Atri, G., Maiolo, M., Pirillo, G.: New algebraic and geometric constructs arising from Fibonacci numbers. In honor of Masami Ito. Soft Comput. **24(23)**, 17497–17508 (2020). https://doi.org/10.1007/s00500-020-05256-1
11. Caldarola, F., d'Atri, G., Maiolo, M., Pirillo, G.: The sequence of carboncettus octagons. In: Sergeyev, Y.D., Kvasov, D.E. (eds.) NUMTA 2019. LNCS, vol. 11973, pp. 373–380. Springer, Cham (2020). https://doi.org/10.1007/978-3-030-39081-5_32
12. Caldarola, F., d'Atri, G., Mercuri, P., Talamanca, V.: On the arithmetic of knuth's powers and some computational results about their density. In: Sergeyev, Y.D., Kvasov, D.E. (eds.) NUMTA 2019. LNCS, vol. 11973, pp. 381–388. Springer, Cham (2020). https://doi.org/10.1007/978-3-030-39081-5_33
13. Caldarola, F., Maiolo, M.: On the topological convergence of multi-rule sequences of sets and fractal patterns. Soft Comput. **24**, 17737–17749 (2020). https://doi.org/10.1007/s00500-020-05358-w
14. Caldarola, F., Maiolo, M., Solferino, V.: A new approach to the Z-transform through infinite computation. Commun. Nonlinear Sci. Numer. Simul. **82**, 105019 (2020). https://doi.org/10.1016/j.cnsns.2019.105019
15. Cococcioni, M., Cudazzo, A., Pappalardo, M., Sergeyev, Y.D.: Solving the lexicographic multi-objective mixed-integer linear programming problem using branch-and-bound and grossone methodology. Commun. Nonlinear Sci. Numer. Simul. **84**, 105177 (2020). https://doi.org/10.1016/j.cnsns.2020.105177
16. D'Alotto, L.: Infinite games on finite graphs using grossone. Soft Comput. **24**, 17509–17515 (2020). https://doi.org/10.1007/s00500-020-0516
17. Frassia, M.G., Serpe, A.: Learning geometry through mathematical modelling: an example with Geogebra. Turk. Online J. Educ. Technol. **2017**, 411–418 (2017)
18. Hastings, H.M., Sugihara, G.: Fractals: A User's Guide for the Natural Sciences. Oxford University Press, Oxford (1994)
19. Hohenwarter, M., Jones, K.: Ways of linking geometry and algebra: the case of Geogebra (2007). http://eprints.soton.ac.uk/id/eprint/50742
20. Hohenwarter, M.: GeoGebra. https://www.geogebra.org
21. Ingarozza, F., Adamo, M.T., Martino, M., Piscitelli, A.: A grossone-based numerical model for computations with infinity: a case study in an Italian high school. In: Sergeyev, Y.D., Kvasov, D.E. (eds.) NUMTA 2019. LNCS, vol. 11973, pp. 451–462. Springer, Cham (2020). https://doi.org/10.1007/978-3-030-39081-5_39
22. Ingarozza, F., d'Atri, G., Iembo, R.: Unimaginable numbers: a case study as a starting point for an educational experimentation. In: Sergeyev, Y.D., Kvasov, D.E. (eds.) NUMTA 2023. LNCS, vol. 14478, pp. xx–yy. Springer, Cham (2024). https://doi.org/10.1007/978-3-031-81247-7_9
23. Karadag, Z., McDougall, D.: Geogebra as a cognitive tool: where cognitive theories and technology meet. In: Bu, L., Schoen, R. (eds.) Model. Learn. Pathways to Math. Underst. Using GeoGebra, pp. 169–181. Sense Publishers (2011). https://doi.org/10.1007/978-94-6091-618-2_12
24. Khali, M., Khalil, U.: Geogebra as a scaffolding tool for exploring analytic geometry structure and developing mathematical thinking of diverse achievers. Int. Electron. J. Math. Educ. **14**, 427–434 (2019). https://doi.org/10.29333/iejme/5746

25. Leonardis, A., d'Atri, G., Caldarola, F.: Beyond Knuth's notation for unimaginable numbers within computational number theory. Int. Electron. J. Algebra **31**, 55–73 (2022). https://doi.org/10.24330/ieja.1058413
26. Lolli, G.: Metamathematical investigations on the theory of grossone. Appl. Math. Comput. **255**, 3–14 (2015). https://doi.org/10.1016/j.amc.2014.03.140
27. Margenstern, M.: Fibonacci words, hyperbolic tilings and grossone. Commun. Nonlinear Sci. Numer. Simul. **21**, 3–11 (2015). https://doi.org/10.1016/j.cnsns.2014.07.032
28. Pepelyshev, A., Zhigljavsky, A.: Discrete uniform and binomial distributions with infinite support. Soft. Comput. **24**, 17517–17524 (2020). https://doi.org/10.1007/s00500-020-05190-2
29. Pirillo, G.: Figure geometriche su un portale del Duomo di Prato. Prato Storia e Arte **121**, 7–16 (2017)
30. Sergeyev, Y.D.: Arithmetic of Infinity. Edizioni Orizzonti Meridionali, Cosenza (2003, 2nd ed 2013)
31. Sergeyev, Y.D.: Lagrange Lecture: Methodology of numerical computations with infinities and infinitesimals. Rendiconti del Seminario Matematico dell'Università e del Politecnico di Torino **68**, 95–113 (2010). https://doi.org/10.48550/arXiv.1203.3165
32. Sergeyev, Y.D.: Using blinking fractals for mathematical modelling of processes of growth in biological systems. Informatica **22**, 559–576 (2011). https://doi.org/10.48550/arXiv.1203.3152
33. Sergeyev, Y.D.: Numerical infinities and infinitesimals: methodology, applications, and repercussions on two Hilbert problems. EMS Surv. Math. Sci. **4**, 219–320 (2017). https://doi.org/10.4171/EMSS/4-2-3
34. Serpe, A.: Geometry of design in high school - an example of teaching with geogebra. In: Gómez Chova, I., López Martinez, A., Candel Torres, I. (eds.) INTED2018, Proceedings 12th International Technology, Education and Development Conference, pp. 463–477. IATED, Valencia, Spain (2018). https://doi.org/10.21125/inted.2018.0668
35. Serpe, A.: Digital tools to enhance interdisciplinary mathematics teaching practices in high school. In: Fulantelli, G., Burgos, D., Casalino, G., Cimitile, M., Lo Bosco, G., Taibi, D. (eds) HELMeTO 2022. CCIS, vol. 1779, pp. 209–218. Springer, Cham (2020). https://doi.org/10.1007/978-3-031-29800-4_16
36. Serpe, A., Frassia, M.G.: Task mathematical modelling design in a dynamic geometry environment: archimedean spiral's algorithm. In: Sergeyev, Y.D., Kvasov, D.E. (eds.) NUMTA 2019. LNCS, vol. 11973, pp. 478–491. Springer, Cham (2020). https://doi.org/10.1007/978-3-030-39081-5_41
37. Serpe, A., Frassia, M.G.: Promote connections between mathematics drawing and history of art in high school through a stem approach. AAPP Atti della Accademia Peloritana dei Pericolanti, Classe di Scienze Fisiche, Matematiche e Naturali **99**, 99S1A5 (2021). https://doi.org/10.1478/AAPP.99S1A5

Game Theory Presented to Italian High School Students in Connection with Infinity Computing

Corrado Mariano Marotta(✉) and Andrea Melicchio

University of Calabria, 87036 Arcavacata di Rende, CS, Italy
corradomariano.marotta@unical.it

Abstract. Game Theory is a rather vast discipline, the purpose of which is to analyze the strategic behaviors of decision-makers (players), or to study the situations in which different players interact pursuing common, different or conflicting objectives. The first purpose of this paper is to discuss the possibility of approaching elementary game theory in high schools. The second objective is to analyze the students' response in terms of learning, but also of liking. Is it possible *"giocare con la teoria dei giochi"* (Engl. transl: play with game theory) in the classroom? The third objective is to highlight the possible links between elementary game theory and the *grossone*-based system introduced by Y.D. Sergeyev. Once again with the didactic aspect of providing high school students with an "easy" and stimulating approach to modern research fields in mathematics. In particular, we describe the approach to a little cycle of lessons of two classes and the performances of a conclusive students' class test. We will see a great students response, above all regarding the connection between game theory and the grossone system.

Keywords: Game theory · Strategic game representation · Mathematical education · Infinity computing · Grossone

1 Introduction

Game Theory is a rather vast discipline, the purpose of which is to analyze the strategic behavior of decision makers (players), that is to study the situations in which different players interact pursuing common, different or conflicting objectives. Players are hypothesized to have rational and intelligent behaviors.

A game is defined as *a situation of strategic interaction between at least two players who behave rationally and intelligently on the basis of rules known to all.*

The term *rational* is directly connected to the neoclassical concept of rationality meaning that each player tries to maximize his final outcome, given a utility function that establishes an order of preference. However, this assumption is however superseded in the sense that it is required that the players, not only are able to solve problems optimally subjected to constraints given, but are aware that their choices influence the behavior of other players.

For this reason, it is preferred to connote the players as *intelligent* and not only as *rational*, in order to indicate the ability both to predict and take into account the behavior of the other players. Game theory can have two different roles:

- The first one (the positive role) is to interpret reality explaining why in certain conflictual situations, the subjects involved (players) adopt certain strategies and certain tactics;
- the second (the prescriptive role) is that one of determining which situations of equilibrium can (or cannot) occur as a result of the interaction of the decision-makers.

In any case, the concepts of solution which are used in the game theory intend to describe the strategies, which the decision-makers should follow, individually or jointly, as a consequence of the rationality hypotheses mentioned above. If then in reality the decision-makers deviate from what is foreseen by the theory, it is, undoubtedly, necessary to ask whether this happens because the model does not capture all the relevant aspects of a situation, or because it is the decision-makers who behave in a non-rational way (or both things). The fundamental difference between the decision theory and the game theory is that in the first, the decision-maker finds himself facing a decision-making problem within some "aleatory states of nature", of which he, maybe, holds a probabilistic characterization; in the second case the player is in front of another decision-maker as a competitor. As a consequence, while in a decision-problem the aim is to achieve an optimal choice (or a sequence of choices), in the second case it is necessary to elaborate a different concept, that is to say that one of equilibrium.

A first classification distinguishes games into *cooperative games* and *non-cooperative games*.

In *cooperative games*, the elementary unit is the coalition of players formed on the basis of binding pacts and opposed to other coalitions. The main problem that arises is not so much on the choice of the moves by the players, but rather on the way to distribute the higher profits from the partnerships.

In *non-cooperative games*, the elementary unit is the player opposed to the others. It is excluded that there can be binding preliminary negotiations, and collateral payments that can take place between the players. It means that each player gains what the outcome of the game attributes to him and that payments between players outside the game are not allowed. Game theory consists of three fundamental blocks:

1. *representation*: it consists of tools and methods to represent a situation of strategic interaction in a formal and graphic way;
2. *decision theory*, which is the representations of individual preferences;
3. *solution theory*: that is methods to understand how players behave.

In turn, the **representation** consists of the following elements:

a. a finite number of **players** $n \geq 2$;

b. the **moves** available to individual players and how they are made. If the players make their moves without knowing the choices made by the others and the game ends after only one move (one shot game) we have static games. If some players, but not all, move first and the other players know the choices made by the first, we have dynamic games;
c. the **outcomes or payoffs** of the game which are the expected profits of the individual players;
d. **preferences** on outcomes.

In this paper we describe a soft approach to game theory proposed to high school students. From many points of view we have taken as a model the experimentation described in [6] concerning an educational approach to chaos theory, but obviously adapting the methodologies to our needs of a high school. In particular, our experimentation involved two classes of the last but one year at the IPSEOA "San Francesco" in Paola (CS), Italy. One has 22 students and the other 21, for a total of 43 students. We organized a cycle of 5 or 6 short lessons for a total of about 6–7 h (depending on the class). The first part of the lesson concerned game theory: a general approach (a little more informal than our introduction above), the difference between strategic and extended forms, the concepts of Pareto optimal, Nash equilibrium and so on. Then, in a second smaller part of the cycle of lessons, we introduced the grossone-based numerical system (see Sect. 2 and Subsect. 2.1 in particular) and we presented to students how it is possible to work together with both of them.

A final class test was administrated to the students, whose results are discussed in details in Sect. 3. Here we report just one of the strongest final conclusion (see Sect. 4): the grossone system seems very appropriate to be taught in high schools, even in connection with (elementary) game theory.

2 Approaching Elementary Game Theory and Infinity Computing in High Schools

2.1 Game Theory via Extended and Strategic Forms

An elementary approach to game theory can be developed with the use of the *strategic form* or the *extended form*.

The representation in extended form takes place through a tree representation, while the strategic (or *normal*) form uses payoff matrices. Both forms have been proposed to students. Questions on the second form appeared in the final test (see Sect. 3), so we give here an example of the first approaches to the strategic form proposed to the students. It deals with the well-known game called *the battle of the sexes* that is explained below.

In separate places, Giulia and Marco choose to spend the evening at the cinema. Giulia prefers to go to the Odeon cinema because they're showing a comedy. Marco wants to go to the Luxe cinema because a thriller is being shown. Both Giulia and Marco would like to spend the evening together but Giulia prefers comedies while Marco prefers thrillers. In this case the players are two:

Giulia and Marco; they make independent and simultaneous choices and have no information when they are making their choice. The moves available to the players are: 1) going to the Odeon cinema to see a comedy (C) or 2) going to the Luxe cinema to see a thriller (T); the outcomes are four: that is, all the possible combinations of the players' moves. The preferences (we indicate with the symbol ">" if an outcome is preferred to another) are given by the ordering on the outcomes.

For Giulia: "Odeon with Marco" > "Luxe with Marco" > "Odeon without Marco" > "Luxe without Marco";

For Marco: "Luxe with Giulia" > "Odeon with Giulia" > "Luxe without Giulia" > "Odeon without Giulia".

Giving the values $3 > 2 > 1 > 0$ to the order of the outcomes above, symmetrically for both the players Giulia and Marco, below there is the strategic representation of the game of sexes and its solution:

$$
\begin{array}{c|cc}
 & \text{Marco} & \\
\text{Giulia} & C & T \\
\hline
C & (3,2) & (1,1) \\
T & (0,0) & (2,3) \\
\end{array}
$$

C obviously stands for comedy, so at the Odeon cinema, T for thriller at the Lux. For instance, the entry at the bottom right, relating to the pair (2,3), is the best for Marco who obtains the maximum possible payoff equal to 3, while Giulia obtains 2. Note that there is not a best solution for both them simultaneously.

Similar examples, representations and exercises have been widely used in class with students. Through them, advanced concepts as *Pareto optimal point*, *Nash equilibrium*, iterated games, etc., have been proposed and explained to the students. Observations and conclusions on the students' response will be given in Sects. 3 and 4.

For general references on basic game theory the reader can see [7] or [23], instead for some examples of applications of game theory in education contexts he/she can see [8] and the references therein.

2.2 Infinity Computing and Game Theory at School

Then it has been developed also a soft approach to infinity computing in the same two classes. Y. Sergeyev proposed a new numerical system at the beginning of this century: it allows to perform computations with infinite and infinitesimal numbers in an easy way, and it is constructed on a new fundamental infinite number called *grossone* and denoted by ①. We refer the reader to the introductory surveys [34,37] or to the book [32]. In the last 20 years Sergeyev's system found a number of successful applications in many areas of mathematics and also other sciences as physics, biology, etc. For instance, [1,5,17,18,22,30,37] contain applications of the grossone-based system to ordinary differential equations, cellular automata, game theory and optimization, [4,9,10,15,16,29,33,35,36] contain applications to fractals, summations and some problems concerning biology. The reader interested to deepen logic foundations of the new system and new solutions to old paradoxes, can see [11,25,37]. We deserve a special mention to a recent research line to which also the present paper belongs: the use of the grossone-based system in high schools for educational purposes (see [2,3,20,21,26–28,31]).

Inside our cycle of class lessons, about 2 h have been devoted to practical class with infinity computing. Such a short time was more than sufficient, by virtue of the great ease of use of the new system for basic calculations. Instead, most of the time was spent working on the connections between game theory and infinity computing. In particular, students met the ideas of sequence (finite or infinite) of games, called tournament, and the possibility to apply grossone-based computations on sequences of games. We will see some easy and basic examples in Sect. 3 discussing the final students test.

3 The Final Test

In this section we discuss the results obtained in the final test, after a short cycle of lessons of about 7–8 h in total, divided in 5 or 6 days (depending on the class). As said in the previous section, about 2 h have been used to teach to the students the very basic fundamentals of the grossone numerical system, also in connection with game theory. The interested reader who wants to learn more deep connections between the two fields can see [18,19] and the references therein.

During the class lectures several examples of questions were proposed to the students, and in particular with open answers. The final students test instead consisted of 8 questions, the first 4 concerning only (basic) game theory and the last 4 dealing with connections between game theory and infinity computing. Furthermore, 7 questions had multiple predefined answers and one containing 5 sub-questions with true/false answers. The details for each question and the results of the students answers are below. The day of the final test 21 students were present in the first class and 19 in the second, for a total number of 40 students.

Question 1. Consider a game with two players (player1, player2), whose strategic representation is given by the following payoff matrix (Fig. 1):

$$\begin{array}{c} \text{player2} \\ \begin{array}{cc} C & T \end{array} \\ \text{player1} \begin{array}{c} C \\ T \end{array} \begin{array}{cc} (15,15) & (5,20) \\ (20,5) & (10,10) \end{array} \end{array} \qquad (1)$$

If player1 makes the first move and chooses strategy C, which one is the best strategy for player2?

(a) Strategy T;
(b) Strategy C;

Table 1. Table of the answers for Question 1. The correct answer is obviously (a), strategy T

Possible answers	Number of students	Percentage
(a) Strategy T	25	62.5%
(b) Strategy C	6	15%
(c) None of the two (T, C)	4	10%
(d) There is no single answer	5	12.5%
No answers	0	0%
Total	40	100%

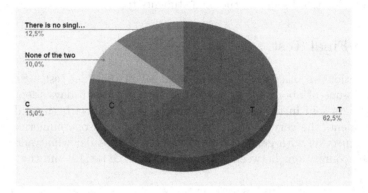

Fig. 1. A pie chart shows the percentages of student responses listed in Table 1 relatively to Question 1. The correct option is (a), strategy T, represented in blue (Color figure online)

(c) None of the two strategies T, C;
(d) There is no single answer.

Question 2. Consider the game of Question 1 with payoff matrix (1). Which one is the best strategy (for both) if the players can cooperate and then decide together the strategies to follow? (Fig. 2).

(a) (C, C): both players choose strategy C;
(b) (C, T): player1 chooses strategy C and player2 strategy T;
(c) (T, C): player1 chooses strategy T and player2 strategy C;
(d) (T, T): both players choose strategy T.

Question 3. Consider the game of Question 1 with payoff matrix (1). Which one is the best strategy (for both) if the players cannot cooperate and therefore they don't know each other's movements? (Table 3 and Fig. 3).

(a) (C, C): both players choose strategy C;
(b) (C, T): player1 chooses strategy C and player2 strategy T;

Table 2. Table of the answers for Question 2. The first option, (C, C), is the correct one

Possible answers	Number of students	Percentage
(C,C): both players choose strategy C	19	47.5%
(C,T): player1 chooses strategy C and player2 T	5	12.5%
(T,C): player1 chooses strategy T and player2 C	9	22.5%
(T,T): both players choose strategy T	7	19.5%
No answers	0	0%
Total	40	100%

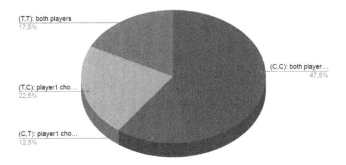

Fig. 2. A pie chart shows the percentages of student responses listed in Table 2 and relative to Question 2. The slice of the pie relating to the correct answer, (C, C), is colored blue (Color figure online)

(c) (T, C): player1 chooses strategy T and player2 strategy C;
(d) (T, T): both players choose strategy T.

Question 4. Consider again the game of Question 1 with payoff matrix (1). Players decide to play 5 times in sequence the game, i.e. a 5-game tournament. Initially, at the first two rounds, they cooperate in the choice of the strategy to pursue and they choose (C, C). But in the round 3, player2 decides to play T. Considering that the first three game strategies are (C, C), (C, C), (C, T), how will the evolution of the tournament be? (Table 4 and Fig. 4).

(a) 4th round (T, T) and 5th round (T, T);
(b) 4th round (C, T) and 5th round (C, T);
(c) 4th round (T, T) and 5th round (C, T);
(d) No definitive answer can be given.

Question 5. Choose true/false (T/F) for each of the following statements (Table 5 and Fig. 5):

(a) A tournament can have 2 ① games;
(b) A tournament can have ① + 1 games;

Table 3. Table of the answers for Question 3. The last option, (T,T), is the correct one

Possible answers	Number of students	Percentage
(C,C): both players choose strategy C	8	20%
(C,T): player1 chooses strategy C and player2 T	9	22.5%
(T,C): player1 chooses strategy T and player2 C	10	25%
(T,T): both players choose strategy T	12	30%
No answers	1	2.5%
Total	40	100%

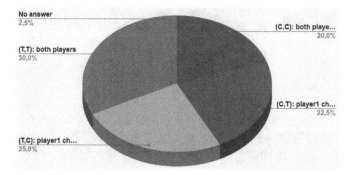

Fig. 3. A pie chart shows the percentages of student responses to Question 3. The green slice is related to the correct answer (T,T) (Color figure online)

(c) A tournament can have a number of games $n \leq$ ①;
(d) A tournament can only have a finite number of games;
(e) There can be no question of a tournament of ① games.

Question 6. Two players decide to play 2 tournaments, one after the other, each consisting of ① rounds. What can I say about the total number of games played? (Table 6 and Fig. 6).

(a) ① games;
(b) 2① games;
(c) Infinite games but we cannot say the number;
(d) 2① − 1 games;
(e) ① × ① games;
(f) No one can give a definite answer.

Question 7. Player1 and player2 play a series of games (possibly several consecutive tournaments). If one of the two players loses ① rounds he leaves the game and player3 enters in his place. Then player3 comes into play (Table 7 and Fig. 7).

Table 4. Table of the answers for Question 4. The first option, i.e. (a), is the correct one

Possible answers	Number of students	Percentage
(a) 4th round (T,T) and 5th round (T,T)	14	35%
(b) 4th round (C,T) and 5th round (C,T)	9	22.5%
(c) 4th round (T,T) and 5th round (C,T)	10	25%
(d) No definitive answer can be given	7	17.5%
No answers	0	0%
Total	40	100%

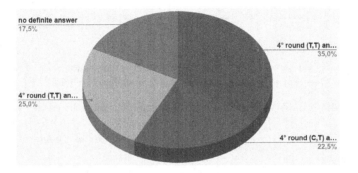

Fig. 4. A pie chart showing the percentages of student responses to Question 4. The blue slice is the one relating to the correct answer (Color figure online)

Table 5. Table of the answers for Question 5. The right answers are, in order, F, F, T, F, F

Possible answers	T	% of T	F	% of F
(a) A tournament can have 2① games	10	25%	30	75%
(b) A tournament can have ①+1 games	15	37.5%	25	62.5%
(c) A tournament can have a number of games $n \leq$ ①	29	72.5%	11	27.5%
(d) A tournament can only have a finite number of games	16	40%	24	60%
(e) There can be no question of a tournament of ① games	15	37.5%	25	62.5%

(a) at the latest after 2①−1 games;
(b) it does not make sense;
(c) after ① games;
(d) never.

Question 8. Player1 and player2 play a series of games (possibly several consecutive tournaments). If one of the two players loses ①/3 rounds he leaves the game and player3 enters in his place. Then player3 comes into play (Table 8 and Fig. 8)

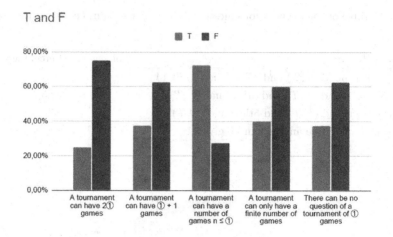

Fig. 5. A histogram showing the percentage of T/F answers for each of the 5 points that make up Question 5. The right answers are, in order, F, F, T, F, F

Table 6. Table of the answers for Question 6. (b) is the correct one

Possible answers	Number of students	Percentage
(a) ① games	6	15%
(b) 2① games	16	40%
(c) Infinite games but we can not say the number	7	17.5%
(d) 2① − 1 games	3	7.5%
(e) ① × ① games	4	10%
(f) No one can give a definite answer	4	10%
No answers	0	0%
Total	40	100%

(a) after ①/3 games;
(b) after 2①/3 games;
(c) at the latest after 2①/3 − 1 games;
(d) never.

4   Conclusions

The response of the students in terms of interest and participation was enthusiastic throughout the cycle of lessons. We also tried to make them play and have fun with the new concepts, exercises in class and more, with very good results and quite high approval from both classes.

The results of the final test have been in general very good. The most difficult point, for the students, was to manage the payoff matrix. It must be said that it is the first time they have seen a matrix in mathematics.

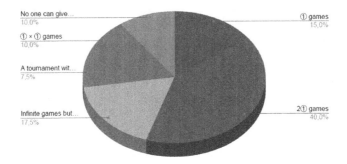

Fig. 6. A pie chart shows the percentages of student responses to Question 6. The red slice is the one relating to the correct answer (Color figure online)

Table 7. Table of the answers for Question 7. The correct answer is (a)

Possible answers	Number of students	Percentage
(a) At the latest after $2① - 1$ games	12	30%
(b) It does not make sense	10	25%
(c) After ① games	10	25%
(d) Never	8	20%
No answers	0	0%
Total	40	100%

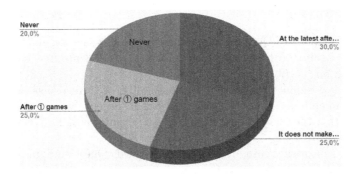

Fig. 7. A pie chart showing the percentages of student responses to Question 7. The blue slice is the one relating to the correct answer (a) (Color figure online)

Quickly analyzing the answers given by the students, we immediately notice how 62.5% answered the (easy) Question 1 correctly and 47.5% Question 2 concerning Pareto optimal. The percentage of correct answers drops to 30% for the more difficult Question 3 concerning Nash equilibrium and to 35% for Question 4.

Table 8. Table of the answers for Question 8. The correct option is (c)

Possible answers	Number of students	Percentage
(a) After ①/3 games	15	37.5%
(b) After 2①/3 games	12	30%
(c) At the latest after 2①/3 − 1 games	10	25%
(d) Never	3	7.5%
No answers	0	0%
Total	40	100%

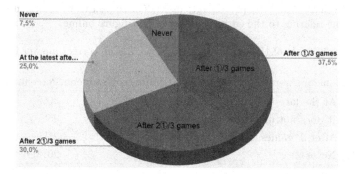

Fig. 8. A pie chart showing the percentages of student responses to Question 8. The yellow slice is the one relating to the correct answer (c) (Color figure online)

The T/F answers of Question 5 were instead a great success, and this demonstrates once again the remarkable ease of use that students find handling the grossone system (cf. [2,3,21]).

A good performance also with Question 6, 40% of correct answers. The last two more difficult questions achieved more than acceptable percentages (also considering that the school in question is not a lyceum): 30% for Question 7 and 25% for Question 8.

The most surprising case, in our opinion, regards Question 5: for each of the 5 items (a)–(e), the majority of the students always gave the correct answers, and in particular from a minimum of 60% to a maximum of 75%. This leads us to believe, in agreement with the conclusions of [2,3,21], that the grossone-based system can be taught with great ease and profit in high schools, also in connection with game theory.

In the near future we intend to carry out more extensive experiments involving a greater number of students, always in the same school or together with other schools. In particular we would like to broaden the experimental horizon in two main directions. In the first we intend to involve unimaginable numbers (see [14,24] and the references within them for definitions and basic properties) which have many points of contact with game theory and the Infinity Computing (see also [2] in this same volume). A triptych formed in this way offers many ideas for work and didactic experimentation. The second direction instead moves towards geometry, and wants to experiment with Infinity Computing in connection with the succession of Carboncettus octagons, combinacorics of words (see [12,13]), Fibonacci numbers and game theory. This direction, towards the combinatorics of (infinite) words and codes, also involves Infinity Computing from a purely theoretical point of view, posing many possible problems to investigate.

Acknowledgments. This research was partially supported by Seeweb s.r.l., Cloud Computing provider based in Frosinone, Italy, and part of DHH Group.

References

1. Amodio, P., Iavernaro, F., Mazzia, F., Mukhametzhanov, M.S., Sergeyev, Y.D.: A generalized Taylor method of order three for the solution of initial value problems in standard and infinity floating-point arithmetic. Math. Comput. Simul. **141**, 24–39 (2017). https://doi.org/10.1016/j.matcom.2016.03.007
2. Antoniotti, L., Astorino, A., Caldarola, F.: Unimaginable numbers and infinity computing at school: an experimentation in Northern Italy. In: Sergeyev, Y.D., Kvasov, D.E., Astorino, A. (eds.) NUMTA 2023. LNCS, vol. 14478, pp. 223–231. Springer, Cham (2025). https://doi.org/10.1007/978-3-031-81247-7_17
3. Antoniotti, L., Caldarola, F., d'Atri, G., Pellegrini, M.: New approaches to basic calculus: an experimentation via numerical computation. In: Sergeyev, Y.D., Kvasov, D.E. (eds.) NUMTA 2019. LNCS, vol. 11973, pp. 329–342. Springer, Cham (2020). https://doi.org/10.1007/978-3-030-39081-5_29
4. Antoniotti, L., Caldarola, F., Maiolo, M.: Infinite numerical computing applied to Peano's, Hilbert's, and Moore's curves. Mediterr. J. Math. **17**, 99 (2020). https://doi.org/10.1007/s00009-020-01531-5
5. Astorino, A., Fuduli, A.: Spherical separation with infinitely far center. Soft. Comput. **24**, 17751–17759 (2020)
6. Bertacchini, F., Bilotta, E., Caldarola, F., Pantano, P.: The role of computer simulations in learning analytic mechanics towards chaos theory: a course experimentation. Int. J. Math. Educ. Sci. Technol. **50**, 100–120 (2019). https://doi.org/10.1080/0020739X.2018.1478134
7. Binmore, K.G.: Game Theory: A Very Short Introduction. OUP Oxford (2007)
8. Burguillo, J.C.: Using game theory and competition-based learning to stimulate student motivation and performance. Comput. Educ. **55**, 566–575 (2010). https://doi.org/10.1016/j.compedu.2010.02.018
9. Caldarola, F.: The exact measures of the Sierpiński d-dimensional tetrahedron in connection with a Diophantine nonlinear system. Commun. Nonlinear Sci. Numer. Simul. **63**, 228–238 (2018). https://doi.org/10.1016/j.cnsns.2018.02.026
10. Caldarola, F.: The Sierpiński curve viewed by numerical computations with infinities and infinitesimals. Appl. Math. Comput. **318**, 321–328 (2018). https://doi.org/10.1016/j.amc.2017.06.024
11. Caldarola, F., Cortese, D., d'Atri, G., Maiolo, M.: Paradoxes of the infinite and ontological dilemmas between ancient philosophy and modern mathematical solutions. In: Sergeyev, Y.D., Kvasov, D.E. (eds.) NUMTA 2019. LNCS, vol. 11973, pp. 358–372. Springer, Cham (2020). https://doi.org/10.1007/978-3-030-39081-5_31

12. Caldarola, F., d'Atri, G., Maiolo, M., Pirillo, G.: New algebraic and geometric constructs arising from Fibonacci numbers. In honor of Masami Ito. Soft Comput. **24(23)**, 17497–17508 (2020). https://doi.org/10.1007/s00500-020-05256-1
13. Caldarola, F., d'Atri, G., Maiolo, M., Pirillo, G.: The sequence of carboncettus octagons. In: Sergeyev, Y.D., Kvasov, D.E. (eds.) NUMTA 2019. LNCS, vol. 11973, pp. 373–380. Springer, Cham (2020). https://doi.org/10.1007/978-3-030-39081-5_32
14. Caldarola, F., d'Atri, G., Mercuri, P., Talamanca, V.: On the arithmetic of knuth's powers and some computational results about their density. In: Sergeyev, Y.D., Kvasov, D.E. (eds.) NUMTA 2019. LNCS, vol. 11973, pp. 381–388. Springer, Cham (2020). https://doi.org/10.1007/978-3-030-39081-5_33
15. Caldarola, F., Maiolo, M.: On the topological convergence of multi-rule sequences of sets and fractal patterns. Soft. Comput. **24**, 17737–17749 (2020). https://doi.org/10.1007/s00500-020-05358-w
16. Caldarola, F., Maiolo, M., Solferino, V.: A new approach to the Z-transform through infinite computation. Commun. Nonlinear Sci. Numer. Simul. **82**, 105019 (2020). https://doi.org/10.1016/j.cnsns.2019.105019
17. Cococcioni, M., Cudazzo, A., Pappalardo, M., Sergeyev, Y.D.: Solving the lexicographic multi-objective mixed-integer linear programming problem using branch-and-bound and grossone methodology. Commun. Nonlinear Sci. Numer. Simul. **84**, 105177 (2020). https://doi.org/10.1016/j.cnsns.2020.105177
18. D'Alotto, L.: Infinite games on finite graphs using grossone. Soft. Comput. **24**, 17509–17515 (2020)
19. Fiaschi, L., Cococcioni, M.: Non-archimedean game theory: a numerical approach. Appl. Math. Comput. **409**, 125356 (2021). https://doi.org/10.1016/j.amc.2020.125356
20. Iannone, P., Rizza, D., Thoma, A.: Investigating secondary school students' epistemologies through a class activity concerning infinity. In: Bergqvist, E., Österholm, M., Granberg, C., Sumpter, L. (eds.) Proceedings of the 42nd Conference of the International Group for the Psychology of Mathematics Education, vol. 3, pp. 131–138. PME, Umeå, Sweden (2018)
21. Ingarozza, F., Adamo, M.T., Martino, M., Piscitelli, A.: A grossone-based numerical model for computations with infinity: a case study in an italian high school. In: Sergeyev, Y.D., Kvasov, D.E. (eds.) NUMTA 2019. LNCS, vol. 11973, pp. 451–462. Springer, Cham (2020). https://doi.org/10.1007/978-3-030-39081-5_39
22. Iudin, D., Sergeyev, Y.D., Hayakawa, M.: Infinity computations in cellular automaton forest-fire model. Commun. Nonlinear Sci. Numer. Simul. **20**, 861–870 (2015)
23. Kolokoltsov, V.N., Malafeyev, O.A.: Understanding Game Theory: Introduction to the Analysis of Many Agent Systems With Competition and Cooperation, 2 edn. World Scientific Publishing Company (2020)
24. Leonardis, A., d'Atri, G., Caldarola, F.: Beyond Knuth's notation for unimaginable numbers within computational number theory. Int. Electron. J. Algebra **31**, 55–73 (2022). https://doi.org/10.24330/ieja.1058413
25. Lolli, G.: Metamathematical investigations on the theory of grossone. Appl. Math. Comput. **255**, 3–14 (2015)
26. Mazzia, F.: A computational point of view on teaching derivatives. Inform. Educ. **37**, 79–86 (2022)
27. Nasr, L.: The effect of arithmetic of infinity methodology on students' beliefs of infinity. Mediterr. J. Res. Math. Educ. **19**, 5–19 (2022)
28. Nasr, L.: Students' resolutions of some paradoxes of infinity in the lens of the grossone methodology. Inform. Educ. **38**, 83–91 (2023)

29. Pepelyshev, A., Zhigljavsky, A.: Discrete uniform and binomial distributions with infinite support. Soft. Comput. **24**, 17517–17524 (2020)
30. Rizza, D.: Numerical methods for infinite decision-making processes. Int. J. Unconv. Comput. **14**, 139–158 (2019)
31. Rizza, D.: Primi passi nell'Aritmetica dell'Infinito. Bonomo Editore (2023, in Italian)
32. Sergeyev, Y.D.: Arithmetic of Infinity. Edizioni Orizzonti Meridionali, Cosenza (2003, 2nd ed 2013)
33. Sergeyev, Y.D.: Blinking fractals and their quantitative analysis using infinite and infinitesimal numbers. Chaos Solitons Fractals **33**(1), 50–75 (2007)
34. Sergeyev, Y.D.: Lagrange lecture: methodology of numerical computations with infinities and infinitesimals. Rendiconti del Seminario Matematico dell'Università e del Politecnico di Torino **68**, 95–113 (2010)
35. Sergeyev, Y.D.: Using blinking fractals for mathematical modelling of processes of growth in biological systems. Informatica **22**, 559–576 (2011)
36. Sergeyev, Y.D.: The exact (up to infinitesimals) infinite perimeter of the Koch snowflake and its finite area. Commun. Nonlinear Sci. Numer. Simul. **31**, 21–29 (2016)
37. Sergeyev, Y.D.: Numerical infinities and infinitesimals: methodology, applications, and repercussions on two Hilbert problems. EMS Surv. Math. Sci. **4**, 219–320 (2017)

Meta Discussion Pedagogical Model to Foster Mathematics Teacher's Professional Development

Antonella Montone[1(✉)], Michele Giuliano Fiorentino[2], and Giuditta Ricciardiello[1]

[1] Department of Science of Education, Psychology and Communication,
University of Bari Aldo Moro, Bari, Italy
{antonella.montone,giuditta.ricciardiello}@uniba.it
[2] Department of Mathematics, University of Bari Aldo Moro, Bari, Italy
michele.fiorentino@uniba.it

Abstract. In this contribution we present an experimental research undertaken with Pre-Service Primary Teachers (PSTs). The aim of this study is to foster the Mathematical Discussion (MD) theoretical framework's learning, in order to give PSTs the opportunity to learn in and from their practice. The MD combined with the Meta Discussion on a Pedagogical model (MDPm), a new theoretical construct that we identified, allow prospective teachers, on the one hand to be introduced to the pedagogical model itself and, on the other hand, to develop in and from practice both theoretical knowledge and practical experience of this model. This means that the MD becomes at the same time the subject of the lecture and the methodology used during the lecture. This new theoretical approach combines Shulman's research about the Pedagogical Content Knowledge (PCK), Ball's Mathematical Knowledge for Teaching (MKT) and further research on what is relevant to learn in and from practice and how this practice can be used for teachers' learning. The experimental research has two different goals: the construction of the MD's characteristics from a theoretical point of view and the way to manage MD in the future professional practice. The analysis of results highlight that the PSTs became aware of how MD works and of its value. In and from their own practice they learned how to manage MD in class.

Keywords: Mathematical discussion · prospective teachers · mathematics teachers' professional development

1 Introduction

This contribution aims to provide new insight on how to train prospective teachers referring to the teachers' professional development research topic. As a matter of fact, knowing mathematics is not enough and it is fundamental to achieve pedagogical competencies.

Current literature on teacher education offers different theoretical frameworks. In particular, [1] provided a model for shared approaches to research in mathematics education, emphasizing the connection between disciplinary knowledge and the pedagogical knowledge which are necessary to teach. [2] elaborated an operational definition of this model by analysing ways for the mathematics teacher's professional development. In this perspective, in this paper we present some results emerging during a larger study, still in progress, concerning prospective teachers' training; in particular, the study concerns introducing prospective teachers to a particular pedagogical model, through their direct experience of its implementation.

In this paper we analyse the data collected during the intervention held with 160 prospective mathematics teachers attending the undergraduate 'mathematics teaching' course in primary education. Data show that experiencing in first person the pedagogical theories could foster the theory's conceptualisation itself. Moreover, this training promotes the future teaching profession's awareness.

In the following, there is an outline of the conceptual background, the definition of the particular pedagogical model in focus, and then, the experimental design, the analysis of some of the data collected and some results obtained.

2 Conceptual Background

The mathematics teachers' training is a fundamental research topic of research in Mathematics Education. During the last years, several research projects have been conducted in this area and several theories emerged. According to [1] to teach a discipline, a specific knowledge is needed, not exclusively a disciplinary one. This happens because disciplinary and pedagogical knowledge are both involved in teaching-learning processes.

Indeed, as Shulman highlights, teachers need an integrated view of disciplinary content and issues related to its teaching, in order to foster effective teaching-learning processes. Furthermore, different conceptualisations and models of specialised mathematical knowledge have been proposed, as result of some research studies: the Mathematical Knowledge for Teaching (MKT) model [3] or the more recent Mathematics Teacher's Specialised Knowledge (MKTS) model [4].

Ball & Even [2] emphasize "the need to focus teachers' education on practice-and the problem of doing it effectively". Taking into account the complex articulation of different contents and their relationships, they suggest us to intervene in practice for teacher training and provide us a prospective for the implementation of educational interventions.

Consequently, to give prospective teachers a starting point for training themselves as future teachers through understanding and interpreting the actions of their future students, it is fundamental to train them through practice and providing them an approach to a given task [2, 5].

Research has problematised teacher training related to these complex contents and then formulated ten key principles of Professional Development [6] among which the following are particularly relevant:

1. Using teachers as participants in classroom activities or students in real situations, model desired classroom approaches during in-service sessions to project a clearer vision of the proposed changes.
2. Solicit teachers' conscious commitment to participate actively in the professional development sessions and to undertake required readings and classroom tasks, appropriately adapted for their own classroom.
3. Allow time and opportunities for planning, reflection, and feedback in order to report successes and failures to the group, to share "the wisdom of practice," and to discuss problems and solutions regarding individual students and new teaching approaches.
4. Recognise that change is a gradual, difficult, and often painful process, and afford opportunities for ongoing support from peers and critical friends.
5. Encourage participants to set further goals for their professional growth.

These principles outline some best practices to promote a clearer vision for changing. In order to put in practice the modalities suggested by Ball & Even, the pedagogical model we have been working on is the Mathematical Discussion (MD) elaborated by [6]. According to Bartolini Bussi [7], the MD is a 'polyphony of voices articulated about a mathematical object (concept, problem, procedure, etc.), which constitutes a motive of the teaching-learning activity'.

The voices (represented by the signs produced by the students) have to be coordinated with the voice of the mathematical culture (witnessed by the teacher herself), and this polyphony is "orchestrated" by the teacher [8].

During a MD, the teacher plays two main roles: mediator and moderator. These roles are enacted by the main following teacher' actions: the "back to the task" action (to reconstruct the context and to foster the (re)emergence of meanings and processes related to the task); "focalizing" action (to focus on aspects consistent with the didactic objective); "request of synthesis" action (to support students in the process of de-contextualisation and generalisation with respect to specific tasks); "offer of synthesis" action (to provide a formulation introducing the desired terms; to ratify the acceptability and mathematical status of a specific meaning).

A mathematical discussion activity is differentiated according to its objective (motive). Different types of discussion can therefore be distinguished. In the following, reference will be made to the Conceptualisation Discussion (MDc) and to a Balance Discussion (MDb). MDc is understood as the process of constructing mathematical concepts, building up suitable connections between already lived experiences and particular mathematical terms [9]. In order to activate a MDc, it is necessary to start by solving an open-ended problem that, on the one hand, addresses and discusses the different solutions provided by the students and, on the other hand, faces mathematical conceptualisation with respect to the mathematical content at stake. MDb is understood as the process of informing, analysing and evaluating the individual solutions proposed to a given problem. It may happen that a Balance Discussion naturally develops into a Conceptualization Discussion (MDb-c).

As a matter of fact, the use of an open-ended problem [10] is crucial for triggering a rich and effective mathematical discussion. Indeed, it is not solved mechanically, but requires different solving strategies. Its solution uses different representations (graphical, symbolic, arithmetical, algebraic, etc.); moreover, it requires mathematical knowledge and skills that are the aim of the educational context in which the open-ended problem is set [11]. Finally, it is necessary for the problem to be interesting, even challenging, for prospective teachers as well as for school students. This will ensure that prospective teachers are involved both cognitively and emotionally, as solvers of the problem. The open-ended problem's characteristic of lending itself to different resolutions and different representations is fundamental in fostering a rich and meaningful MD both on the level of mathematical conceptualisation and on the level of teacher-training.

The mathematical content of the open-ended problem is that of equi-extension. It is proposed not in a procedural way through measurement, but through direct conceptualisation of the extension and to verify an equivalent extension through the Mathematical Quantity.

3 The Proposed Model and the Research Hypotheses

Considering the general aim of developing suitable knowledge for prospective teachers we elaborated a new methodology. It aims at enchaining the didactical experience lived by the students in participating to a MD orchestrated by their instructor and focussing on a specific mathematical content, with a new experience, still involving them in a discussion, still orchestrated by their instructor, but with a new aim: reflecting on their previous lived experience, identifying the key theoretical aspects characterizing the pedagogical model constituted by the Pedagogical model of MD. In other words, we elaborated a new theoretical construct that we named Meta Discussion on a Pedagogical model (MDPm). As said, this constitutes a new level of discussion that has as its motive the conceptualization of MD as a Pedagogical model.

Indeed, the MDPm, aims to define the characteristics of MD itself and in particular to recognise the teacher's actions in and from practice [2].

Within such a complex conceptual framework the main hypothesis of our research study can be synthesized as follows: through the combination of MD and MDPm, prospective teachers can be introduced to the pedagogical model of MD, and develop in and from practice both theoretical knowledge and practical experience of this model. Consistently with this hypothesis we designed a teaching experiment with the aim of exploring and deepening such hypothesis. The following section is devoted to describing the research methodology.

4 The Research Methodology and Experimental Setting

In this paper we present a specific experimental path of a general project that is design-based research [12]. In particular, we design operational phases in which we foster the relationship between the participation of prospective teachers in the two kinds of discussion: the MD (both of balance and of conceptualization) and MDPm, in this way we intend to foster the development in and from practice of both practical experience and theoretical knowledge of this model. From the MDPm transcript, data on the evolution and construction by the prospective teachers of the pedagogical model's theoretical knowledge is collected. Particularly relevant are the data focused on the trainer's role and his/her actions during the orchestration of the discussion.

To validate the specific experimental path, a qualitative analysis of the MDPm's transcriptions was carried out, according to the criteria of credibility, dependability, transferability and confirmability [13], to ensure trustworthiness. In this section we describe the general setting of the specific experimental path in focus. The research involved 160 prospective teachers in mathematics, fourth-year students in Primary Education, and attending Mathematics Education Course. The experimental path and in particular both MDb-c and MDPm take place on online platforms [14]. These discussions have been recorded and subsequently transcribed for the analysis. The training mode and in particular both MDb-c and MDPm take place synchronously on Microsoft Teams platform. Both the discussions take place with the whole class group. These discussions have been recorded and then transcribed for the analysis.

5 The Experimental Design

The overall aim of the broader study we have focused on, is to foster theoretical learning from situational learning. We want to investigate how through the combination of MDb-c and MDPm, prospective teachers can be introduced to the pedagogical model of MD, and develop in and from practice both theoretical knowledge and practical experience of this model.

This hypothesis is at the base of the design of the experimental path. It consists of a succession of phases identified by activities informed by our hypothesis on the educational relationship between experiencing a MD and participating in a MDPm on such an experience. To validate our research study, the educational path experimentation, already carried out and analysed in a pilot study [15], has been conducted again with a new group of prospective teachers.

The organisation of the educational intervention is described below according to the different phases:

Phase 0: introduction of the MD theoretical model. In a preparatory phase to the experimental path, a lecture on MD was given, in which the theoretical elements characterising the pedagogical model and the teachers' actions orchestrating the discussion were presented.

Phase 1: For the start of the experimental activities, an open-ended mathematics problem sufficiently elementary to be understood by anyone and included in school curricula was chosen. The topic of equi-extension is a complex one, generally based on the procedural activity of calculating the measure of equi-extended areas, and not on the idea of extension as a quantity related to equi-decomposability. The following problem was proposed to be solved in groups.

> *Two brothers receive a rectangular piece of land as inheritance. In order to divide it into two parts of the same size, one of them suggests planting a stake anywhere on the land and joining it to the four stakes driven into the four vertices of the rectangular land. One of the brothers will take two non-adjacent triangles, the other the remaining part. Are the two parts really equal? Justify your answer.*

The students solved the above problem firstly in groups and did a preliminary discussion among them, without the instructor, on the solution to share with the whole class and the instructor.

Starting from the solutions of the problem in the following lesson, a MDb-c was developed by the instructor. The aim of the MDb-c is to bring out all the problems' solutions strategies, in converging towards a shared solution, and to bring out and make explicit the equi-extension's concept. In the design and then in the realisation of the MDb-c, we referred to the patterns described by the MD pedagogical model. In particular, the instructor who orchestrates the discussion performed the actions expected from the Pedagogical model. In so doing we intended to make the prospective teacher observe these actions as performed by the 'teacher' and at the same time, experience the effect of these actions as students.

Phase 2: MDPm on the previous MDb-c. In this phase, engaged in a MDPm, prospective teachers are asked to reflect on the activity carried out in phase 1: recognise the elements characterising a MDb-c, identifying its main steps; identify the actions performed by the teacher and characterising the pedagogical model. Mathematics in this MDPm remains in the background, although it is not ignored by the participants. MD, and in particular the teacher' actions, as a pedagogical model become the subject of discussion. A key aspect for the prospective teachers' professional development lies in the transition from the role of student to that of teacher, projecting himself/herself in the actions performed by the instructor.

6 Analysis Criteria and Analysis

In the following we report the analysis concerning the MDPm carried out in phase 2. The following analysis points out how the MDPm let the characteristics of the MD pedagogical model emerge. These characteristics are made explicit by the prospective teachers. The analysis of MDPm's transcription, highlights how the prospective teachers recognise the teacher's actions during a MDb-c. These actions enable them to re-construct MD's general characteristics. Then, the analysis of the same transcript was aimed at highlighting the experience in practice of prospective teachers during MD. Moreover, we report how the MDPm let some characteristics of the discussion that the students have done in small groups without the instructor emerge. We believe that the comparison between the first discussion in small groups and the next MDb-c all together with the instructor, highlight the value of the instructor's presence during the MDb-c.

After solving a mathematical open-ended problem on equi-extension by students, the teacher starts the MDPm, referring to the lesson in which students had participated in the MD related to the previous mathematical problem's solution.

Below an example drawn from an excerpt from the MDPm transcript:

(0:00) Instructor: What did we do in the last lesson? What was the teacher's aim?

(0:35) Francesca: in my opinion, when the problem has been proposed to us, when we have been divided into groups and when we had to discuss firstly with our peers, and then with the teacher... for all this I suppose we did a Mathematical

Discussion, [...] as a pedagogical model and on the solutions of the problem [...] in addition your role was especially that of mediator of the discussion

The instructor starts the MDPm in a new way, which differs from a MD which follows the solution of a problem. Indeed, the instructor does a "back to the task" not evoking the problem' solutions carried out by the groups, but proposing to discuss on the MD in which the students participated. The aim is to bring out the MD's characteristics. For this reason, we could define this action carried out by the instructor as a new action: "Back to the experience". This action aimed at going back over the discussion experience in order to recognise the MDb-c as a pedagogical model. The question is formulated by the teacher in a different way: there is no longer a reference to the problem to be solved but an attempt is made to problematise the MDb-c as a pedagogical model.

Francesca's answer shows a recognition of the MDPm's aim to make prospective teachers discuss about MDb-c. She recognizes the MD as a pedagogical model and her answer probably follows a question that induces shifting the discussion's focus from the mathematical content to the pedagogical model. Furthermore, she detects the specific role of the teacher, as mediator. The MDPm proceeds and the teacher invites Ornella to speak.

(3:02) Ornella: Another thing I'd like to add is that... you had another role, specifically that of moderator. In particular when you gave the opportunity to most of us to intervene during the discussion... you never stopped our speech or you never came to a conclusion and we also came to an uncertainty... Let's say you never said "this is the right solution", "this is another right solution" ... you replicated our words...

(5:07) Daniele: I agree with Ornella, in the last lesson we did a MD and, furthermore, I had the impression that ... that mathematics can be discussed, precisely that through reasoning guided by you teachers...the various ideas emerged...

Ornella refers to the experience related to the concept of moderator as characteristic of the teacher's roles in a MD. From Ornella's words it emerges that, at the moment when the student talks about his/her solution, the teacher intervenes to let most of them intervene during the MD and she never says if the solution is correct/incorrect. Ornella also recognises the action of replicating students' words which we can classify as the action of "mirroring" as a characteristic teacher's action of a MD.

Daniele acknowledges that the activity in the previous lesson was a MD; he also states that a discussion can evolve a mathematical meaning, which is a particular aspect that characterises MD.

The discussion proceeds and the instructor offer a synthesis:

(5:48) Instructor: Daniele said we discussed Mathematics, Francesca said, the role of the teacher is seen to be one of mediator, Ornella added that it is also that of moderator... but can you tell me what were the very elements that allowed you to recognise this activity as a discussion?

(6:17) Francesca: your intervention sometimes was offering synthesis and sometimes was aimed at investigating more deeply what we were saying... these aspects

made me recognise the teacher's role as mediator and moderator... you also picked up the threads of Daniele's speech to... try to understand better what he said... and the words that you used, made me understand this is a MD.

(07:35) Alessia: Yes, following what Francesca said, I think that the language has been a very important element. I mean, we used an appropriate way that we all noticed to talk about mathematics.

Here the instructor performs a "meta-focusing" action (which is a new version of the focusing action characteristic of a MD), which aims to make explicit the characterising elements that made it possible to recognise a MD in the experienced activity. In Francesca's intervention the concept of mediator and moderator as characteristic of the teacher's roles in a MD emerged.

From Francesca's words it emerges that the teacher intervenes to take up and reformulate what the student said while talking about his/her solution. Moreover, Francesca recognises the action that we call "offering synthesis", as a characteristic teacher's action of a MD, as a linguistic mediation.

Recalling what Francesca said, Alessia highlights the language's function in the evolution of the mathematical concept at stake. In other words, Alessia emphasises the function of language in the evolution of the mathematical concept as characteristic of the MD pedagogical model.

(8:04) Instructor: So... synthesising... Francesca says... I recognise the discussion, that is, I recognise the elements characterising the MD because there is the role of the teacher who becomes mediator and moderator. The teacher intervened to offer synthesis, sometimes intervened to investigate. Francesca says that the teacher never gave us theories, but always waited for us to define the theoretical aspects and then, as Alessia said, it seemed to us that in this talking of mathematics, the new thing was the language used by the teacher.

(9:57) Anna: What impressed me is that you always asked why, when one of us intervened... by proposing a solution. You tried to repeat the solution with other words so that it was clear to the whole class and you tried to involve all of us in the discussion.

Here the teacher synthesised what emerged from the three previous interventions and repeated some words said by the students. She also recalled through Francesca's words a further characteristic of the MD pedagogical model, that is the teacher's role as mediator and moderator.

In her intervention Anna highlights some specific teacher's actions as mediator in always asking why referring to her experience during MDb-c. Furthermore, Anna recognises the MD as a pedagogical model in the action by the teacher to involve the whole class.

Referring to the summary intervention and Anna's answer, we can observe the evolution from direct reference to the lived experience - what was specifically done and said - to reference to the pedagogical model.

In the following excerpts, we report the comparison between the discussion in small groups and the MDb-c.

(13:06) Mary: ... in my opinion, there's something about the time administration, we didn't have a time deadline... Indeed, we are still talking about it and this let us think that the discussion has not finished yet... In addition, you didn't give us a hurry in the research of a solution... This allowed us to express our thoughts and ideas...

Mary recognises the moderator's role and she adds another element useful to characterize the DM's pedagogical model, referring to the time administration.

(15:22) Ornella: I'd like to add that another element is the interaction itself... I mean to express agreement or disagreement... but also, as you said, the difficulty into interpret someone else's point of view... when we highlighted some differences in words' use...Is it correct? Is it completely wrong?... so, just the interaction and the conflict... just trying to justify one or more alternatives... because the aim was not to give a solution that was more correct than others, but the real aim was to let us interact. The main difference between the discussion we had in the small groups and the other one in the big class group with you is the presence of the teacher as mediator, I think...

Ornella with her words highlights her experience as student which allows prospective teachers to notice some typical aspects of peer discussion. In this kind of discussion, it may happen to be in agreement or disagreement with others and therefore it is essential to understand someone else's point of view.

Moreover, the distinction between the discussion in small groups and the MD with the instructor in the big class group, allows her to recognise the teacher's action. In the first discussion, the students have the aim to solve an open-ended problem, in the MD the students have the aim to share all together the solutions emerged during the work in small groups. The instructor has a different aim, that is to foster the interaction and bring out the construction of mathematical concepts involved.

(20:06) Sara: I think that both the discussions were aimed at <u>let us become the main actors of our own learning... you didn't give us a knowledge to directly be used in solving an exercise but, on the contrary, you gave us a problem to build knowledge...</u> In my opinion the difference between the small group discussion and the MD is the teacher's role as mediator. Moreover, the MD helped those students who in the first workgroup didn't find a complete solution.

Sara also recognises the teacher's role of mediator through the comparison between the development of a discussion in small groups and the one of a MD. Indeed, this is aimed at the conceptualization of a mathematical knowledge to be built as highlighted in the underlined text.

(22:10) Daniele: in my opinion there are two essential differences... I'd like to recall what Francesca and Alessia already said, in a previous intervention.... They were talking about language... Isn't it? The language used in the MD with you, is different from that we used in the small group discussion... I mean... I see a difference between the contexts in which the two discussions took place...

The informal context of the first discussion... it seemed more informal, because the discussion was among peers... on the contrary in the MD the attention was focused also on the language... we paid attention to use the correct words and also you were requesting us more attention in the use of language when you said "you have not convinced me...". This expression stimulated our use of language in searching for a justification comprehensible for you and the whole class... The differences are the context and the language used in both the contexts... So, the MD's aim was to discuss all together and let all our points of view emerge...

In this excerpt is evident how the language role emerges again, also in the comparison between the two different discussions. Daniele recalls the previous interventions by Francesca and Alessia and distinguishes on the one hand the aim of the discussion in small groups to give a solution and on the other hand the MD's aim to let different points of view emerge and share them with the peers. In this way MD's role is synthesised in understanding and persuading.

As in the following intervention by Stefania, which is quoted below, the prospective teachers become increasingly aware of the teacher's role through a distancing from the lived situation.

(31:55) Stefania: So, in the MD the cognitive conflict emerged... there was a constant constructing and deconstructing of what we were saying, going to investigate why, through your questions "But I didn't understand... but are you sure? But can you make me understand? No, but I don't see it...". At one point when I was describing my solution I distanced myself from what I was saying; that is, I realised that I was able to treat the subject, the solution... with more detachment; when I have to justify it, I have to demonstrate something and I am asked certain questions that derive from an a priori analysis (designed by the teacher who knows what to ask) I realised that I was treating it with distance and therefore could be more aware and impartial in the discussion. Distancing oneself from oneself and cognitive conflict are the two fundamental aspects of mathematical discussion realised thanks to the teacher's mediation and moderation. Even now in this discussion, I realise that I distanced myself from the fact that I was involved in the discussion in the previous lesson, thanks to your questions and the interventions of my peers.

Stefania states that she experienced a situation in which it was possible for her, thanks to the teacher's intervention, to implement the distancing hypothesis. Moreover, as she describes, she becomes aware of the teacher's role. The teacher's guidance enabled a process of decontextualisation of the solution with respect to the problem. Finally, in the underlined expression, Stefania makes explicit reference to the teachers' roles in the pedagogical model.

7 Preliminary Results and Concluding Remarks

The new theoretical construct MDPm and the training intervention on the MD pedagogical model made it possible for prospective teachers to reflect on previous lived experience MD. We assume that, in this way, they could identify the key theoretical aspects characterising the MD pedagogical model. According to Ball & Even [2], through the MDPm, prospective teachers recognise the teacher's actions performed in this model, in and from practice, after having personally experienced them.

Analysis of the transcripts revealed evidence of the recognition of theoretical aspects from the experienced situation. It happens particularly in the identification of the teacher's actions recognised through a detachment from the student's role and assuming the role of future teacher. Indeed, Francesca in her intervention (00:35) identifies MD as an appropriate pedagogical model that is constructed and/or recognised. In particular, during the MDPm, the students recognise and appropriate the patterns of the model, recognise the effectiveness of the model with respect to its function and finally, distinguish two specific functions: 1) solving the problem and constructing the mathematical concept; 2) recognising the nature of mathematics as something that can be discussed. In this research study, the implementation of the direct experience, reflects the suggestions declared by Clarke's ten key principles of Professional Development, concerning the involvement of prospective teachers in active participation for their professional development.

Data analysis shows that the training activity carried out by experiencing in first person the pedagogical theories fostered the theory's conceptualisation itself and the future teaching profession's awareness.

References

1. Shulman, L.S.: Those who understand: knowledge growth in teaching. Educ. Res. **15**(4), 4–14 (1986)
2. Ball, D.L., Even, R.: Strengthening practice in and research on the professional education and development of teachers of Mathematics: next steps. In: Even, R., Ball, D.L. (eds.) The Professional Education and Development of Teachers of Mathematics, pp. 255–259. Springer, New York (2009)
3. Ball, D.L., Thames, M.H., Phelps, G.: Content knowledge for teaching: what makes it special? J. Teach. Educ. **59**(5), 389–407 (2008)
4. Carrillo-Yañez, J., et al.: The mathematics teacher's specialised knowledge (MTSK) model. Res. Math. Educ. (2018). https://doi.org/10.1080/14794802.2018.1479981
5. Fiorentino, M.G., Montone, A., Rossi, P.G., Telloni, A.I.: A digital educational path with an interdisciplinary perspective for pre-service mathematics primary teachers' professional development. In: Fulantelli, G., Burgos, D., Casalino, G., Cimitile, M., Lo Bosco, G., Taibi, D. (eds.) Higher Education Learning Methodologies and Technologies Online. HELMeTO 2022. Communications in Computer and Information Science, vol. 1779, pp. 663–673. Springer, Cham (2023). https://doi.org/10.1007/978-3-031-29800-4_50
6. Clarke, D.M.: Ten key principles from research for the professional development of mathematics teachers. In: Aichele, D.B., Coxfors, A.F. (eds.) Professional Development for Teachers of Mathematics (Yearbook of the National Council of Teachers of Mathematics). Reston, VA: NCTM, pp. 37–48 (1994)

7. Bartolini Bussi, M.G.: Verbal interaction in mathematics classroom: a Vygotskian analysis. In: Steinbring, H., Bartolini Bussi, M.G., Sierpinska, A. (eds.) Language and Communication in Mathematics Classroom, NCTM, Reston, Virginia, pp. 65–84 (1998)
8. Bartolini Bussi, M.G., Mariotti, M.A.: Semiotic mediation in the mathematics classroom: artefacts and signs after a vygotskian prospective. In: English, L., Bartolini Bussi, M., Jones, G., Lesh, R., Tirosh, D. (eds.) Handbook of International Research in Mathematics Education, second revised edition. Mahwah, NJ: Lawrence Erlbaum, pp. 746–783 (2008)
9. Bartolini Bussi, M.G., Boni, M., Ferri, F.: Social interaction and knowledge at school: The Mathematical Discussion. Technical report no. 21 NRD of Modena, Municipality of Modena, pp. 11–12 (1995). https://www.comune.modena.it/memo/prodotti-editoriali/saperi-e-discipline/allegati/interazione_sociale_e_conoscenza_a_scuola.pdf
10. Pehkonen, E.: Introduction to the concept "open-ended problem." In Use of open-ended problems in mathematics classroom (Issue 7) (1997). http://coreylee.me/en/publications/2001_self-efficacy_change.pdf%5Cnhttp://files.eric.ed.gov/fulltext/ED419714.pdf
11. Serpe, A., Frassia, M.G.: Task mathematical modelling design in a dynamic geometry environment: archimedean spiral's algorithm. In: Sergeyev, Y., Kvasov, D. (eds.) Numerical Computations: Theory and Algorithms. NUMTA 2019. Lecture Notes in Computer Science, vol. 11973, pp 478–491. Springer, Cham (2020). https://doi.org/10.1007/978-3-030-39081-5_41
12. Swan, M.: Design research in mathematics education. In: Lerman, S. (ed.) Encyclopedia of Mathematics Education, pp. 148–152. Springer, Cham (2020)
13. Guba, E.: Criteria for assessing the trustworthiness of naturalistic inquiries. Educ. Tech. Res. Dev. **29**(2), 75–91 (1981)
14. Serpe, A.: Digital tools to enhance interdisciplinary mathematics teaching practices in high school. In: Fulantelli, G., Burgos, D., Casalino, G., Cimitile, M., Lo Bosco, G., Taibi, D. (eds.) Higher Education Learning Methodologies and Technologies Online. HELMeTO 2022. Communications in Computer and Information Science, vol. 1779, pp 209–218. Springer, Cham (2022). https://doi.org/10.1007/978-3-031-29800-4_16
15. Fiorentino, M.G., Mariotti, M.A., Montone, A.: Prospective mathematics teachers' professional development through meta discussion on a pedagogical model. In: Drijvers, P., Csapodi, C., Palmér, H., Gosztonyi, K., Kónya, E. (eds.), Proceedings of the Thirteenth Congress of the European Society for Research in Mathematics Education (CERME13), pp. 3435–3442. Alfréd Rényi Institute of Mathematics and ERME (2023)

A Variational Quantum Soft Actor-Critic Algorithm for Continuous Control Tasks

Antonio Policicchio[✉], Alberto Acuto, Paola Barillà,
Ludovico Bozzolo, and Matteo Conterno

NTT DATA Italia S.p.A., Via Ernesto Calindri 4, 20143 Milan, Italy
antonio.policicchio@nttdata.com
https://it.nttdata.com/

Abstract. Quantum Computing promises the availability of computational resources and generalization capabilities well beyond the possibilities of classical computers. An interesting approach for leveraging the near-term, Noisy Intermediate-Scale Quantum Computers, is the hybrid training of Parameterized Quantum Circuits (PQCs), i.e. the optimization of a parameterized quantum algorithms as a function approximation with classical optimization techniques. When PQCs are used in Machine Learning models, they may offer some advantages over classical models in terms of memory consumption and sample complexity for classical data analysis. In this work we explore and assess the advantages of the application of Parametric Quantum Circuits to one of the state-of-art Reinforcement Learning algorithm for continuous control - namely Soft Actor-Critic. We investigate its performance on the control of a virtual robotic arm by means of digital simulations of quantum circuits. A quantum advantage over the classical algorithm has been found in terms of a significant decrease in the amount of required parameters for satisfactory model training, paving the way for further developments and studies. A quantum advantage over the classical algorithm has been found in terms of a significant decrease in the amount of required parameters for satisfactory model training, paving the way for further developments and studies.

Keywords: Reinforcement Learning · Parametrized Quantum Circuits · Continuous Control

1 Introduction

Deep Reinforcement Learning (DRL) [1], the combination of Deep Learning and Reinforcement Learning (RL), has emerged as an interesting approach for tackling with continuous control systems that require measuring and adjusting the controlled quantities in continuous-time. In continuous control, compared to RL, DRL allows to fix critical issues relative to the dimensionality and scalability of data in tasks with sparse reward signals. However, despite recent improvements, the challenges of learning complex control tasks with DRL are still far from being solved for real-world applications. This is mainly due to some well know issues with DRL: sample efficiency, generalization, and computing resources for training the learning algorithms [2]. Sample efficiency means how much data are needed to be collected in order to build an optimal policy to accomplish the designed task. According to [2], several issues in continuous control prevent an effective sample efficiency: the agent cannot receive a training set provided by the environment unilaterally, but rather information which is determined by both the actions it takes and the dynamics of the environment; although the agent aims at maximizing the long-term reward, it can only observe the immediate reward; there is no clear boundary between training and test phases, since the time the agent spends trying to improve the policy often comes at the expense of utilizing this policy, which is often referred to as the exploration-exploitation trade-off [3]. On the other hand, generalization refers to the capacity to use previous knowledge from a source environment to achieve a good performance in a target environment, and applicability for flexible long-term autonomy. This is widely seen as a necessary step to produce artificial intelligence that behaves similar to humans. Moreover, it has to be noted that, given the large amount of data to reach optimal results, DRL is computationally intensive, and it requires high-performance computers for model training and fastening the learning process. More progress is required to overcome such limitations, as both gathering experiences by interacting with the environment and collecting expert demonstrations for RL are expensive procedures.

Quantum Computing (QC) promises the availability of computational resources and generalization capabilities well beyond the possibilities of classical computers [4]. Even if fault-tolerant quantum devices are still far to come, near-term devices – Noisy Intermediate-Scale Quantum Computers (NISQ) [5], limited in the number of qubits, coherence times and operations fidelity – can already be utilized for a variety of problems. One promising approach is the hybrid training of Variational or Parameterized Quantum Circuits (VQCs or PQCs), i.e. the optimization of a parameterized quantum algorithm as a function approximation with classical optimization techniques [6]. PQCs are typically composed of fixed gates, e.g. controlled NOTs, and adjustable gates, e.g. qubit rotations parameterized by a set of free parameters. The main approach is to formalize problems of interest as variational optimization tasks, then using a hybrid quantum-classical hardware setup to find approximate solutions. By implementing some subroutines on classical hardware, the requirement of quantum resources is significantly reduced, particularly in the number of qubits, circuit depth, and coherence time.

Quantum Machine Learning (QML) typically involves training a VQC in order to analyze either classical or quantum data [7]. QML models may offer some advantages over classical models in terms of memory consumption and sample complexity for classical data analysis. Moreover, a recent research presented a comprehensive study of generalization performance in QML after training on a limited number of training data points, showing that a good generalization is somehow guaranteed from few training data [8]. All such aspects look promising to overcome the DRL issues discussed above for continuous control.

Starting from recent utilization of VQCs in RL problems [9–11], the possible application of quantum-classical hybrid algorithms to the continuous control task of a robotic arm has been investigated. Specifically, the advantages of the application of VQCs to one of the state-of-art Reinforcement Learning techniques for continuous control – the Soft Actor-Critic (SAC) – have been explored and assessed by means of digital simulations of quantum circuits. The work demonstrates that continuous control is a promising field of application for Quantum Reinforcement Learning (QRL).

2 Quantum Reinforcement Learning for Continuous Control

2.1 Reinforcement Learning for Continuous Control

Reinforcement Learning [12] is a Machine Learning technique in which an agent learns to behave in an environment by performing actions and assessing the results of those actions. For each good action, the agent gets positive reward, and for each bad action, the agent gets negative reward or penalty. In a standard RL setup consisting of an agent interacting with an environment E in discrete time steps, at each time step t the agent receives an observation x_t, takes an action a_t, and receives a reward r_t. A schematic representation is shown in Fig. 1.

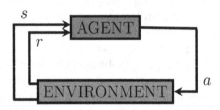

Fig. 1. Schematic model of Reinforcement Learning algorithm: s is the environment state, r is the reward, and a is the agent action [13].

In general, the environment may be partially observed, so that the entire history of the observation and action pairs, $s_t = (x_1, a_1, ..., a_{t-1}, x_t)$, is required to describe interaction environment-actor. Here, it is assumed the environment is fully observed, so that the environment state corresponds to the observation, i.e. $s_t = x_t$. The environment can be stochastic or deterministic and, in order to model it, a Markov decision process [14] is used with a state space S, an action space $A \in \mathbb{R}^N$, an initial state distribution $p(s_1)$, a transition dynamics $p(s_{t+1}|s_t, a_t)$, and a reward function $r(s_t, a_t)$. An agent's behavior is defined by a policy, π, which maps states to a probability distribution over the action space given the state space, $\pi : S \to P(A)$. The return from the current state is defined as the sum of discounted future rewards, $R_t = \sum_{i=t}^{T} \gamma^{i-t} r(s_{i,a_i})$, with a discounting factor $\gamma \in [0, 1]$. Note that the return depends on the action taken, therefore the policy π affects it directly. The goal in RL is to maximize the return, and this is strictly correlated to learning the best policy able to reach this goal. A useful function to describe the expected return combined with the state s_t and the action a_t following the policy π, is the action-value Q-function which depends only on the environment [13]. The action-value function is fundamental in RL, as it underlies one of the first learning algorithm: the Q-learning algorithm [15] and its extension to Deep Q-learning in the context of Deep Learning. Q-learning is the main algorithm of the Value-Based approach to RL. Value-Based approaches try to find or approximate the optimal value function, which is a mapping between an action and a value. The higher the value, the better the action. The robotic control typically requires a continuous action space, and Q-learning cannot be applied due to the enormous number of possible configurations that need to be explored and the difficulty to converge to an optimal solution for the given task. Policy-Based algorithms have been proposed for continuous control [13]. Such algorithms try to find the optimal policy directly without leveraging the Q-value [16]. The Policy-Based approach has its most suitable applications in continuous and stochastic environments, where faster convergence is guaranteed. On the other hand, Value-Based approaches are more sample-efficient and steady [17].

The Actor-Critic DRL algorithm merges together Policy-Based and Valued-Based approaches [16]. The core idea is to split the model in two components: the Actor, that defines an action based on the state, and the Critic, that produces the state-values. The actor essentially controls how the agent behaves by learning the optimal policy, and the Critic evaluates the action by computing the value function. The training of the neural networks representing the Actor and the Critic is performed separately, using gradient descent [18]. As training proceeds, the Actor learns to produce better actions to reach the goal, and the Critic gets better at evaluating those actions. The training process of the model begins with a starting state of the environment being observed by the agent. Then the agent takes an action, leading the environment into a new state, and letting the agent achieve a reward (step). When a terminal state is encountered, a learning episode ends – an episode being a collection of all subsequent steps. In this process the model optimization procedure is required, in order to update model parameters according to a specified loss function. For the Actor-Critic algorithm, the update of neural network weights occurs at every step, and not at the end of every episode. The Actor-Critic approach suffers from high sensitivity to the values of hyperparameters. To tackle this problem and improve performance, stability and proficiency, the Soft Actor-Critic has been proposed by Haarnoja et al. in [19]. SAC introduces multiple elements to the Actor-Critic approach:

- A large buffer memory to efficiently update the weights using the previous history, thus increasing learning stability.
- A target neural network of the Critic is introduced to further improve the stability and efficiency of learning the approximation of the value-state. To update the target Critic network, the soft function update procedure is used: values include a factor to combine new and old weights obtained from the

updated and previous steps, and to this purpose a loss value of state and a loss of action must be defined.
- The maximum entropy of RL is introduced to increase the environment exploration, improve learning, and reduce sensitivity to hyperparameters. The loss of this maximum entropy was transformed for this case from the "classical" one considering the use of a neural network, and introducing an input noise vector sampled from a fixed spherical Gaussian distribution to calculate the action. In this context, maximum entropy represents the measure of chaos or disorder, strictly linked to the fact that the policy acts as randomly as possible to make the model more robust to variations.

2.2 Variational Quantum Circuits for Reinforcement Learning

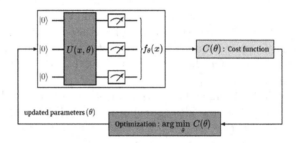

Fig. 2. Schematic representation of a Variational Quantum Algorithm.

A Variational Quantum Circuit consists of various quantum gates and wires, with learnable parameters to adjust these gates [7]. Like the parameters in a neural network, these learnable parameters in VQCs can be optimized to approximate complex continuous functions. So, VQCs are also known as quantum neural networks. A typical VQC has three main components: encoder, variational circuit, and measure. The encoder is usually the first operation of any quantum circuits. Given a classical input x, the encoder encodes it into a quantum state. The variational circuit often follows an encoder circuit. Usually, it consists of alternate layers of rotation gates and entanglement gates (e.g., a closed chain or ring of CNOT gates). After the variational circuit, the states of qubits are measured. Like with neural networks, a VQC is trained through gradient descent. Given a quantum simulator, backpropagation [20] can be applied to compute gradient analytically and optimize the VQC parameters. However, backpropagation is not applicable for a physical Quantum Computer, since it is impossible to measure and store intermediate quantum states during computation without impacting the whole computation process. Instead, parameter-shift is used to compute gradients for any multi-qubit VQCs. This is done by evaluating the circuit when shifting the parameter by a specific quantity, and calculating the difference before and after the shifting [21]. After getting gradients, learnable parameters are optimized with traditional optimizers, such as RMSprop and Adam [22]. A schematic for this approach is shown in Fig. 2. Unlike neural networks, VQCs do not express non-linearity, except for measurement, unless a particularly encoding strategy is applied. Additionally, it can be observed how the measurement operations lead to non-linearity. To introduce more non-linearity, depolarizing gates could be added, but there is no theorem able to quantify how many of them are needed. This may be a problem in DRL, due to the significant presence of non-linear functions in the Q-values and Actor components. To introduce this non-linearity and deal with the no-cloning property of quantum computing, a new ansatz called *data re-uploading* has been defined. The concept has been introduced in [23], and is meant to introduce non-linearity by reapplying the encoding layer multiple times, to have a more composite expressiveness of the function. The strategy is therefore to use more quantum gates without increasing the number of required qubits.

3 A Virtual Robotic Arm as an Environment for Continuous Control Task

The task of robotic arm operation requires continuous control, due to the continuous-valued observations coming from sensors and actions for accurate movement control. The DRL promises robots autonomously acquiring complex behaviors from sensor observations [2,24–26], enabling robots to move towards unstructured environments, to handle unknown objects, and to learn a state representation suitable for several tasks. A robotic arm can be described as a chain of links that are moved by joints containing motors to change the link position and orientation. Figure 3 shows the schematic diagram of a simple 2-dimensional, four-joints robotic arm mounted and fixed on the first joint. The arm is able to move the links using the joints on the 2-dimensional plane and can independently move each link clockwise and counter-clockwise up to a given velocity. Last joint is referred to as the end effector.

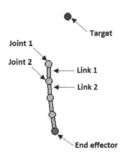

Fig. 3. Simplified schematic diagram of mechanical components of a four-joint robotic arm.

Such environment has been created using Box2D [27], by adapting parts of OpenAI Gym environment Acrobot-v1 [28]. Support for continuous action and state spaces is enabled, and the dynamics of a planar manipulator with an arbitrary number of links can be simulated. For the scope of this work, a two-link structure has been considered, with link 1 connecting the center (i.e. joint 1) to the so called middle effector (i.e. joint 2), and link 2 connecting the middle effector to the end effector. Physical parameters of the environment are summarized in Table 1. As with standard custom Gym environments, all parameters are customizable. Let $\bm{x}_m = (x_m, y_m)$, $\bm{x}_e = (x_e, y_e)$ and $\bm{x}_t = (x_t, y_t)$ be the positions of the middle effector, end effector and target respectively. The environment allows for accessing the following states: target position $\bm{x}_t$; end effector position $\bm{x}_e$; rotation angles (θ, ϕ) of the two links with respect to vertical axis. Based on the state of the environment, an algorithm $\mathcal{A}$ defines the actions to perform, applying torques (a_m, a_e) to the middle effector and to the end effector, with intensity ranging between $[-max\_torque, max\_torque]$. Once an arbitrary number of episodes N and a maximum number of steps per episode (250 in this case) are chosen, the logic flow shown in Algorithm 1 begins. When an episode starts, the target position is randomly initialized, while the effector positions are always initialized so that the robotic arm is in the direction in which gravity pulls. Then, for each step in the episode, the agent gets state s of the environment and computes, according to it and through the algorithm $\mathcal{A}$, the actions to take. After that, the agent computes the Euclidean distance d between $\bm{x}_e$ and $\bm{x}_t$. Once a distance threshold T is defined, reward r for each step is set to $-d$ if $d > T$, otherwise it is set to $+5$; r is then used to update the episode return R (i.e., the sum of all rewards obtained in it). Episode terminates when $d \leq T$, or when maximum number of steps is reached.

Table 1. Two-links robotic arm environment parameters.

Parameter	Description	Used Values
Link mass	Mass of links connecting the joints	0.01 Kg
Link height	Vertical length of links	0.5 m
Link width	Horizontal length of links	0.1 m
Episode length	Maximum number of steps for every episode	250
Frames per second	Refresh rate for animation	50
Distance threshold	Distance for which the target is considered reached	0.25 m
Max joint velocity	Maximum velocity of rotation of link	2.5 rad/s
Max joint torque	Maximum torque that can be applied on joint	1000 Nm

Algorithm 1. Environment workflow

Require: number of episodes N
Require: algorithm $\mathcal{A}$
  for N episodes **do**
    init target position (x_t, y_t)
    init effectors positions $(x_m, y_m), (x_e, y_e)$
    init episode reward $r = 0$
    for 250 steps **do**
      get environment state s
      perform actions $(a_m(\mathcal{A}, s), a_e(\mathcal{A}, s))$
      get new end effector position (x_e, y_e)
      compute $d \leftarrow \sqrt{(x_t - x_e)^2 + (y_t - y_e)^2}$
      if $d > Threshold$ **then**
        $r \leftarrow r - d$
      else
        $r \leftarrow r + 5$
        break
      end if
    end for
  end for

4 Implementation of Classical and Quantum Soft Actor-Critic Algorithms

4.1 Soft Actor-Critic Algorithm

Soft Actor-Critic has been proposed in [19]. Its implementation replicates the algorithm structure as proposed in [22] and is summarized in the pseudo-code shown in Algorithm 2. Indeed, the present work has started from the quantum-classical hybrid RL model proposed in [22], to solve the Pendulum-v0 OpenAIGym environment [28]. The main components of the Soft Actor-Critic algorithm are: one Actor, two Critics and two target Critics. The Critic and target Critic networks share the same architecture. Due to the deep learning nature of this algorithm, all of these components are neural networks. Inspired by the Deep Q-Learning approach, there are target Critic components since the data need to be independent and identically distributed ("i.i.d.") to correctly model the Q-functions: unfortunately, any two consecutive states do not have this property, and an independent copy of the Critic - the target - is necessary to meet this condition. Moreover, recent advances have shown that using a copy for both Critic and target Critic components improves the performance and the stability of the algorithm. The Critic networks are used to calculate the Q-functions of the actual state, while the target Critics are used to calculate the Q-functions of the next state, by choosing the minimum of the target outputs. These functions, as can be seen in Algorithm 2, are used for optimization of the Critic components.

Algorithm 2. Soft-Actor-Critic algorithm

Require: initial actor parameters θ, initial critics parameters ϕ_1 and ϕ_2, γ, α, ρ, empty experience replay D.
Initialize the actor network with θ.
Initialize two critics networks with ϕ_1 and ϕ_2 respectively.
Set target critics: $\phi_{targ,1} \leftarrow \phi_1$ and $\phi_{targ,2} \leftarrow \phi_1$.
for each time-step **do**
  Observe state S, select action $A \sim \pi_\theta(\cdot|S)$ and execute A in the environment.
  Observe next state S', reward R, and binary done signal d to indicate whether S' is terminal state or not.
  Store (S, A, R, S', d) in D.
  Reset the environment if $d = 1$.
  Sample a batch of transitions $B = (S, A, R, S', d)$ from D randomly.
  Compute target values $y(R, S', d) = R + \gamma(1-d)(min_{i=1,2}Q_{\phi_{targ,i}}(S', A') - \alpha \log \pi_\theta(A', S'))$ where $A' \sim \pi(\cdot, S')$.
  Update ϕ_i by minimizing: $\mathbb{E}_B[(Q_{\phi_i}(S, A) - y(R, S', d))^2]$ for $i = 1, 2$.
  Update θ by maximizing: $\mathbb{E}_B[min_{i=1,2}Q_{\phi_i}(S, \tilde{A}_\theta) - \alpha \log \pi_\theta(\tilde{A}_\theta|S)]$.
  Do a soft update for target action-value networks: $\phi_{targ,i} \leftarrow \rho\phi_{targ,i} + (1-\rho)\phi_i$ for $i = 1, 2$.
end for

The Actor components outputs two values for this algorithm: the mean and the variance of a Gaussian distribution that will be used to sample the actions used for each state. This means that the approach is statistical, but it can be converted to deterministic by using the mean of the distribution.

4.2 Variational Quantum Soft Actor-Critic Algorithm

In the Quantum Soft Actor-Critic (QSAC) algorithm some neural network layers have been replaced with a VQC. There is a major issue when using VQCs, namely the difficulty to approximate non-linear functions. In order to address it, two solutions already tested in [22] have been used:

- data re-uploading: a technique introduced in [23] where every VQC layer, except the last one, reapplies data encoding to have a more composite function expressiveness.
- neural networks: layers of neural networks are added before (for quantum Critic only) and after (for quantum Actor and Critic components) to leverage the high capacity of non-linearity using activation functions.

The used structure for the Actor component is shown in Fig. 4a. The VQC replaces the entire input layer in the hybrid structure, and neural network layers are later used. This allows for dimensional flexibility and the capacity to exploit potential quantum advantages. The encoding layer of the VQC implements the angle encoding technique, and every gate is parametrized to increase capacity to find a useful representation. The ansatz, that is repeated $n - 1$ times, is composed by: a generic rotation over the three axis controlled by three different

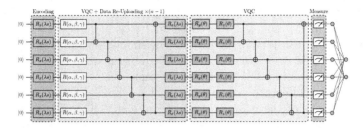

(a) The architecture of quantum-classical hybrid Actor component of the QSAC

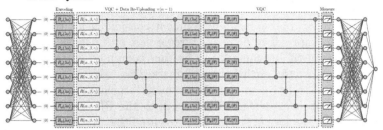

(b) The architecture of quantum-classical hybrid Critic and target Critic components of the QSAC

Fig. 4. The architecture of quantum-classical hybrid Actor, Critic and target Critic components of the QSAC.

parameters, the CNOT gates to introduce entanglement, and a further encoding layer using data re-uploading. The last ansatz is composed by only y and z-axis rotations and CNOT gates. Finally, measurement is applied using the Pauli-Z gate.

Figure 4b shows the structure of the hybrid circuit of the Critic component. As it can be seen the VQC, apart from the number of qubits, is equal to the one used for the actor. The real difference is the presence of multiple neural network layers before and after the VQC.

5 Experiments

Both classical and quantum SAC algorithms have been implemented using the Python programming language [29]. Also the robotic arm environment has been developed in the same language. TensorFlow Quantum [30] has been used as the development framework for Quantum Machine Learning. This library works by simulating the quantum circuit using Cirq [31] and distributing the computational workload through the multi-threading features of TensorFlow [32]. All classical components of the quantum-classical hybrid algorithm have been implemented using the TensorFlow library, in order to improve performance, reduce memory consumption, reduce dependencies, and avoid possible conflicts and errors during execution. Due to the simulation nature and the long time required to train this model, it was chosen not to explore how the introduction of noise may affect this kind of algorithm. The simulation and training processes have been executed on a Virtual Machine equipped with 16 GB RAM, Intel Xeon CPU and Ubuntu 22.04 Operating System.

The QSAC architecture presented in Sect. 4.1 has been tested against several configurations of the neural networks and the VQC learnable parameters and hyperparameters. Each configuration has been compared to a classical SAC implementation of equivalent complexity in terms of the learnable parameters, using whenever possible the same exact hyperparameter values, optimizer and activation functions.

The structure of robotic arm environment considered for this work allows for precise resolution rules to be defined, regardless of the initial position of the target. A deterministic benchmark can be thus computed as described in [33]. For the experiments, the environment is marked as "solved" according to two criteria: the first is relative to the deterministic benchmark, while the second is considered in absolute terms with respect to the RL and QRL algorithms considered. About the former, it has been required that the mean of returns for the last 1000 episodes of training falls within the range $[\mu - \sigma, \mu]$, where μ is the average mean of the return distribution of the deterministic benchmark, and σ its standard deviation. Regarding the latter, it has been required for a maximum of 1% of episodes to fail in the last 1000 episodes of the training. In particular, an episode is considered failed when the end effector does not reach the target within 250 steps. Both criteria have to be satisfied before 5000 episodes are reached.

All presented plots in the next sub-sections refer to an average of 20 episodes over 10 runs, with the shaded area representing one standard deviation. All networks have been optimized through Adam [34], and the hyperparameters have been chosen by a grid search mechanism.

Figure 5 Left shows a comparison of three training curves and the deterministic benchmark: in red representing deterministic benchmark, green QSAC with 2700 total learnable parameters, blue SAC with 3000 parameters (i.e. about the same amount of the QSAC) and orange SAC with 100 times the number of parameters of the QSAC (i.e. 274k parameters). A clear quantum advantage has been found in the number of learnable parameters. In particular, the classical SAC with the same number of parameters as the QSAC does not converge. The SAC requires 100 times the amount of parameters compared to the quantum algorithm to solve the environment with the same learning curve.

Other checks have been done to verify if this QSAC algorithm presents more interesting properties. Indeed, Fig. 5 Right contains a comparison between the reward frequency distribution of the SAC (red histogram) and the QSAC (blue histogram), where advantages in terms of model stability cannot be noticed (Table 2).

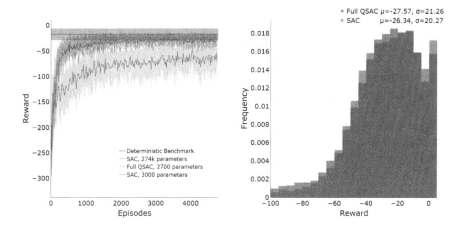

Fig. 5. Left: learning curves of classical and quantum-classical SAC architectures tested on the robotic arm environment. Right: return distributions of QSAC and SAC. (Color figure online)

Table 2. Architectures and hyperparameters of the SAC and QSAC configurations.

Hyperparams	Model		
	SAC 3000 total parameters	SAC 270000 total parameters	QSAC 2700 total parameters
γ	0.99	0.99	0.99
α	0.2	0.2	0.2
Learning rate	0.0003	0.0003	0.0003
Memory size	1000000	1000000	1000000
Optimizer	Adam	Adam	Adam
Actor neurons	(6,7)(8,(1,1))	(6,7)(8,(1,1))	(6,VQA(5 layers),8,1)
Actor act. func	(linear,relu,linear)	(linear,relu,linear)	(linear,relu,linear)
Actor params	149	149	112
Critic neurons	(8,22)(22,21)(21,1)	(8,256)(256,256)(256,1)	(8,VQA(20 layers),8,1)
Critic act. func.	(linear,relu,relu,linear)	(linear,relu,relu,linear)	(linear,relu,relu,linear)
Critic params	719	67840	650
Total params	3025	271509	2712

6 Conclusions and Future Works

A quantum variational approach to the Soft Actor-Critic algorithm has been studied, looking for and evaluating possible quantum advantages over its classical counterpart when applied to a continuous control task, in particular that of a simulated robotic arm. Several quantum-classical hybrid network architectures of the Soft Actor-Critic algorithm have been experimented by making use of digital simulation of quantum circuits. For all architectures, a comparison with the classical implementation of the algorithm with comparable number of learnable parameters has been carried out. Quantum SAC, with both Actor and Critic components leveraging Variational Quantum Circuits, exhibits a huge quantum advantage with respect to its classical counterpart. Namely, it requires 100 times fewer learnable parameters to reach the same performance as the classical SAC model.

As a future work, an amplitude encoding of the state action will be tested in order to introduce more non-linearity to the quantum circuits, possibly avoiding the use of classical dense neural network layers in the Critic and Actor components of the algorithm. Moreover, the evaluation of the quantum algorithm with a physical quantum computer will be carried out. Indeed, while data re-uploading seems to work particularly well for quantum reinforcement learning using simulators, further experiments are required in order to understand if such an approach could be exploited on real quantum devices, dealing with decoherence and noise, and suffering from potential design limitation.

Acknowledgements. This work was fully funded by NTT DATA Corporation (Japan) and supported by NTT DATA Italia S.p.A. (Italy).

References

1. Wang, H., et al.: Deep reinforcement learning: a survey. Front. Inf. Technol. Electron. Eng. **21** (2020)
2. Liu, R., Nageotte, F., Zanne, P., Mathelin, M., Dresp-Langley, B.: Deep reinforcement learning for the control of robotic manipulation: a focused mini-review. Robotics **10**, 22 (2021). https://doi.org/10.3390
3. Sutton, R., Barto, A.: Reinforcement Learning: An Introduction. The MIT Press (2018). http://incompleteideas.net/book/the-book-2nd.html
4. Nielsen, M., Chuang, I.: Quantum Computation and Quantum Information: 10th Anniversary Edition. Cambridge University Press (2011)
5. Preskill, J.: Quantum computing in the NISQ era and beyond. Quantum **2**, 79 (2018). https://doi.org/10.22331
6. Benedetti, M., Lloyd, E., Sack, S., Fiorentini, M.: Parameterized quantum circuits as machine learning models. Quantum Sci. Technol. **4**, 043001 (2019). https://doi.org/10.1088
7. Schuld, M., Petruccione, F.: Machine Learning with Quantum Computers. Springer, Cham (2021). https://doi.org/10.1007/978-3-030-83098-4. https://books.google.it/books?id=-N5IEAAAQBAJ
8. Banchi, L., Pereira, J., Pirandola, S.: Generalization in quantum machine learning: a quantum information standpoint. PRX Quantum **2** (2021). https://doi.org/10.1103
9. Heimann, D., Hohenfeld, H., Wiebe, F., Kirchner, F.: Quantum deep reinforcement learning for robot navigation tasks. arXiv (2022). https://arxiv.org/abs/2202.12180
10. Chen, S., et al.: Variational quantum circuits for deep reinforcement learning. IEEE Access **8**, 141007–141024 (2020)
11. Valdez, F., Melin, P.: A review on quantum computing and deep learning algorithms and their applications. Soft Comput. (2022)
12. Nian, R., Liu, J., Huang, B.: A review on reinforcement learning: introduction and applications in industrial process control. Comput. Chem. Eng. **139**, 106–886 (2020). https://www.sciencedirect.com/science/article/pii/S0098135420300557

13. Lillicrap, T., et al.: Continuous control with deep reinforcement learning. arXiv (2015). https://arxiv.org/abs/1509.02971
14. Puterman, M.: Markov Decision Processes: Discrete Stochastic Dynamic Programming. Wiley (2005). https://cds.cern.ch/record/1319893
15. Watkins, C., Dayan, P.: Q-learning. Mach. Learn. **8**, 279–292 (1992)
16. Konda, V., Tsitsiklis, J.: Actor-critic algorithms. Soc. Ind. Appl. Math. **42** (2001)
17. Li, H., Lau, T.: Reinforcement learning: prediction, control and value function approximation. arXiv (2019). https://arxiv.org/abs/1908.10771
18. Han, M., Zhang, L., Wang, J., Pan, W.: Actor-critic reinforcement learning for control with stability guarantee. arXiv (2020). https://arxiv.org/abs/2004.14288
19. Haarnoja, T., Zhou, A., Abbeel, P., Levine, S.: Soft actor-critic: off-policy maximum entropy deep reinforcement learning with a stochastic actor. arXiv (2018). https://arxiv.org/abs/1801.01290
20. Rumelhart, D., Hinton, G., Williams, R.: Learning representations by back-propagating errors. Nature **323**, 533–536 (1986)
21. Crooks, G.: Gradients of parameterized quantum gates using the parameter-shift rule and gate decomposition. arXiv (2019). https://arxiv.org/abs/1905.13311
22. Lan, Q.: Variational quantum soft actor-critic. arXiv (2021). https://arxiv.org/abs/2112.11921
23. Pérez-Salinas, A., Cervera-Lierta, A., Gil-Fuster, E., Latorre, J.: Data re-uploading for a universal quantum classifier. Quantum **4**, 226 (2020). https://doi.org/10.22331
24. Gu, S., Holly, E., Lillicrap, T., Levine, S.: Deep reinforcement learning for robotic manipulation with asynchronous off-policy updates. arXiv (2016). https://arxiv.org/abs/1610.00633
25. Ibarz, J., Tan, J., Finn, C., Kalakrishnan, M., Pastor, P., Levine, S.: How to train your robot with deep reinforcement learning: lessons we have learned. Int. J. Rob. Res. **40**, 698–721 (2021). https://doi.org/10.1177
26. Kilinc, O., Montana, G.: Reinforcement learning for robotic manipulation using simulated locomotion demonstrations. Mach. Learn. **111**, 465–486 (2021). https://doi.org/10.1007
27. Catto, E.: Box2D, a 2D physics engine for games (2011). https://box2d.org/
28. Brockman, G., et al.: OpenAI Gym. arXiv (2016). https://arxiv.org/abs/1606.01540
29. Van Rossum, G., Drake, F.: Python 3 reference manual. CreateSpace (2009)
30. Broughton, M., et al.: TensorFlow quantum: a software framework for quantum machine learning. arXiv (2020). https://arxiv.org/abs/2003.02989
31. Gidney, C., et al.: Cirq, Zenodo (2022). https://doi.org/10.5281/zenodo.6599601
32. Abadi, M., et al.: TensorFlow: large-scale machine learning on heterogeneous systems (2015). https://www.tensorflow.org/. Software available from tensorflow.org
33. Acuto, A., Barillà, P., Bozzolo, L., Conterno, M., Pavese, M., Policicchio, A.: Variational quantum soft actor-critic for robotic arm control (2022)
34. Kingma, D., Ba, J.: Adam: a method for stochastic optimization. arXiv (2014). https://arxiv.org/abs/1412.6980

Named Entity Recognition to Extract Knowledge from Clinical Texts

Ileana Scarpino[1]((✉)), Rosarina Vallelunga[1], and Francesco Luzza[2]

[1] Department of Medical and Surgical Sciences, University of Catanzaro, Catanzaro, Italy
{ileana.scarpino,rosarina.vallelunga}@unicz.it
[2] Department of Health Science, University of Catanzaro, Catanzaro, Italy
luzza@unicz.it

Abstract. Clinical texts encompass a wide range of information such as patient's history, disease diagnosis and prescribed drugs, reflecting details and nuances that are valuable in providing knowledge.

In the present study, a Natural Language Processing approach, specifically Named Entity Recognition (NER), is applied to extract important concepts from gastroenterology clinical texts.

NER is a task of text analytics to identify, in written documents, named entities ranging from general concepts to information in specific fields.

The application is performed through freely available Python packages of the spaCy library adapted to the English language. Although the spaCy's NER is generic, models trained in the clinical domain are used to identify categories belonging to the medical sector of interest. In particular, the main goal is to find as much entities as possible by paying attention to major bowel diseases, such as Ulcerative Colitis and Crohn's disease. We performed two experiments, the first one involves the use of the ScispaCy package for scientific text processing. The second one applies Med7 that is a spaCy NER model for labeling drug information trained for identification of seven medication-related concepts, dosage, drug names, duration, form, frequency, route of administration, and strength.

The results show that two approaches applied for NER analysis perform well to extract knowledge from gastroenterology clinical texts.

The results obtained will allow to evaluate the efficiency of the proposed methodology and to analyze, through the extracted entities, the profile and aspects of the diseases considered and the associated drugs.

Keywords: Named Entity Recognition · Natural Language Processing · Clinical Report

1 Introduction

Named Entity Recognition (NER) is a method within Natural Language Processing (NLP) that falls under the domain of Information Extraction, aiming to automatically identify, extract, and classify key information within documents, resulting in structured data.

NER made its initial appearance in 1995 [1], delineating three primary categories (Entity, Name, and Number). The primary objective behind NER implementation was the examination of text to discern proper nouns and subsequently assign them classifications. Collobert et al. [2] introduced a NER approach based on neural networks, marking the first instance of domain independence.

This methodology has become widespread, with numerous adaptations proposed over the past decade leveraging technologies such as Recurrent Neural Networks (RNN) and word embedding. [3].

In many fields, such as medicine, NLP applications can be useful for textual data for research purposes, making efficient use of the vast resources available with the goal of improving documentation, quality, and efficiency of health care [4].

For example, the chemical literature is very rich in information on chemical entities; extracting relationships about molecules and their properties can help scientists expand knowledge about drug characteristics.

This huge amount of data, also deriving from clinical texts, has motivated the development of new information extractions; it is, therefore, clear that medicine needs a breakthrough and technological revolution, which makes it possible to identify new interesting markers that cannot be identified with statistical methods. With the evolution of NLP techniques, doctors can be supported in the interpretation of clinical information, giving added value to textual data that would otherwise remain not exploited.

Our work discusses a NLP approach, specifically NER, applied to extract important concepts from Inflammatory Bowel Disease (IBD) clinical text, where a literature overview shows that little research of this kind has been done.

Crohn's Disease (CD) and Ulcerative Colitis (UC) are both IBD that impact the gastrointestinal tract and may lead to the development of polyps, influenced by genetic factors. Despite significant advancements over the last two decades, the intricate nature of IBD presents substantial challenges. Traditional scientific approaches have struggled to tackle crucial research inquiries, including the development of future targeted and personalized diagnostic strategies. [5–7].

The paper is organized as follows. Section 2 presents a description of IBD and the structure of clinical reports, offering an overview of NER, as a method for recognizing information and entities in the medical field.

Section 3 describes materials and methods, considering the construction of the data set and an overview of the models considered for NER.

Section 4 discusses the results of the analysis. Finally, Sect. 5 concludes the paper and presents possible future work.

2 Background

IBD encompasses two distinct conditions, namely UC and CD, with a small subset of patients exhibiting an intermediate form. UC manifests as an inflammatory process localized to the colon, characterized by clinical symptoms such as bloody diarrhea, the presence of mucus in stools, abdominal pain, and weight loss.

On the other hand, CD has the potential to affect any segment of the digestive tract, extending from the mouth to the anus. Its symptoms and complications parallel those observed in UC. [8].

The barriers present in the processes of diagnosis and treatment of patients, such as those suffering from IBD, can be overcome through Machine Learning (ML) methodologies, efficiently and effectively managing the flow of clinical data and speeding up interpretation of information [9].

The existing data sources serve as inputs for analytical techniques, spanning from conventional statistical methods to sophisticated approaches like data mining, machine learning, clustering, text analysis, and image analysis. These methods collectively contribute to enhancing our understanding of IBD and addressing existing gaps within this field. [10].

The use of computer techniques favors an accurate and rapid diagnosis and an effective treatment to prevent and predict side effects and complications with a consequent improvement in the patient's quality of life which represents one of the objectives of scientific research. In the field of IBD and in the medical field in general, potential sources of NLP analysis have numerous strengths and limitations, including medical records, clinical trial data, and e-health applications.

Given the dispersion of clinical information across numerous electronic records, the application of NLP approaches can streamline the organization of this data, effectively mitigating the constraints associated with permanent records. [11].

Enhancing the efficacy of NLP is crucial for the efficient organization, interpretation, and identification of patterns within textual data, as highlighted in [12], particularly in unstructured clinical reports. These advancements can potentially yield valuable discoveries by extracting pertinent information from medical records and leveraging NLP techniques. In particular, the use of relation extraction and named entity recognition approaches can favor a better knowledge of the properties of the drugs prescribed to patients with IBD, with a consequent characterization of the disease itself.

2.1 Information Extraction from IBD Clinical Texts

The medical record is the set of data and information (personal, health, social, environmental, legal) collected by health personnel during a patient's hospitalization to document diagnoses and therapies. Clinical reports generally have no defined structure and are usually written in a free format, which makes them difficult to analyze.

The utilization of ML in the context of IBD serves as a research avenue aimed at enhancing patient health outcomes. It offers patients increased access to care, facilitates comprehension of their health status, supports preventive measures, and enables early diagnoses.

The identification of medical concepts and extraction of information presents a challenging yet crucial task. It acts as a vital component for transforming unstructured data into a structured and tabulated format, facilitating subsequent analytical activities. [13].

IBD have created new challenges that traditional scientific methods have failed to address, NLP performance is important for organizing, interpreting, and recognizing patterns from textual data [10], such as unstructured clinical reports.

Information mining allows discoveries from medical records to characterize the diseases themselves, including IBD. For example, pharmacovigilance can be improved using text mining, and obtaining drug adverse event data from medical notes [14]. A benefit lies in the fact that clinical reports, which include endoscopy reports, are frequently accessible not only in PDF format but also in plain text. This availability enables the application of NLP approaches, including NER and Relationship Extraction (RE). These methods serve as integral components for tasks involving information extraction within the clinical domain. [3].

2.2 Clinical Named Entity Recognition

NER constitutes a broad and extensively explored field within NLP [15]. Most research publications approach NER as a supervised task, involving the training of a model on annotated data and subsequent application to new text [16].

NER plays a pivotal role in extracting crucial concepts and entities of interest, such as disease names, drug names, and lab tests, from detailed clinical narratives containing patient information. Machine learning methods frame the clinical NER task as a labeling problem for a given input sequence, typically individual words derived from clinical text [17].

Over the past decades, numerous automatic NER systems have emerged and been utilized for the recognition of chemical entities. These systems can be categorized into four distinct groups:

- NER systems based on dictionaries utilize collections of vocabulary tailored for specific domains, typically sourced from archives within the relevant domain. Instances of dictionaries within the realms of chemistry and biomedicine include the Jochem dictionary [18]. This dictionary is applied for the identification of small molecules and drugs within textual content. Another notable example is the DrugBank dictionary, specifically designed for cataloging drugs;
- NER systems based on rules operate on a set of manually crafted rules for extracting entity names, as outlined in [19]. These rule-based models are constructed using sets of rules that incorporate grammatical elements (such as parts of speech) and syntactical aspects (like word precedence). In some cases, these rule sets may be combined with dictionaries to enhance their performance;
- NER systems employing the ML approach, as described in [20–22], utilize statistical models to identify particular entity names. These models generate characteristic-based representations of observed data, which are dependent on annotated documents;

- The hybrid NER system. NER approach uses the positive features of each approach [19].

In the field of biomedicine, NER addresses various classes, including proteins, genes, diseases, drugs, organs, DNA sequences, RNA sequences, and potentially others [23]. Drugs, particularly in their pharmaceutical context, represent a specialized category of chemicals that holds significant relevance in biomedical research. An elementary and straightforward approach to NER involves directly comparing textual expressions found in a pertinent lexical repository with the raw text.

Recognition of drug-related entities is a crucial step in intricate biomedical Natural Language Processing (NLP) tasks, such as the extraction of pharmacogenomic, pharmacodynamic, and pharmacokinetic parameters. A key challenge in chemical text mining revolves around identifying chemical entities mentioned in texts. Extracting molecules and their properties from mined data aids research scientists, particularly in the domain of drug development.

A specific instance is Chemical NER, which autonomously identifies instances of chemical entities within a given text. [19].

In [24] An entity recognition tool called biomedical neural and multi-type normalization (BERN) has been proposed that recognizes known entities and discovers new entities and identifies overlapping entity types. BERN uses the BioBERT NER models of Lee et al. [25] to label genes/proteins, diseases, drugs, chemicals, and species. BioBERT NER models can be used to discover new entities from the most recent biomedical literature.

Biomedical-named entity recognition (BioNER) is a fundamental step for mining COVID-19 literature. Pre-trained supervised BioNER models, such as SciSpacy, are used to identify highly domain-specific entity types (e.g., animal models of diseases) or emerging ones (e.g., coronaviruses) for COVID-19 studies [26].

3 Materials and Methods

In the present study, NLP techniques were applied, in particular NER approaches, for the extraction of clinical and pharmacological information starting from gastroenterology reports relating to colonoscopy instrumental examinations with a diagnosis of Crohn's disease and ulcerative colitis. The analysis shows the objective to apply approaches of NER to IBD, using the ScispaCy package for word processing and Med7 for the identification of 7 medication-related concepts, dosage, drug names, duration, form, frequency, route of administration, and strength.

The description of the models is presented to offer an overview of the objective of our work by showing the results obtained in the following sections.

3.1 Gastroenterology Dataset

The dataset consists of 230 medical reports written in Italian by specialists in the field of Gastroenterology, concerning Crohn's disease and ulcerative colitis.

The clinical reports are structured as follows: a first section is dedicated to the way in which the patient booked the visit which can be outpatient or from the national health system; the other section includes the reason for the visit and any instrumental tests, the instrument used, any biopsy taken, the response of the instrumental investigation and the operator who performs the analysis. The patient's personal data have been obscured from the reports for privacy by extracting the reason why the patient underwent diagnostic visits.

We extracted the motivation written by the doctor and for which the patient was subjected to instrumental investigation. The extracted texts were collected for the creation of the dataset in *CSV* file format. The longest document contains 105 words and the shortest 39 words.

To allow for analysis, the texts were collected and automatically translated by the Google API into English to allow the use of the original packages.

The *Googletrans* module has been imported, which is used to facilitate the translation of languages, in this case from Italian to English.

3.2 NER Using ScispaCy

In the year 2019, the Allen Institute for Artificial Intelligence (AI2) introduced *ScispaCy*[1], an extensive open-source spaCy pipeline tailored for Python. This tool is specifically crafted for the analysis of biomedical and scientific text through Natural Language Processing (NLP). ScispaCy stands out as a robust resource, particularly for Named Entity Recognition (NER), which involves identifying keywords, known as entities, and organizing them into distinct categories.[2].

From the literature, it appears that this is a new Python library for the practical processing of biomedical and scientific texts [27].

ScispaCy contains two major packages released: en_sci_sm[3] and $en\_core\_sci\_md$[4]. Models in the $en\_core\_sci\_sm$ package have a larger vocabulary and include word vectors, while those in $en\_core\_sci\_md$ have a smaller vocabulary and do not include word vectors [28]; $en\_ner\_bc5cdr\_md$[5] [28] is an important spaCy NER model trained on *bc5cdr* corpus[6] to recognize two types of entities as *disease* and *chemical* [27].

To provide users with more specific requirements on entity types, there exist four additional packages $en\_ner\_[bc5cdr-craft-jnlpba-bionlp13cg]\_md$ with finer-grained NER models trained on *bc5cdr* (for chemicals and diseases), *craft* (for cell types, chemicals, proteins, genes), *jnlpba* (for cell lines, cell types, DNAs, RNAs, proteins) and *bionlp13cg*(for cancergenetics), respectively [28]. ScispaCy models are fast, easy to use, and scalable.

---

[1] https://spacy.io/universe/project/scispacy.
[2] https://towardsdatascience.com/using-scispacy-for-named-entity-recognition-785389e7918d.
[3] https://allenai.github.io/scispacy/.
[4] https://allenai.github.io/scispacy/.
[5] https://allenai.github.io/scispacy/.
[6] https://www.biocreative.org/tasks/biocreative-v/track-3-cdr/Accessedon08/19.

3.3 NER Using Med7

Med7[7] is a spaCy NER model for labeling drug information. The model is trained to recognize seven categories: *drug, dosage, duration, form, frequency, route*, and *strength*. Med7 is recognized as a transformer-based model that has been trained on electronic health records (EHR) to extract key concepts in clinical settings [29].

Identifying medical concepts and extracting information is a challenging task that is important for analyzing unstructured data in structured and tabulated formats.

It is noted that frequently the annotations present within the medical records are intrinsic, making it more challenging to train models such as Med7 useful for accurately identifying entities relating to drugs [13].

In many studies, the initial training of the Med7 model involved predicting the next word and utilized a dataset comprising 2 million free-text patient records. Subsequently, fine-tuning was performed specifically for the named-entity recognition task.

Importantly, the Med7 model is designed to operate efficiently without the need for costly infrastructure. It can be utilized on standard machines equipped with CPU, making it a practical and accessible solution. [13]. NER models in different applications help to save a lot of time and effort to read textual data manually [30].

4 Results

This section is dedicated to the discussion of the results obtained by applying NER to our clinical dataset through two known libraries ScispaCy and Med7. It is possible to visualize the results through the graphical representations that we obtained by executing the code with the Python language.

The IBD dataset underwent NLP techniques to generate a Python word cloud to show in a first step insights into the latent aspects mentioned in the dataset. The most frequent words are represented through the word cloud which allows the exploration of textual data. In the word cloud, each term has different sizes according to the frequency in the corpus, as shown in Fig. 1.

The word cloud allows the immediate identification of the main concepts of the explored texts. Terms such as "Therapy", "Illness", "Endoscopy", "Colonoscopy", "Ulcerative colitis", "Crohn's", and "Diagnosis" represent the possibility of the presence of intestinal disease and any neoplasms, identified through instrumental tests.

---

[7] https://github.com/kormilitzin/med7

Fig. 1. Word cloud showing the most frequent words in the IBD dataset (230 documents) after preprocessing. Tokens with the largest font size are the most frequent.

Terms such as "Ileum", "Rectus", and "Sigmoid" are the affected segments, while words such as "Abdominal", "Pain", "Diarrhea", "Bowel Habit", "Abscess", "Rettoragy", represent the symptoms and effects associated with the disease of interest.

Of particular interest for the analysis in question, we highlight terms such as "Infliximab", "Mesalazine", "Deltacortene", "Adalimumab", "Beclometasone", "Benefit", "Remission" which allow us to know which are the prescribed drugs and the possibility for patients to find benefit thanks to them.

4.1 NER Analysis Using ScispaCy

After a preliminary data exploration, the package ScispaCy have been installed which is a Python package containing spaCy models for biomedical, scientific, or clinical text processing and en_ner_bc5cdr_md which represents a spaCy NER model trained on the *bc5cdr* corpus, which recognizes entities such as *disease* and *chemical*.

Figure 2 shows an extract of the results in the NER using ScispaCy package and *en_ner_bc5cdr_md* model.

Table 1 represents all the terms that have been grouped into the two entities recognized by the imported module. The *chemical* column contains all the drugs associated with and prescribed in the presence of IBD, while the *disease* column

```
abdominal pain diarrhea DISEASE
fistula DISEASE
crohn disease DISEASE
intestinal occlusion DISEASE
crohn ileo disease DISEASE
azathioprine CHEMICAL
crohn DISEASE
abdominal pain DISEASE
budesonide CHEMICAL
colon crohn disease DISEASE
salazopyrin CHEMICAL
sideropenic anaemia DISEASE
colon crohn disease DISEASE
abdominal pains fever diarrhea DISEASE
weight loss DISEASE
steroid CHEMICAL
crohn ileo colic phenotype inflammatory disease DISEASE
azathioprine CHEMICAL
intestinal inflammatory disease DISEASE
crohn like ileo colic DISEASE
dex CHEMICAL
ulceration DISEASE
pain DISEASE
diarrhoea nausea vomiting DISEASE
fever DISEASE
mesalazine CHEMICAL
crohn ileo colic disease DISEASE
colic DISEASE
crohn ileo colic disease fistulizing spine DISEASE
perisigmoid abscess DISEASE
```

Fig. 2. Example of extracted output from NER analysis through *en_ner_bc5cdr_md* model from ScispaCy package.

contains all the terms relating to the diagnosed pathology and all the possible symptoms associated with it and interesting tracts.

4.2 NER Analysis Using Med7

In the second experiment, the analysis was developed using Med7 which is a transferable clinical natural language processing model for medical records to recognize entities in the clinical context. The *en_core_med7_lg* model trained on medical records, capable of recognizing seven categories, was imported.

At this stage, the model helped identify us drug-associated characteristics such as *drug, dosage, duration, form, frequency, route,* and *strength*.

The results of this phase allowed us to confirm that the Med7 modules can extract the entities of interest also starting from clinical reports of patients with IBD.

Figure 3 shows an extract of the report in which the entities of interest have been identified in the dataset.

Table 1. Entities recognized with *en_ner_bc5cdr_md* model.

DISEASE	CHEMICAL
– Crohn Disease	– Steroid
– Rectocolitis Ulcerosa	– Prednisone
– Abdominal pain	– Azathioprine
– Diarrheal	– Salazopyrin
– Epigastralgia	– Mesalazine
– Perianal abscess	– Mercaptopurine
– Fistula	– Adalimumab
– Ileo colic	– Ciprofloxacin
– Leukopenia	– Methylprednisone
– Pneumonia	– Metronamine
– Overinfection	– Prednisolone
– Intestinal occlusion	– Methotrexate
– Sideropenic anemia	– Golimumab
– Fever	– Infliximab
– Intestinal Inflammatory	– Beclomethasone
– Ulceration	– Urbason
– Nausea vomiting	– Betamethasone
– Colon cancer	– Topster
– Dyspepsia	– Medrol
Achalasia	– Tapazole
– Ileal resection	– Telmisartan
– Stenosis	– Rifaximin
– Cramps	– Ketorolac
– Acne	
– Neoplasm	
– Diabetes Mellitus	
– Haematuria	
– Hyperaemia	
– Bleeding	
– Duodenal ulcer	
– Pancolitis	
– Rectorrhagia	
– Erosions	
– Polyps	
– Clostridium	

patient with sclerosing mesenteritis diagnosed on the surgical site in 10 2017 after resection of paraortic lymphadenopathic mass multiple loops of the small intestine tangential resection of the blind duodenum part of the ascending colon Histological exam proliferation of fused cellular elements arranged in bundles with interposition of keloid-type myxoid stroma In 2019 subocclusive episodes treated with medical therapy Since then under therapy with `Azathioprine DRUG` `capsules FORM` `100 mg STRENGTH` to `daily FREQUENCY` `for 5 months DURATION` `PO ROUTE` soministration is recommended

```
[('Azathioprine', 'DRUG'),
 ('capsules', 'FORM'),
 ('100 mg', 'STRENGTH'),
 ('daily', 'FREQUENCY'),
 ('for 5 months', 'DURATION'),
 ('PO', 'ROUTE')]
```

Fig. 3. Example of extracted output from NER analysis through *en_core_med7_lg* model from Med7 package.

5 Discussion and Conclusions

In this work, anonymized gastroenterology medical reports (IBD medical reports) were analyzed by using NER methods, with the aim of extracting as many entities as possible from the main intestinal diseases, such as Ulcerative Colitis and Crohn's disease.

The study was performed in two phases. The first one involves the use of the model *en_ner_bc5cdr_md* imported from ScispaCy package to recognize two specific entities such as *chemical* and *disease*. The model has allowed us to group in the *chemical* set the names of the drugs associated with intestinal diseases, while in the *disease* set, we find the characterization of the disease through the symptoms associated with it and the possible presence of tumor lesions. The second experiment involves the use of the model *en_core_med7_lg* imported from Med7 package which made it possible to extract properties related to drugs, identifying specific entities such as *drug, dosage, duration, form, frequency, route,* and *strength*.

This work underlined the importance of Natural Language Processing methodologies to analyze clinical documents. The results have made it possible to broaden the knowledge of aspects that are often negligible due to the enormous flow of data present in clinical contexts, such as in the field of gastroenterology. We can confirm that the NER models taken into consideration were useful to extract as much information as possible from the clinical dataset.

This represents a preliminary approach for future developments, for which an increase of available data and better model performance are expected.

References

1. Grishman, R., Sundheim, B.M.: Message understanding conference-6: a brief history. In: COLING 1996: The 16th International Conference on Computational Linguistics, vol. 1 (1996)
2. Collobert, R., Weston, J., Bottou, L., Karlen, M., Kavukcuoglu, K., Kuksa, P.: Natural language processing (almost) from scratch. J. Mach. Learn. Res. **12**, 2493–2537 (2011)
3. Bose, P., Srinivasan, S., Sleeman, W.C., IV., Palta, J., Kapoor, R., Ghosh, P.: A survey on recent named entity recognition and relationship extraction techniques on clinical texts. Appl. Sci. **11**(18), 8319 (2021)
4. Hou, J.K., Imler, T.D., Imperiale, T.F.: Current and future applications of natural language processing in the field of digestive diseases. Clin. Gastroenterol. Hepatol. **12**(8), 1257–1261 (2014)
5. Bernstein, C.N.: Treatment of ibd: where we are and where we are going. Off. J. Am. Coll. Gastroenterol.— ACG **110**(1), 114–126 (2015)
6. Actis, G.C., Pellicano, R., Rosina, F.: Inflammatory bowel diseases: current problems and future tasks. World J. Gastrointest. Pharmacol. Therapeut. **5**(3), 169 (2014)
7. Perera, N., Dehmer, M., Emmert-Streib, F.: Named entity recognition and relation detection for biomedical information extraction. Front. Cell Dev. Biol. 673 (2020)
8. Szymańska, S., et al.: Inflammatory bowel disease–one entity with many molecular faces. Gastroenterol. Rev./Przegląd Gastroenterologiczny **14**(4), 228–232 (2019)
9. Luo, Y., et al.: Natural language processing for EHR-based pharmacovigilance: a structured review. Drug Saf. **40**, 1075–1089 (2017)
10. Hashimoto, R.E., Brodt, E.D., Skelly, A.C., Dettori, J.R.: Administrative database studies: goldmine or goose chase? Evid.-Based Spine-Care J. **5**(02), 074–076 (2014)
11. Scarpino, I., Vallelunga, R., Luzza, F., Cannataro, M.: Machine learning approaches in inflammatory bowel disease. In: Computational Science–ICCS 2022: 22nd International Conference, London, UK, 21–23 June 2022, Proceedings, Part II, pp. 539–545. Springer, Heidelberg (2022)
12. Khurana, D., Koli, A., Khatter, K., Singh, S.: Natural language processing: state of the art, current trends and challenges. Multimedia Tools Appl. **82**(3), 3713–3744 (2023)
13. Kormilitzin, A., Vaci, N., Liu, Q., Nevado-Holgado, A.: Med7: a transferable clinical natural language processing model for electronic health records. Artif. Intell. Med. **118**, 102086 (2021)
14. Menti, E., et al.: Bayesian machine learning techniques for revealing complex interactions among genetic and clinical factors in association with extra-intestinal manifestations in ibd patients. In: AMIA Annual Symposium Proceedings, vol. 2016, p. 884. American Medical Informatics Association (2016)
15. Nguyen, G., Dlugolinský, Š, Laclavík, M., Šeleng, M., Tran, V.: Next improvement towards linear named entity recognition using character gazetteers. In: van Do, T., Thi, H.A.L., Nguyen, N.T. (eds.) Advanced Computational Methods for Knowledge Engineering. AISC, vol. 282, pp. 255–265. Springer, Cham (2014). https://doi.org/10.1007/978-3-319-06569-4_19
16. Korkontzelos, I., Piliouras, D., Dowsey, A.W., Ananiadou, S.: Boosting drug named entity recognition using an aggregate classifier. Artif. Intell. Med. **65**(2), 145–153 (2015)

17. Wu, Y., Jiang, M., Xu, J., Zhi, D., Xu, H.: Clinical named entity recognition using deep learning models. In: AMIA Annual Symposium Proceedings, vol. 2017, p. 1812. American Medical Informatics Association (2017)
18. Tang, B., Cao, H., Wu, Y., Jiang, M., Xu, H.: Recognizing clinical entities in hospital discharge summaries using structural support vector machines with word representation features. In: BMC Medical Informatics and Decision Making, vol. 13, pp. 1–10. BioMed Central (2013)
19. Tang, D., Duan, N., Qin, T., Yan, Z., Zhou, M.: Question answering and question generation as dual tasks. arXiv preprint arXiv:1706.02027 (2017)
20. Sutskever, I., Vinyals, O., Le, Q.V.: Sequence to sequence learning with neural networks. Adv. Neural Inf. Process. Syst. **27** (2014)
21. LeCun, Y., Bengio, Y., Hinton, G.: Deep learning. Nature **521**(7553), 436–444 (2015)
22. Mikolov, T., Chen, K., Corrado, G., Dean, J.: Efficient estimation of word representations in vector space. arXiv preprint arXiv:1301.3781 (2013)
23. Nadeau, D., Sekine, S.: A survey of named entity recognition and classification. Lingvisticae Investigationes **30**(1), 3–26 (2007)
24. Kim, D., et al.: A neural named entity recognition and multi-type normalization tool for biomedical text mining. IEEE Access **7**, 73729–73740 (2019)
25. Lee, J., et al.: Biobert: a pre-trained biomedical language representation model for biomedical text mining. Bioinformatics **36**(4), 1234–1240 (2020)
26. Wang, X., Song, X., Li, B., Zhou, K., Li, Q., Han, J.: Fine-grained named entity recognition with distant supervision in covid-19 literature. In: 2020 IEEE International Conference on Bioinformatics and Biomedicine (BIBM), pp. 491–494. IEEE (2020)
27. Tarcar, A.K., Tiwari, A., Dhaimodker, V.N., Rebelo, P., Desai, R., Rao, D.: Healthcare ner models using language model pretraining. arXiv preprint arXiv:1910.11241 (2019)
28. Neumann, M., King, D., Beltagy, I., Ammar, W.: Scispacy: fast and robust models for biomedical natural language processing. arXiv preprint arXiv:1902.07669 (2019)
29. Sezgin, E., Hussain, S.A., Rust, S., Huang, Y.: Extracting medical information from free-text and unstructured patient-generated health data using natural language processing methods: Feasibility study with real-world data. JMIR Format. Res. **7**, e43014 (2023)
30. Surana, S., Chekkala, J., Bihani, P.: Chatbot based crime registration and crime awareness system using a custom named entity recognition model for extracting information from complaints. Int. Res. J. Eng. Technol. (IRJET) **8**(4) (2021)

Applied Mathematical Modelling in the Physics Problem-Solving Classroom

Annarosa Serpe(✉)

Department of Mathematics and Computer Science, University of Calabria, Rende, CS, Italy
`annarosa.serpe@unical.it`

Abstract. This paper aims to contribute to the literature on planning and implementing mathematical modelling tasks in the higher education curriculum.

Specifically, we propose and discuss a lesson plan carried out within the course of Mathematics Education II - Master's Degree in Mathematics, to train and qualify students on how to capture the interplay between Physics and Mathematics in the perspective of mathematical modelling and problem-solving.

The goal is to make transparent what is meant by modelling by showing how it is possible to develop integrated learning environments beyond the traditional boundaries between subjects both at the subject level and at the level of required pedagogy, illustrating this characterisation with an example regarding a theme that 'subsumes' the two disciplines. It is important in the educational training of future teachers to explore ways based on influential theoretical expositions of the concept of interdisciplinarity. This is especially true when examples of didactic transposition are provided where Mathematics and Physics are understood as forms of knowledge that have relationships and differences, which are deeply intertwined and which co-evolve, mutually generating new problems that lead to permeating the labile boundaries between them to generate new knowledge.

These examples of didactic transposition, therefore, are rich in interdisciplinarity connections, which show students the relevant aspects of the two disciplines from a historical and epistemological point of view and their connections.

Keywords: Higher education · Mathematical modelling · Physics problem-solving

1 Introduction

Mathematics and Physics are two intrinsically connected disciplines, and their mutual influences have played an essential role in their development. The role of Mathematics in relation to Physics is not only technical, but also structural, meaning that one is linked to the definition and interpretation of concepts, relationships, and problems of the other.

Describing the relationship between Mathematics and Physics from an educational perspective is not simple. This awareness is relevant also in the context of teaching Mathematics and Physics in secondary school. From research literature (see i.e. [8]), it is evinced that students often have difficulty using mathematical tools and methods in a physical context and combining mathematics and physics in problem-solving.

The literature demonstrates that the main problem lies not in the lack of purely mathematical skills, but in the use of Mathematics in a different context, such as Physics.

Mathematics modelling is a special type of problem-solving as it allows one to determine the solution of many problems of an applied nature and beyond. For this reason, today learning to construct mathematical models of simple situations drawn from real-life contexts turns out to be very important for students [4]. Teachers play an important role in successfully implementing mathematical modelling because in their lessons they must develop specific strategies and attitudes, which must be supported by concrete actions through appropriate choices, strategies, and teaching methodologies. Mathematics Modelling is a cognitive strategy in which an object or situation is replaced by a model, and by examining this model, information about the object or situation originally given can be obtained. Consequently, modelling activities in an interdisciplinary context between Physics and Mathematics helps students think about the bonds between these two disciplines, and in particular, what mode of action they recognise for Mathematics in providing a model of a physical situation. Even in the pre-service teacher education and training, it is necessary to favour approaches that effectively promote the interrelationships between Mathematics and Physics in order to encourage and sustain the scientific thinking of the future students of these teachers. This paper focuses on implementing teaching interdisciplinary practice centred on applied modelling as a special type of problem-solving concerning relationships between Mathematics and Physics. The teaching practice was proposed and discussed in a lesson within the course of Mathematics Education II - Master Degree in Mathematics, where the students are introduced to the study of the didactical frameworks aimed at coordination and mutual interaction between the two disciplines. The paper is structured as follows. Section two describes the theoretical framework. Section three presents the methodology used for developing didactic practice, and shows its implementation via a problem-solving example, within a theme that 'subsumes' the two disciplines. Finally, in section four conclusions are drawn and recommendations are suggested.

2 Theoretical Framework

In Italian high schools, the synergy between the interdisciplinary dimension of scientific sub-jects is still weak; this prevents the students from developing critical skills, comparison and interconnections of knowledge, especially in the application of Mathematics to Physics. According to Niss [15], the student's conception of a mathematical concept is determined by the set of specific domains in which that concept has been introduced to the student. When a concept is introduced in a narrow mathematical domain, the student may see it as a formal object with arbitrary rules. This results in the recognised difficulty of application of the concept in new settings. The use of Mathematics in Physics is not simple because the main teaching problem consists in formalising a situation already analysed phenomenologically with the aim of finding a theoretical model suited to the facts [16].

If the objective is to teach students to find the relationships between physical meaning and mathematical meaning, how should teachers structure their teaching?

A first step is to question ourselves more about the physical meaning of mathematical formulas, that is, to highlight the importance of not only asking ourselves what a formula looks like, but also why its shape is exactly what it is. Didactically, it means shifting attention from "describing and calculating" to "understanding and explaining".

For example, the same mathematical structure of the equation can be applied to diverse physical phenomena, but the semantics can be very different. Additionally, one should not forget that problem-solving in Physics requires the students to question, reason, and refine their knowledge [2].

Problem-solving requires students to think and that is why it forms an important part of teaching. There are many functions to which a problem can be used in teaching, not all of which are obvious; these must be clear and explicit in the mind of the teacher who proposes the problem. The blending of Mathematics and Physics also implies attention to the different "representations" with which a physical phenomenon or situation can be described (e.g. various formal languages including equations, verbal language, graphs, and other types of diagrams) [12]. It is therefore essential that the teacher helps students develop the ability to use different representations and move easily from one to another. From this perspective, it obviously becomes important to work at an educational level on one of the main mathematical tools that allows for the movement between the two disciplines: mathematical modelling in Physics, the use of which is also strongly advocated in the National Guidelines[1].

Problem-solving centred on model-eliciting tasks offers a rich and a highly authentic view of activities in a classroom as the students have the opportunity to express, test, modify, revise, and refine their own ways of thinking during the process of designing powerful conceptual tools that embody constructs that students are intended to develop [18]. In Physics problem-solving, the mathematisation process involves multilevel, layered reasoning. Its importance is not only for the decisive role that it plays within the study of Physics and for the consequences it shows in learning, but also because it is a methodology that provides key concepts and rational procedures that are valid far beyond Physics itself, as they provide the students with intellectual tools for analysis and synthesis of the complex reality in which they live. Its methodological basis consists –at least at the beginning– in understanding the need to always operate at different levels, such as making abstractions on the phenomenon, abstractions on the data and subdivision and hierarchies of mathematical relationships. For example, we have modelling as a domain-specific practice in Physics, which often involves the production of a mathematical equation from measurements of a physical system.

Currently, digital tools can support modelling activities in interdisciplinary teaching [7, 19], especially when dealing with realistic problems, such as processing models with complex function terms or reducing the effort of calculation. Digital tools can perform a range of tasks in teaching applications and modelling because they allow you to experiment and explore [14, 18]. The simulations are very similar to experimentations because they use models that provide insights into the real system presented in the model or into the model itself [10]. Coming from an applied mathematics perspective, computer simulation through manipulating representations can help model-based learning as students may understand mathematics and physics concepts more elaborately by observing

---

[1] Indicazioni Nazionali per il Curriculum. Ministero dell'Istruzione, dell'Università e della Ricerca (MIUR), 2010.

the direct consequences of the changes they make [1, 6]. Working on modelling problems with digital tools requires two translation processes. After translating mathematical terms into the computer's language, the results obtained then have to be interpreted into mathematical language. Finally, the initial problem will be solved when the mathematical results are applied to the real situation [20]. This process is represented in literature in the extended Modelling cycle (see Fig. 1) of Greefrath [9], and Blum & Leiß [3].

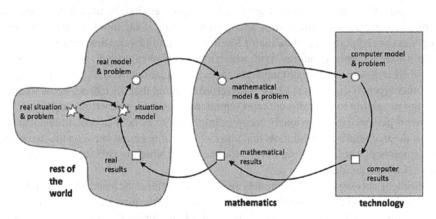

Fig. 1. Extended Modelling cycle by Greefrath [10], and Blum & Leiß [3].

3 Methodology

For teachers, it is not easy to implement modelling in the classroom, much less know how to manage things while students work on modelling problems. Therefore, it is important for teachers that students gradually acquire these competences [11].

In the case of a physics problem, the importance of the methodology lies in providing key concepts and rational procedures that are valid far beyond physics itself. In fact, so that the teaching practice leads to good results these important elements must be kept in mind: define the context of use, the strategies to be used, the forms of analysis, and the representation of the system. The problem-solving strategy is very significant because it not only functions as a field of application but also provides an experience of the actions that creates the need for the application of mathematising. Indeed from a pragmatic perspective, Mathematics is defined not only as purely instrumental but a complex and multifaceted interweaving of instrumental, structural functions and linguistics [5]. In promoting the translation process between the real-world and the mathematical world, the teacher's role is crucial because s/he provides a teaching and learning model, where attention is focused on the different representation registers. Therefore, the teacher is seen also as a semiotic mediator. Coming from the theoretical framework, the didactical methodology emphasises the perspective of mathematical modelling and problem-solving as ideas that cut across Mathematics and Physics and encompass both contextual and conceptual aspects.

3.1 A Simple Mathematical Model

Modelling in an interdisciplinary context means tackling an interactive process among multiple disciplines, highlighting the different forms of possible representation with a view to recognising and valorising the plurality of solutions. For example, the mathematical model of Newton's second law of motion[2] is the well-known equation

$$F = ma \tag{1}$$

where:
F = net force acting on the body in (N);
m = mass of the object in (kg);
a = acceleration in (m/s^2).
Equation (1) can be recast also in the following form[3]:

$$a = \frac{F}{m} \tag{2}$$

where a is the dependent variable reflecting the system's behaviour, F is the forcing function, and m is a parameter.

Equation (2) describes a natural process in mathematical terms and represents a simplification of reality because it yields reproducible results and can be used for predictive purposes. Equation (1) has a simple algebraic form, and its solution can be obtained easily, but there are mathematical models of physical phenomena, that require more complex mathematical techniques. The following example shows an example of this kind.

3.2 Illustrative Example

Problem: *A parachutist of mass 68.1 kg jumps out of a stationary hot air balloon. Compute velocity prior to opening the chute. The drag coefficient is equal to 12.5 kg/s.*

In this problem, Newton's second law can be used to determine the terminal velocity of a free-falling body near the Earth's surface. The mathematical model that describes the dynamic behaviour of the system is a differential equation. Often, finding the solution of a differential equation is not easy. In this case, however, we can make use of an approximate solution. Numerical calculation allows us to find an approximate solution of the differential equation which becomes useful if the margin of error is acceptable.

To solve the parachutist problem *Euler's method* will be used, which is a first-order method for solving differential equations.

Mathematical Model. Expressing a as $\frac{dv}{dt}$ into Eq. (2) we get:

$$\frac{dv}{dt} = \frac{F}{m} \tag{3}$$

where:

---

[2] Newton's second law of motion.
[3] Dependent variable = f(independent variables, parameters, forcing functions).

v = velocity in (m/s);
t = time in (s);
F = net force in (N);
m = mass in (kg).

Since the parachutist is moving, we recognise this as a dynamic problem; thus, F is:

$$\sum F = F_g + F_u \tag{4}$$

where:
F_g = gravitational force;
F_u = upward force of air resistance.

Considering F_u proportional to the square of the velocity,

$$F_u = -c_d v^2 \tag{5}$$

where c_d = drag coefficient in (kg/s). The relationship between the acceleration of the falling parachutist to the forces acting on her/him can be expressed by the differential equation:

$$\frac{dv}{dt} = g - \frac{c_d}{m} v^2 \tag{6}$$

Numerical Approach. The velocity can be approximated because Δt is finite[4],

$$\frac{dv}{dt} \cong \frac{\Delta v}{\Delta t} = \frac{v(t_{i+1}) - v(t_i)}{(t_{i+1} - t_i)} \tag{7}$$

where:
Δv = the difference in velocity;
Δt = the difference in time;
$v(t_i)$ = velocity at an initial time t_i;
$dv(t_{i+1})$ = velocity at some later time t_{i+1}.

Equation (7) is called a *finite divided difference approximation* of the derivative at the time t_i. Thus, Eq. (6) can be written as

$$v(t_{i+1}) = v(t_i) + \left[g - \frac{c_d}{m} v(t_i)^2 \right] (t_{i+1} - t_i) \tag{8}$$

and rearranged to yield

$$v_{i+1} = v_i + \frac{dv_i}{dt} \Delta t \tag{9}$$

where v_i designates velocity at time t_i and $\Delta t = t_{i+1} - t_i$. So, Eq. (6) has been transformed into an equation that can be used to determine the velocity algebraically at t_{i+1} using the slope and previous values of v and t. Given an initial value for the velocity at some time t_i, we can easily compute:

New value = old value + slope × step size.

---

[4] $\frac{dv}{dt} = \lim_{\Delta t \to 0} \frac{dv}{dt}$.

Applied Mathematical Modelling in the Physics Problem-Solving Classroom

This numerical approach considers the entire time interval t during which the motion is studied and divides it into very small intervals Δt. It is therefore a question of finding a solution in a succession of moments in time, where each moment is separated from the next by Δt. Starting from the initial values of quantities known ($t_0 \equiv t$; $v_0 \equiv v$), or as in the case of F_0, easily computed, we get:

$$\frac{F_0}{m} = g - \frac{c_d}{m} v_0^2.$$

At the end of the first interval, $t = t_1 = \Delta t$; $v_1 \cong v_0 + a_0 \Delta t$;

$$\frac{F_1}{m} = g - \frac{c_d}{m} v_1^2.$$

At the end of the second interval, $t_2 = t_1 + \Delta t = 2\Delta t$; $v_2 \cong v_1 + a_1 \Delta t$;

$$\frac{F_2}{m} = g - \frac{c_d}{m} v_2^2.$$

For the ith interval, we have: $t_i = t_{i-1} + \Delta t = i\Delta t$; $v_i \cong v_{i-1} + a_{i-1}\Delta t$;

$$\frac{F_i}{m} = g - \frac{c_d}{m} v_i^2.$$

This procedure approximating a continuous dynamical system with a discrete one is called *Euler's Method*, which is one of the oldest and simplest methods to find the numerical solution of ordinary differential equations or the initial value problems.

Euler's method was applied to our problem because it is a first-order approximation to the derivative where 'i' is the iterator for time (t).

Numerical Solution. Now we can compute the falling parachutist's velocity with *Euler's Method* employing a step size of 2 s for the calculation. The velocity of the parachutist at time $t_i = 0$ is zero. At the time $t_{i+1} = 2$ s substituting all problem data in Eq. (8) to yields:

$$v = 0 + \left[9.8 - \frac{12.5}{68.1}(0)^2\right](2) = 19.60\,\text{m/s}$$

From $t = 2$ to 4 s, we get:

$$v = 19.60 + \left[9.8 - \frac{12.5}{68.1}(19.60)^2\right](2) = 32.00\,\text{m/s}$$

Ultimately, starting from an initial condition, Eq. (8) can be applied repeatedly to calculate the velocity as a function of time. This calculation can be easily carried out on the computer, without neglecting the fact that to obtain good accuracy it is necessary to take many small steps.

Euler's Method Algorithm. The demonstrative function now emerges with the algorithmic-procedural superstructure of Mathematics applied to Physics. The pseudocode is then developed which strengthens the argumentative and interpretative function of the process in a more abstract way. Therefore, the algorithm that helps us solve

Eq. (8) is based on a loop that uses an IF/THEN structure to test whether adding $t+dt$ will take us beyond the end of the interval. The pseudocode drawn up to plan the algorithm for the parachutist problem is shown in Fig. 2. Before entering the loop, the variable h is used to indicate the value of the time step, dt, this to make sure that our routine does not change the given value of dt if and when we make smaller the time step.

```
            g = 9.81;
input       cd, m
input       ti, vi, tf, dt
            t = ti
            v = vi
            h = dt
            DO
                IF t + dt > tf THEN
                    h = tf - t
                END IF
                dvdt = g - (c_d/m) * v
                v = v + dvdt * h
                t = t + h
                IF t ≥ tf EXIT
            ENDDO
            DISPLAY v
```

Fig. 2. Pseudocode to solve Eq. (8) using *Euler's method*.

It is easy, in this algorithm to make mistakes: to enter a step size greater than the calculation interval, for example, $tf - ti = 5$ and $dt = 20$.

To prevent this from happening we can use the modular programming approach that, also, favours the development of new modules to perform additional tasks and which can then be easily incorporated into the existing organised scheme. In addition, we should now recognise the generic nature of the routine which does not present specific references to the problem of the parachutist.

Furthermore, the derivative is not computed within the function by an explicit equation, but another function, *dy*, is necessary, and it must be invoked to compute it, we will see this shortly.

Computer Simulation. To enable greater clarity and understanding of concepts from an interdisciplinary perspective, the algorithm represents the evolution between code and implementation of the descriptive process. An implemented algorithm is, on the one hand, an intellectual gesture, and, on the other, a functioning system that incorporates in its structure the material assumptions about perception, decision, and communication

[17]. Different software can be used for implementation. MATLAB was chosen because it has a variety of functions and operators that allow convenient implementation of many of the numerical methods.

The students in the Mathematics Education II - Master's degree in Mathematics course already know and use this software.

Alternatively, the EXCEL software can be used, as with this is it easier to implement the algorithm in an upper secondary school final class.

Using MATLAB, we can write M-files program to implement numerical calculations. Now, to compute velocity of the parachutist using the following *function file*

$$[y, yplot] = \text{Euler}(dt, ti, tf, vi, m, cd, cmax)$$

where the inputs $dt, ti, tf, vi, m, cd, cmax$ are passed into the *function file Euler* via its argument list (dt,ti,tf,yi) and the output is returned via the assignment statement y= Euler(dt,ti,tf,yi).

The following MATLAB Main1-file is developed from the pseudocode (Fig. 2) executing the calculation from $ti = 0$ to 20 s with a step size of 2 s.

```
m=68.1;
cd=12.5;
ti=0;
tf=20;
vi=0;
dt=2;
cmax=tf/dt;
[y,yplot]=Euler(dt,ti,tf,vi,m,cd,cmax)
plot(0:dt:dt*(length(yplot)-1),yplot,'.r','MarkerSize',10)
hold on
plot(0:dt:dt*(length(yplot)-1),yplot,'b')
```

The results displayed are the following:

```
>> main1
y = 52.8938
yplot =  19.6200    32.0374    39.8962    44.8700    48.0179
         50.0102    51.2711    52.0691    52.5742    52.8938
```

The results are plotted in Fig. 3.

However, utilising straight-line segments to approximate a continuously curving function creates some discrepancy between the two results. To decrease such discrepancies a smaller step size can be used.

In Eq. (8) if we use $1 - s$ intervals the straight-line segments track closer to the true solution; therefore, the error committed is smaller.

Using smaller and smaller step sizes, without the help of the computer, would make such numerical solutions impractical to accurately model the velocity (see Fig. 4).

Knowing that a *method* is convergent, is necessary from a theoretical point of view but may not be sufficient from a practical point of view, given that the calculations must,

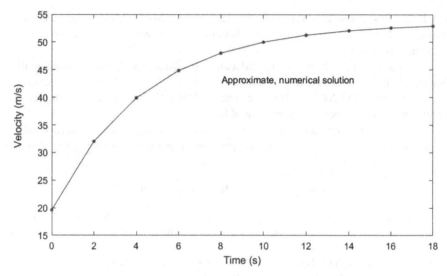

Fig. 3. Plot of falling parachutist's velocity.

in any case, be performed with an *h* finished. In Fig. 4, it can be seen that the error from which the solution calculated by *Euler's Method* is affected is the approximation of the derivative with the incremental ratio.

At each step, therefore, two different errors occur: the one produced locally (*local error*) in that step and the one accumulated in the previous steps (*global error*).

In the last step of this development, the MATLAB `Main1-file` is converted into a proper function. This can be done in the following `Euler.m-file`.

```
function [yy,yplot]=Euler(dt,ti,tf,yi,m,cd,cmax)
t = ti;
y = yi;
h = dt;
c=1;
while c<=cmax
  if t + dt > tf
  h = tf - t;
  end
  dydt = dy(t, y, m, cd);
  y = y + dydt * h;
  yplot(c)=y;
  t = t + h;
  if t >= tf
        c=c+1;
        break
  end
  c=c+1;
end
  yy = y;
```

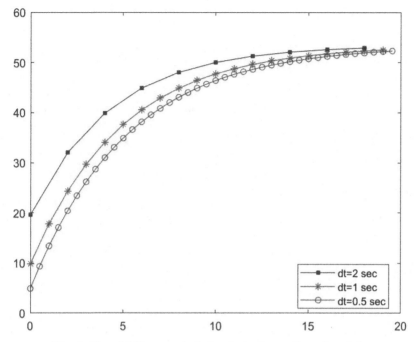

Fig. 4. Plot of falling parachutist's velocity for smaller value of *dt*.

Now, to compute the derivative we can write the following M-file:

```
function dv= deriv(v)
dv=9.81-(12.5/68.1)*v^2;
end
```

that will be saved as dy.m and then returned to the command mode. To call back the function and see the result, we can use the following commands:

```
>> m=68.1;
>> cd=12.5;
>> ti=0;
>> tf=20.;
>> vi=0;
>> dt=0.1;
>> euler(dt,ti,tf,vi,m,cd)
```

The development of the numerical model for the computerised solving of the mathematical model is extremely important. *Euler's Method* was used to solve the parachutist problem. The goal was to study it, understand it and predict its evolution.

The conclusions provide the approximate solutions that depend on the simplification of the real problem, a procedure necessary to introduce the mathematical language that gradually helps to improve the approximation by refining the model with increasingly precise initial data.

4 Conclusions

Favouring teaching practices in which the process of building models is alternated and integrated with experimental procedures makes the study of Mathematics and Physics - usually considered difficult and complex - rich and stimulating. In fact, the benefits brought by the use of models and computer simulations are multiple: they provide alternative collaborative and interactive environments, they arouse students' curiosity by enhancing passion and intellectual liveliness, imagination and taste for abstraction, they guarantee a better quality of learning because they encourage personal discovery by helping students to act and think, allowing them to develop new knowledge.

The use, then, of the IT tool offers support for exploration, experimentation and modelling activities with many advantages including: the visualisation of the structure, the interaction processes and the evolution of the models; easing the burden of difficulties related to Mathematics; the search for models that make Physics/Mathematics more accessible; and last, but not least, the advantage of focusing qualitative reasoning with models, facilitating (and motivating) the transition to quantitative interpretations.

Despite all these advantages revealed by this pedagogical approach - which is contextualised in the epistemological framework of constructionism - its use in teaching practice in Italian high schools is still very discontinuous.

This lack of continuity can be attributed - perhaps - to the fact that most teachers are not adequately trained to didactically transpose the two disciplines from an application point of view, that is, to design teaching practices linked to the types of models that are usually used to represent a system, the specificities necessary to represent them and the degree of complexity of the modelling and simulation tools. For this reason, it is important to practically prepare the future teacher to implement didactical activities in this regard. A computer simulation model enables teachers to create more motivating, enriching educational environments and have access to content and activities that would otherwise be difficult or impossible to do. For some complex problems, computer modelling is not only less expensive than building a physical model or experimenting with the actual system, but it may also be the only way to attempt to understand the system.

The Italian high school curriculum includes the introduction of mathematical modelling and computer simulation and also the use of educational programming languages for writing simulation models. Providing examples of didactic transpositions where Mathematics and Physics are understood as disciplines helps future teachers to develop modelling and representational skills.

The parachutist problem integrates the two disciplines and so gives real meaning to interdisciplinarity and demonstrates how applied mathematics can be inserted into the curriculum, as expected by the Italian National Guidelines.

The future Physics and Mathematics teachers have to focus especially on the methodology aimed at making students understand how Physics provides Mathematics with interesting problems to investigate, and Mathematics provides Physics with powerful tools for analysing data and theorising. The use of computers with the MATLAB program for modelling and simulation enables the creation of interdisciplinary educational environments that are highly accessible because they provide meaningful ways to construct, describe, explain, manipulate, or predict patterns and regularities associated with complex situations. The example proposal in the Mathematics Education II course wants to be a foundation for the development of interdisciplinary teaching practices in pre-service higher secondary education teacher training.

According to Mason [13], a major contribution to effective teaching of modeling lies in enculturating students into what it is like to perceive the world as a modeller. The method an example exposed in this paper aims to do exactly this.

References

1. Bertacchini, F., Bilotta, E., Caldarola, F., Pantano, P.: The role of computer simulations in learning analytic mechanics towards chaos theory: a course experimentation. Int. J. Math. Educ. Sci. Technol. **50**(1), 100–120 (2019). https://doi.org/10.1080/0020739X.2018.1478134
2. Bing, T.J., Redish, E.F.: Analyzing problem solving using math in physics: epistemological framing via warrants. Phys. Educ. Res. **5**, 1–15 (2009). https://doi.org/10.1103/PhysRevST PER.5.020108
3. Blum, W., Leiß, D.: How do students and teachers deal with modelling problems? In: Haines, C., Blum, W., Galbraith, P., Khan, S. (eds.) Mathematical Modelling: Education, Engineering and Economics–ICTMA 12, pp. 222–231. Horwood Publishing, Chichester (2007). https://doi.org/10.1533/9780857099419.5.221
4. Borromeo Ferri, R.: Learning How to Teach Mathematical Modelling in School and Teacher Education. Springer, Switzerland (2018). https://doi.org/10.1007/978-3-319-68072-9
5. Caccamo, M.T., Serpe, A.: Mathematics in physics problem-solving. A kinematics study in high school. Atti della Accademia Peloritana dei Pericolanti-Classe di Scienze Fisiche, Matematiche e Naturali **1**(1), 2 (2023). https://doi.org/10.1478/AAPP.1011A2
6. Clark-Wilson, A., Robutti, O., Thomas, M.: Teaching with digital technology. ZDM Int. J. Math. Educ. **52**(7), 1223–1242 (2020). https://doi.org/10.1007/s11858-020-01196-0
7. Fiorentino, M.G., Montone, A., Rossi, P.G., Telloni, A.I.: A digital educational path with an interdisciplinary perspective for pre-service mathematics primary teachers' professional development. In: Fulantelli, G., Burgos, D., Casalino, G., Cimitile, M., Lo Bosco, G., Taibi, D. (eds.) HELMeTO 2022. CCIS, vol. 1779, pp. 663–673. Springer, Cham (2023). https://doi.org/10.1007/978-3-031-29800-4_50
8. Galbraith, P., Stillman, G.: A framework for identifying student blockages during transitions in the modelling process. Zentralblatt für Didaktik der Mathematik **38**, 143–162 (2006). https://doi.org/10.1007/BF02655886
9. Greefrath, G.: Using technologies: new possibilities of teaching and learning modelling–overview. In: Kaiser, G., Blum, W., Borromeo Ferri, R., Stillman, G. (eds.) Trends in Teaching and Learning of Mathematical Modelling. International Perspectives on the Teaching and Learning of Mathematical Modelling, vol. 1, pp. 301–304. Springer, Dordrecht (2011). https://doi.org/10.1007/978-94-007-0910-2_30
10. Greefrath, G., Siller, H.S.: Modelling and simulation with the help of digital tools. In: Stillman, G., Blum, W., Kaiser, G. (eds.) Mathematical Modelling and Applications. International Perspectives on the Teaching and Learning of Mathematical Modelling, pp. 529–539. Springer, Cham (2017). https://doi.org/10.1007/978-3-319-62968-1_44
11. Kaiser, G., Schwarz, B., Tiedemann, S.: Future teachers' professional knowledge on modeling. In: Lesh, R., Galbraith, P., Haines, C., Hurford, A. (eds.) Modeling Students' Mathematical Modeling Competencies. International Perspectives on the Teaching and Learning of Mathematical Modelling, pp. 433–444. Springer, Dordrecht (2013). https://doi.org/10.1007/978-94-007-6271-8_37
12. Lopez-Gay, R., Martinez Saez, J., Martinez Torregrosa, J.: Obstacles to mathematization in physics: The case of the differential. Sci. Educ. **24**, 591–613 (2015). https://doi.org/10.1007/s11191-015-9757-7

13. Mason, J.: Modelling modelling: where is the centre of gravity of-for-when teaching modelling? In: Matos, J.F., Blum, W., Houston, K., Carreira, S.P. (eds.) Modelling and Mathematics Education, ICTMA 9 - Applications in Science and Technology, pp. 39–61. Horwood Publishing (2001). https://doi.org/10.1533/9780857099655.1.39
14. Montone, A., Fiorentino, M.G., Mariotti, M.A.: Learning translation in geometric transformations through digital and manipulative artefacts in synergy. In: Zaphiris, P., Ioannou, A. (eds) Learning and Collaboration Technologies. Designing Learning Experiences. HCII 2019. LNCS, vol. 11590, pp. 191–205. Springer, Cham. (2019). https://doi.org/10.1007/978-3-030-21814-0_1
15. Niss, M.: Aspects of the nature and state of research in mathematics education. Educ. Stud. Math. **40**, 1–24 (1999)
16. Redish, E.F., Kuo, E.: Language of physics, language of math: disciplinary culture and dynamic epistemology. Sci. Educ. **24**, 561–590 (2015). https://doi.org/10.1007/s11191-015-9749-7
17. Serpe, A.: A computational approach with MATLAB software for nonlinear equation roots finding in high school maths. In: Sergeyev, Y., Kvasov, D. (eds.) Numerical Computations: Theory and Algorithms. NUMTA 2019. LNCS, vol. 11973, pp. 463–477. Springer, Cham (2020). https://doi.org/10.1007/978-3-030-39081-5_40
18. Serpe, A.: Digital tools to enhance interdisciplinary mathematics teaching practices in high school. In: Fulantelli, G., Burgos, D., Casalino, G., Cimitile, M., Lo Bosco, G., Taibi, D. (eds.) Higher Education Learning Methodologies and Technologies Online. HELMeTO 2022. CCIS, vol. 1779, pp. 209–218. Springer, Cham. (2023). https://doi.org/10.1007/978-3-031-29800-4_16
19. Serpe, A., Frassia, M.G.: Task mathematical modelling design in a dynamic geometry environment: Archimedean Spiral's Algorithm. In: Sergeyev, Y., Kvasov, D. (eds.) Numerical Computations: Theory and Algorithms. NUMTA 2019. LNCS, vol. 11973, pp. 478–491. Springer, Cham (2020). https://doi.org/10.1007/978-3-030-39081-5_41
20. Vorhölter, K., Greefrath, G., Borromeo Ferri, R., Leiß, D., Schukajlow, S.: Mathematical modelling. In: Jahnke, H., Hefendehl-Hebeker, L. (eds.) Traditions in German-Speaking Mathematics Education Research. ICME-13 Monographs. Springer, Cham (2019). https://doi.org/10.1007/978-3-030-11069-7_4

AIR SAFE: Leveraging IoT Sensors and AI Models to Foster Optimal Indoor Conditions

Mariangela Viviani[1], Simone Colace[1], Daniele Germano[1], Sara Laurita[1], Giuseppe Papuzzo[2], and Agostino Forestiero[2](✉)

[1] WISH s.r.l, Rende, Italy
[2] ICAR-CNR, Rende, CS, Italy
agostino.forestiero@icar.cnr.it

Abstract. The need of safe and livable indoor environments has intensified recently, given the great amount of time people spend indoor. In addition, the recent COVID-19 pandemic has moved the interest from the outdoor to indoor spaces. To guarantee an optimal indoor environmental quality, monitoring and regulating many variables (such as indoor and outdoor temperature, pollutants concentration, noise, and brightness) is necessary. In this context, we have developed AIR SAFE, an IoT and AI based infrastructure to monitor and control environmental quality in closed spaces. AIR SAFE uses Machine Learning models to make predictions of temperature, relative humidity, and CO_2 concentration. These predictions, together with data from a network of IoT sensors, are used to take actions on windows and the air conditioning system with the aim of modifying for the better the room environment. We show the results of the AI model we have developed for predicting indoor concentration of CO_2, relative humidity, and temperature. Our Long Short Term Memory (LSTM) model has been tested against literature models using simulated data at first, and then testing the best models on real data. Using real data, the LSTM network performs best at forecasting temperature and relative humidity, while Random Forest is the best CO_2 concentration predictor.

Keywords: Air quality · Internet of Things · Smart Environment · IoT sensors · Machine Learning · Neural Networks

1 Introduction

Nowadays, people spend most of their time in closed environments, in offices, or at home. Therefore, secure and highly livable environmental conditions are needed, also to reduce the probability of aerial viruses spreading.

The wide spreading of Internet of Things (IoT) in various fields has allowed the emergence of Smart Buildings, that are capable of acting on the state of the systems, such as windows, air conditioning or heating, to modify the indoor environment on the base of rules decided and thresholds fixed by the user or the Building Manager. In general, safe and comfortable states or situations are a combination of objective (e.g., air quality) and subjective (such as thermal comfort) elements.

Air quality, indoor or outdoor, is determined by the concentration of pollutants in the environment. Pollutants include carbon dioxide (CO_2), ozone (O_3), nitrogen dioxide (NO_2) and also particulates (such as $PM_{2.5}$). Occupancy, number of windows, and Heating, Ventilation and Air Conditioning system (HVAC) also contribute to determine air quality in indoor environments. Air quality is regulated by the World Health Organization (WHO) air quality guidelines [12] and by the American Society of Heating, Refrigerating and Air Conditioning Engineers (ASHRAE) Standard 62.1 [3]. Thermal comfort is a subjective feeling of satisfaction with the room temperature. It is evaluated by the ASHRAE Standard 55 [2]. The latest indoor temperature standard in Italy has been set by the Ministry Decree 386 [4] and by ENEA (the Italian National Agency for New Technologies, Energy and Sustainable Economic Development) [6] in October 2022. Thermal comfort depends on environmental factors, such as temperature, humidity, wind speed and thermal radiation, but also on personal aspects, like clothing and metabolic rate. It ranges between -3 (Cold) and $+3$ (Hot), with 0 being the optimal state.

HVAC systems account for the major part of energy consumption in buildings. Energy efficiency and the climatic crisis we are living in demand for an optimal design of the systems controlling and regulating the airflow. Also, to keep an eye on people's well-being and maintain high productivity, optimal conditions for thermal comfort and air quality are essential. For these reasons, models that are able to predict indoor conditions and provide alerts or take actions to re-establish the optimal state are increasing in demand.

The first attempts to solve this problem were using statistical models, as, for example, multi-linear regression (MLR). For these kind of models, a detailed knowledge of the environment was required and they were difficult to generalize. With the advent of Artificial Intelligence new horizons have opened and smart algorithms, implementing Machine Learning (ML) or Deep Learning, have been developed. These algorithms do not need explicit programming, but are able to extrapolate the features of the system.

Our aim is to design a system that, using real-time data and predictions from AI models, is able to determine the state of windows, thermostats and lights, in order to ensure the healthiest and most livable environments to occupants in a room. To implement the best AI models suitable in our case, we consider a comparison between different models for the prediction of indoor temperature, relative humidity and CO_2 concentration on a simulated dataset. In particular, we compare Random Forest (RF), convolutional neural network (CNN), multilayer perceptron (MLP) and long-short term memory (LSTM) for CO_2 concentration; RF, support vector regression (SVR), MLR and LSTM for relative humidity prediction; RF, MLP and LSTM for temperature prediction. We test these models at first using two years of simulated data, and then using a month of real data collected in our demonstrators.

Our results show that, in the simulated data testing phase, RF is the optimal model for predicting air quality and thermal comfort, while LSTM is better at predicting relative humidity. When testing the best models on real data, RF is still the best model for predicting CO_2 concentration, while LSTM performs better at the prediction of temperature and relative humidity.

The paper is organized as follows: in Sect. 2 we review the recent models for the prediction of CO_2 concentration, temperature and relative humidity, we then illustrate the models whose performances are compared in Sect. 3, introduce the dataset and the metrics used for the comparison in Sect. 4 and report the results in Sect. 5. Finally, we discuss the results and talk about some future developments.

2 Related Works

In 2016, [21] used a RF algorithm to predict CO_2 concentration in indoor environments using internal and external temperatures, relative humidity and date and time. They concluded that the prediction is good enough to avoid the installation of CO_2 sensors and derive the quantity indirectly. A study on the prediction of indoor temperature in [11] reported that LSTM can reduce the prediction error of 50% with respect to a MLP model. The authors of [1] compared different models for indoor air prediction and relative humidity prediction for livestock animal barns. In particular, they compared MLR, MLP, RF, decision tree regression (DR) and SVR models. They found that RF was the best model to predict both indoor temperature and relative humidity. In the study of [16] it has been proposed a CNN algorithm for predicting indoor CO_2 concentration. The novelty of the work consists in the constant update of the code with data collected by means of a sliding window, ensuring high performances over time. In their study, [17] developed a framework to evaluate indoor air quality, estimating CO_2 and $PM_{2.5}$ concentration. Their model is made of a modified LSTM network, where the forget gate is absent which provides lighter computation. Statistical and ML models, with LSTM and RF among them, were compared in [7]. The LSTM model outperforms all the others in the prediction of CO_2 emissions. Latest works employ more sophisticated models. A review of the recent developments in the field of indoor environmental quality monitoring can be found in [20]. They list some recent works using Deep Learning techniques to control and predict energy consumption and thermal comfort in indoor environments. As an example, [19] combined k-means with CNN and LSTM to predict energy consumption and reported that this model is able to capture spatio-temporal features to improve the forecasting of buildings energy consumption. Finally, [5] use also a CNN-LSTM model but for predicting indoor temperature. The comparison with Multi-Layer Perceptron and LSTM models shows that the addition of a CNN part to the model can extract spatial features to correct for unknown patterns in the data.

The latest frontier in the field of ML prediction with IoT is Artificial Intelligence of Things (AIoT), in which Artificial Intelligence and IoT are used in synergy to obtain efficient decision-making. The work of [14] develops an innovative framework, Edge MLOps, that uses Edge Computing for ML inference on data gathered by the IoT network, and Cloud Computing for the continuous training of ML models and to gather, aggregate, and visualize data. In [15], the authors propose EdgeMiningSim, a simulation-driven methodology that aims to overcome the challenges posed by IoT applications in the data mining domain.

They do so by taking into account the orthogonal and interdependent requirements of IoT applications and the features of Edge/Cloud deployments. The methodology is specifically tailored to the IoT context and is flexible enough to be applied to any IoT application. It simultaneously considers manifold (algorithmic, infrastructural, and contextual) aspects that are only partially or individually analyzed in the narrow literature of IoT data mining. It is inspired by software engineering principles and supported by the simulation activity for simultaneously assessing the overall impact of a given IoT data mining task along with the computing, communication, and energy dimensions. By adopting a distributed EC paradigm, EdgeMiningSim can effectively and efficiently perform IoT data mining even in those demanding IoT application domains that involve resource-constrained IoT devices. The authors in [10], focus on the urgent problem of air pollution attributed to traditional energy sources and the health hazards linked to PM2.5.

The work introduces an Internet of Things (IoT) system that employs a hybrid machine learning (ML) architecture to forecast PM2.5 levels on both cloud and edge computing devices. The system's versatility was assessed through testing on a personal computer (PC) for cloud-based predictions and a Raspberry Pi for edge device forecasts, making it suitable for use in remote, low-connectivity environments. The study results underscore the superior performance of the NARX/LSTM algorithm in terms of precision, and the NARX/XGBRF combination in achieving a trade-off between accuracy and processing speed on edge devices such as the Raspberry Pi.

On the base of the models discussed above, we test different techniques, such as RF, CNN and LSTM, on simulated and real data, in order to decide which model performs best for air quality and thermal comfort forecasting in our case. The complete system is using Cloud Computing to work, but could evolve in a hybrid Edge/Cloud computing system in a next step, in a manner similar to the papers cited above.

3 ML Models

To predict CO_2 concentration, relative humidity and indoor temperature we compared the performances of different models: MLR, RF, SVR, MLP, CNN, and LSTM.

MLR is an extension of a simple linear regression in which the response variable, y, depends on multiple explanatory variables, x_i. The purpose of the model is to determine the coefficients of the curve of degree n, where n is the number of independent variables, that can best represent the data.

Decision Trees (DT) are a commonly used ML technique. A decision tree is essentially a flow chart made of decision blocks, or nodes, connected by branches and terminating in leaves, which constitute the final decisions. The algorithm can extract a set of rules from data to determine the most probable result, i.e., the value of y. A more efficient prediction can be obtained using an ensemble of DTs, in what is called random forest (RF) algorithm. RF takes an ensemble of DTs, selecting a sub-sample of variables using bootstrap, i.e., random selection with replacement. The final output is obtained by averaging the output of each DT. The error is estimated through the out-of-bags (OOB) observations, namely the observations left out from the training set.

SVR is a class of Support Vector Machine (SVM) models, originally designed to solve classification problems. SVM finds the hyperplane that separates best the data in classes. A reasonable choice for the hyperplane is the one that defines the largest separation, or margin, between the classes. SVR are the SVM extension to solve non-linear regression problems. Given the continuous nature of the response variable it is necessary to define an additional threshold, ε, to construct the best hyperplane, i.e., the one that does not intercept any of the explanatory variable points. Then, linear regression is employed to minimize the error between the hyperplane points and the response variable points.

A MLP is a neural network consisting of at least three layers of nodes: input, hidden, and output layers. Data flows from the input layer to the output layer through the hidden layers. Each layer receives the information regarding the prediction error, in order to adjust the weights of the nodes (*backpropagation*). All the nodes have a non-linear activation function, that helps in capturing the non-linearities of the problems. MLPs belong to the class of fully connected networks, where each node in every layer is connected to all other nodes through arcs with certain weights.

CNNs are widely used for image recognition (see [8,13]) but have been employed lately also for time series forecasting (e.g., [16,19]). The basic units of a CNN are the kernels, small-size matrices that are multiplied convolutionally with all the blocks of data. Usually, after every convolutional layer a Pooling layer is applied. Such a layer is able to maximize the feature extraction achieved by the kernels. At the end of the convolutional block, the CNN is flattened in a fully connected network.

Recurrent Neural Networks (RNN) are a kind of neural networks where cycles are present. This means that the output of certain nodes can be the input on the same layer or on previous layers. These kinds of neural networks come particularly handy in analyzing time series, as they can retain information from previous timesteps. LSTMs are a class of RNN where the information flux passes through three gates: input, forget and output gate. Each of these gates can decide if the information is to be carried on or to be discarded, allowing the network to retain the information it needs and forget what is useless.

4 Setup and Models

In this Section we detail the models we used, explaining the parameters choice for every model. At first we describe how we created the synthetic dataset used for testing the models, then we define the parameter choice for the models and, finally, we explain how we collected the data for testing the models on a real dataset and illustrate the results of the best models on it.

4.1 Synthetic Data

We use synthetic data calculated through a simulation of realistic values for a closed environment under internal and external factors. The data consists of two years of data with a ten minutes cadence for eight variables: internal and external temperature, relative humidity, CO_2 concentration, heating and cooling system, occupation and window status (open/closed). To keep the temporal information in the data, we transformed the variable Datetime, splitting it into time and month columns and assigning to each value in the columns a combination of sine and cosine, where the trigonometric circle is split into twelve for the month column and into twenty-four for the time column [9]. We want to predict the values for the next 10, 20 and 30 min (i.e., $t+1, t+2, t+3$).

We use MLR for relative humidity prediction following [1]. It does not require parameter optimization, but we scale the data to the range [0, 1].

We use RF for CO_2 concentration, relative humidity and temperature prediction [1]. The optimization is conducted using grid search. For the CO_2 prediction model the final number of estimators (number of trees) is $n_{estimators} = 280$ and the maximum depth for the trees is $max\ depth = 9$; for the temperature we obtain $n_{estimators} = 220$ and $max\ depth = 16$; for relative humidity, grid search gives $n_{estimators} = 130$ and $max\ depth = 16$.

SVR is used for predicting relative humidity [1]. Given the complexity of the data and the high number of regressors, a 3rd degree polynomial kernel is selected, with $\varepsilon = 0.1$.

MLP is employed for CO_2 [18,21] and temperature forecasting. For the forecast of CO_2 concentration our model is made of an input layer with 12 nodes and an hidden layer with 32 nodes and $tanh$ as activation function. For temperature we use again an input layer of 12 nodes, but now implement three hidden layers with 32 nodes and using $relu$ activation function.

CNN is used for CO_2 [16] and temperature prediction. The model is the same in both cases and it is made of an input using 12 kernels and 1 channel, a convolutional layer of size 64 × 64 with 5 kernels, a second convolutional layer 32 × 32 with 4 kernels, a MaxPooling layer using 2 kernels. After flattening, a dense layer with 64 nodes is added and, finally, a dense layer with 1 node closes the network. *Relu* is used as activation function. Padding is set to zero.

The LSTM model employed has a single LSTM layer. We select a learning rate of 0.001 and set an EarlyStopping function to stop the training phase when the validation loss does not improve for five epochs to avoid overfitting, and save the weights of the best epoch. The tests are conducted using dropout ∈

$\{0.0, 0.02, 0.03, 0.05\}$ and we select the best models for the case with dropout $\neq 0.0$. The LSTM models selected for each quantity and forecast to predict are shown in Table 1.

Table 1. Best LSTM models for the prediction of CO_2 concentration, relative humidity and temperature.

	LSTM units	Activation function	Batch size	Learning rate	Dropout
$CO_2(t+1)$	256	relu	36	0.001	0.20
$CO_2(t+2)$	256	relu	64	0.001	0.30
$CO_2(t+3)$	256	relu	128	0.001	0.20
$RH(t+1)$	128	tanh	256	0.001	0.30
$RH(t+2)$	256	relu	64	0.001	0.20
$RH(t+3)$	256	relu	64	0.001	0.30
$Temp(t+1)$	128	relu	128	0.001	0.20
$Temp(t+2)$	256	relu	36	0.001	0.30
$Temp(t+3)$	128	relu	36	0.001	0.20

4.2 Real Data

To test the models on real data, we collect one month of data from our offices, located in Southern Italy, for the quantities CO_2 concentration, indoor and outdoor temperature, relative humidity, heating and cooling system, window status (open/closed) and passive infrared (PIR) data. The last variable is a measure of the room occupation.

The setup and the sensors used are showed in Fig. 1. External temperature data are collected by a MCF-LW12TERPM Enginko sensor mounted on the roof of the building, while indoor data are collected by a Multi-sensor AM319 9 in 1 Milesight. The changes in the state of windows, lighting and air conditioning system are recorded through a control panel connected to our database.

We correct the data for missing values, outliers and resample to a 10 min cadence, to have the same resolution as the simulated dataset.

5 Results and Discussion

In this Section we compare the performances of the different models for the ten, twenty and thirty minutes prediction.

To compare the results of the different models we look at the mean absolute error (MAE) and at the root mean square error (RMSE) defined as:

$$MAE = \frac{\sum_{i=1}^{N} |y_i - \hat{y}_i|}{N}, \quad (1)$$

$$RMSE = \sqrt{\frac{1}{N} \sum_{i=1}^{N} (y_i - \hat{y}_i)^2}, \quad (2)$$

where y are the real data to predict, $\hat{y}$ are the model predictions, N is the dimension of the test set, $y_{\max}$ and $y_{\min}$ are the maximum and minimum value for the variable y. To test our result on the simulated data, we split the dataset keeping the last six months as test set and use the rest for training.

Fig. 1. Above: office chosen for the sensors installation. Below (from left to right): control panel, window motor, multi-sensor and external sensor.

5.1 Synthetic Data

The errors for the prediction of CO_2 concentration, relative humidity and temperature on simulated data are shown in Tables 2, 3 and 4, and Fig. 2.

RF is the best model for predicting CO_2 concentration for all the prediction horizons. CNN1D and MLP perform second and third best at 10 and 20 min forecast, while LSTM is the second best models for the 30 min forecast. In the case of relative humidity, LSTM performs best with the ten minutes prediction, while RF is again better on the twenty and thirty minutes predictions. For temperature, again RF gives the best predictions, with LSTM second best model.

Table 2. Errors on CO_2 prediction for the synthetic data.

Model	t+1		t+2		t+3	
	MAE	RMSE	MAE	RMSE	MAE	RMSE
RF	7.3679	15.9521	13.1696	27.9710	18.7597	39.4285
CNN1D	8.9078	17.5078	16.9219	32.3626	25.2171	47.1168
MLP	8.8654	17.3964	17.0949	32.2291	24.1223	46.6313
LSTM	8.1945	18.1488	16.4475	35.0303	19.9654	44.1167

Table 3. Errors on relative humidity prediction for the synthetic data.

Model	t+1		t+2		t+3	
	MAE	RMSE	MAE	RMSE	MAE	RMSE
RF	0.3331	0.8017	0.5117	1.0207	0.6827	1.2115
SVR	0.2914	0.7810	0.5417	1.2033	0.7669	1.5078
MLR	0.3081	0.7774	0.5955	1.2319	0.8553	1.5862
LSTM	0.3057	0.7746	0.5685	1.1764	0.7803	1.4339

Table 4. Errors on indoor temperature prediction for the synthetic data.

Model	t+1		t+2		t+3	
	MAE	RMSE	MAE	RMSE	MAE	RMSE
RF	0.0405	0.1515	0.0773	0.2367	0.0985	0.2840
MLP	0.0960	0.2747	0.1623	0.4254	0.2393	0.5348
LSTM	0.0806	0.2642	0.1300	0.4046	0.1578	0.4749

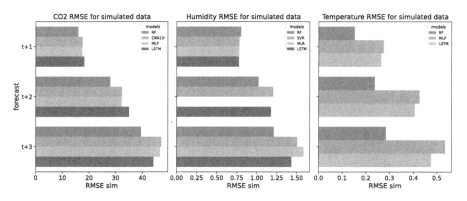

Fig. 2. RMSE error for the simulated data for CO_2 concentration (left), relative humidity (center) and temperature (right) predictions.

5.2 Real Data

We show MAE and RMSE for the prediction of CO_2 concentration, relative humidity and temperature on real data in Tables 5, 6 and 7, and Fig. 3.

Table 5. Errors on CO_2 prediction for real data.

Model	t+1		t+2		t+3	
	MAE	RMSE	MAE	RMSE	MAE	RMSE
RF	19.8272	53.4199	34.8013	85.9956	48.4585	111,311
CNN1D	22.4626	54.9991	42.9299	89.8207	55.6355	113,790
MLP	33.4062	78.5365	79.0392	158.078	76.3828	157,000
LSTM	21.2043	56.7998	37.0268	92.5422	50,8780	115,996

Table 6. Errors on relative humidity prediction for real data.

Model	t+1		t+2		t+3	
	MAE	RMSE	MAE	RMSE	MAE	RMSE
RF	0.1869	0.3323	0.4494	0.7660	0.6417	1.0091
SVR	1.1585	2.9350	1.4001	3.3671	1.5777	3.6818
MLR	0.2503	0.3963	0.4614	0.6964	0.6345	0.9349
LSTM	0.1454	0.3192	0.2787	0.5318	0.3723	0.6827

Table 7. Errors on indoor temperature prediction for real data.

Model	t+1		t+2		t+3	
	MAE	RMSE	MAE	RMSE	MAE	RMSE
RF	0.1936	0.3428	0.2267	0.3897	0.5148	1.008
MLP	0.4703	0.9759	0.4715	0.9675	0.7571	1.1816
LSTM	0.1455	0.2507	0.2352	0.4288	0.2917	0.5353

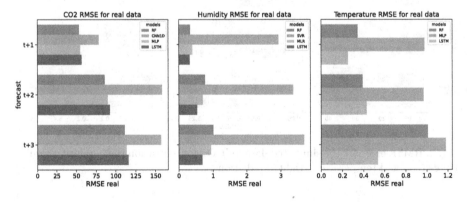

Fig. 3. RMSE error for real data for CO_2 concentration (left), relative humidity (center) and temperature (right) predictions.

Also in the case of real data RF is the model that can predict CO_2 concentration with the highest accuracy at all prediction horizons. MLP and LSTM are, respectively, second and third best models. For relative humidity the best model is LSTM, as it was the case for the simulated data. The worst performance comes from SVR: the RMSE in Fig. 3 is much larger than the one from the other models. Regarding temperature prediction, using real data, the model which gives the lowest error with the 10 min forecast is LSTM, while RF has the smallest error for the prediction at 20 min, and again LSTM shows the lowest error in the 30 min forecast.

In order to check whether the results of the models on real data are statistically different, we apply the Mann-Whitney U test to the distributions of the predictions from RF and LSTM models, for all the three quantities to predict and at all forecast horizons. The Mann-Whitney U test is a non-parametric test. The null hypothesis tests that, given two distributions, they are statistically the same. We fix the confidence level, $\alpha = 0.05$, and we calculate the p-values for the test. The results are reported in Table 8.

Table 8. p-values for the Mann-Whitney U test.

Forecast	CO_2	RH	Temperature
t+1	0.37	0.34	$2.99 \cdot 10^{-4}$
t+2	$3.72 \cdot 10^{-4}$	0.02	$2.65 \cdot 10^{-6}$
t+3	$1.04 \cdot 10^{-8}$	0.19	0.01

The distributions of the predictions of RF and LSTM models for relative humidity are statistically the same. This means that we can use one model or the other and the predictions from each of them will be statistically the same. Regarding the CO_2 concentration predictions, at 10 min the distributions from RF and LSTM are statistically the same, while for the 20 and 30 min predictions the p-values tell us to reject the null hypothesis. For temperature prediction, at 10 and 20 min forecast we reject the null hypothesis of statistically equal distributions, but the distributions are the same for the 30 min forecast. We decide to use RF for the prediction of CO_2 concentration, and LSTM for predicting temperature and relative humidity.

6 Conclusions and Future Work

In this work we compare different models for the prediction of room temperature, CO_2 concentration and relative humidity, three quantities to assess the wellness in indoor environments. We compare different models for each of the quantities we want to predict, i.e., random forest (RF), convolutional neural networks (CNN1D), multi-layer perceptron (MLP) and long-short term memory (LSTM) for CO_2 concentration prediction, RF, support vector regression (SVR), multi-linear regression (MLR) and LSTM for relative humidity prediction, and RF, MLP and LSTM for temperature prediction. We test the models at first on a simulated dataset, created using physical laws, and then on real data, taken from the instrumentation installed in our offices in Southern Italy.

On the synthetic data, RF gives the best predictions for CO_2 concentration and temperature, while LSTM performs best on relative humidity predictions. When testing on real data, again RF and LSTM result in the lowest errors. This time, while RF remains the best model to predict CO_2, LSTM results to be the best at relative humidity and temperature forecast. We note here that the fact that the RF model performance on temperature prediction gets worse when dealing with real data, can mean that RF has the best guess at the function generating the synthetic data and, therefore, the performance on real data gets worse.

With these results at hand, we include the best models, RF for CO_2 and LSTM for temperature and relative humidity, in the AIR SAFE system. Together with real-time data and the current state of windows and heating and lighting systems, the results of the prediction models are part of the automation process that decides the next state of windows, heating and lighting system. AIR SAFE is currently being tested in our headquarters offices, in Naples and Rende (CS), and in a schoolroom located in the South of Italy.

Acknowledgements. This work is part of the project Artificial Intelligence Reporting system for SAFe indoor and efficient Energy (AIR SAFE), funded by POR Campania FESR 2014–2020 Asse 3 OS 3.1 Azione 3.1.1 - CUP B67H22002620007.

References

1. Arulmozhi, E., Basak, J.K., Sihalath, T., Park, J., Kim, H.T., Moon, B.E.: Machine learning-based microclimate model for indoor air temperature and relative humidity prediction in a swine building. Animals **11**(1) (2021). https://doi.org/10.3390/ani11010222. https://www.mdpi.com/2076-2615/11/1/222
2. ASHRAE_55 (2020). https://www.ashrae.org/technical-resources/bookstore/standard-55-thermal-environmental-conditions-for-human-occupancy
3. ASHRAE_62.1 (2022). https://www.techstreet.com/ashrae/standards/ashrae-62-1-2022?product_id=2501063
4. DM_383_6.10.2022. https://www.mite.gov.it
5. Elmaz, F., Eyckerman, R., Casteels, W., Latré, S., Hellinckx, P.: CNN-LSTM architecture for predictive indoor temperature modeling. Build. Environ. **206**, 108327 (2021). https://doi.org/10.1016/j.buildenv.2021.108327. https://www.sciencedirect.com/science/article/pii/S0360132321007241
6. ENEA dal D.lgs: https://www.efficienzaenergetica.enea.it/component/jdownloads/?task=download.send&id=542&Itemid=101
7. Kumari, S., Singh, S.K.: Machine learning-based time series models for effective CO2 emission prediction in India. Environ. Sci. Pollut. Res., 1614–7499 (2022). https://doi.org/10.1007/s11356-022-21723-8
8. Liu, Q., et al.: A review of image recognition with deep convolutional neural network. In: Huang, D.-S., Bevilacqua, V., Premaratne, P., Gupta, P. (eds.) ICIC 2017. LNCS, vol. 10361, pp. 69–80. Springer, Cham (2017). https://doi.org/10.1007/978-3-319-63309-1_7

9. Mary Jasmine, E., Milton, A.: The role of hyperparameters in predicting rainfall using n-hidden-layered networks. Nat. Hazards **111**, 489–505 (2022). https://doi.org/10.1007/s11069-021-05063-3
10. Moursi, A.S., El-Fishawy, N., Djahel, S., Shouman, M.A.: An IoT enabled system for enhanced air quality monitoring and prediction on the edge. Complex Intell. Syst. **7**(6), 2923–2947 (2021)
11. Mtibaa, F., Nguyen, K.K., Azam, M., Papachristou, A., Venne, J.S., Cheriet, M.: LSTM-based indoor air temperature prediction framework for HVAC systems in smart buildings. Neural Comput. Appl. **32**(23), 17569–17585 (2020). https://doi.org/10.1007/s00521-020-04926-3
12. World Health Organization: WHO global air quality guidelines: particulate matter (PM2.5 and PM10), ozone, nitrogen dioxide, sulfur dioxide and carbon monoxide. World Health Organization (2021)
13. Pak, M., Kim, S.: A review of deep learning in image recognition. In: 2017 4th International Conference on Computer Applications and Information Processing Technology (CAIPT), pp. 1–3. IEEE (2017)
14. Raj, E., Buffoni, D., Westerlund, M., Ahola, K.: Edge MLOps: an automation framework for AIoT applications. In: 2021 IEEE International Conference on Cloud Engineering (IC2E), pp. 191–200. IEEE (2021)
15. Savaglio, C., Fortino, G.: A simulation-driven methodology for IoT data mining based on edge computing. ACM Trans. Internet Technol. **21**(2) (2021). https://doi.org/10.1145/3402444
16. Segala, G., Doriguzzi-Corin, R., Peroni, C., Gazzini, T., Siracusa, D.: A practical and adaptive approach to predicting indoor CO2. Appl. Sci. **11**(22) (2021). https://doi.org/10.3390/app112210771. https://www.mdpi.com/2076-3417/11/22/10771
17. Sharma, P.K., et al.: IndoAirSense: a framework for indoor air quality estimation and forecasting. Atmos. Pollut. Res. **12**(1), 10–22 (2021). https://doi.org/10.1016/j.apr.2020.07.027. https://www.sciencedirect.com/science/article/pii/S130910422030218X
18. Skön, J.P., Johansson, M., Raatikainen, M., Leiviskä, K., Kolehmainen, M.: Modelling indoor air carbon dioxide (CO_2)concentration using neural network. Int. J. Environ. Ecol. Eng. **6**(1), 37–41 (2012). https://publications.waset.org/vol/61
19. Somu, N., Raman, M.R.G., Ramamritham, K.: A deep learning framework for building energy consumption forecast. Renew. Sustain. Energy Rev. **137**, 110591 (2021). https://doi.org/10.1016/j.rser.2020.110591. https://www.sciencedirect.com/science/article/pii/S1364032120308753
20. Tien, P.W., Wei, S., Darkwa, J., Wood, C., Calautit, J.K.: Machine learning and deep learning methods for enhancing building energy efficiency and indoor environmental quality - a review. Energy AI **10**, 100198 (2022). https://doi.org/10.1016/j.egyai.2022.100198. https://www.sciencedirect.com/science/article/pii/S2666546822000441
21. Vaňuš, J., Martinek, R., Bilik, P., Žídek, J., Dohnalek, P., Gajdo, P.: New method for accurate prediction of CO2 in the smart home (2016). https://doi.org/10.1109/I2MTC.2016.7520562

Short Papers

Unimaginable Numbers and Infinity Computing at School: An Experimentation in Northern Italy

Luigi Antoniotti[1], Annabella Astorino[2], and Fabio Caldarola[3](✉)

[1] IIS Palladio, via Tronconi 22, Treviso, Italy
luigiantoniotti@gmail.com
[2] Department of Computer Engineering, Modeling, Electronics and Systems Engineering, University of Calabria, 87036 Arcavacata di Rende, CS, Italy
annabella.astorino@dimes.unical.it
[3] Department of Environmental Engineering, University of Calabria, 87036 Arcavacata di Rende, CS, Italy
fabio.caldarola@unical.it

Abstract. In this paper we describe an experimentation carried out in a high school in northern Italy. The focus is to present to students new concepts of very big numbers and the infinite, studying their response, approaches, intuit and the suitability to apply in larger scale. In particular the new concepts and notations regard unimaginable numbers and the infinity computing. Several exercises have been suggested to students arousing much interest. Also a final test has been proposed and it is discussed in part in this paper. Among many observations and conclusions, we confirm a great ease of use of infinity computing by students and an almost immediate and intuitive degree of reception. Also unimaginable numbers, hyperoperations and Knuth's powers proved highly educational value, but they are more difficult to master.

Keywords: Mathematical education · Infinity computing · Unimaginable numbers · Student learning

1 Introduction

In the literature, it is possible to find several projects that concern the development of new learning approaches in the context of high schools. These have different purposes, such as promoting greater motivation for learning, developing skills and propensities for critical and creative thinking.

In this paper we describe an experimentation carried out in 2023 in a high school in Treviso, northern Italy. This study was conducted in parallel and simultaneously with a similar one carried out in a second high school in Crotone, southern Italy (see [19] in this same volume for details). The aim of this study in two schools was to investigate the students' response to some teaching activities concerning new approaches to very big numbers and infinity. In particular we employed infinity computing similarly as in 2019 in three Italian schools (see

[2,18]), but now, for the first time to our knowledge, employing in a mathematical education research also the so-called *unimaginable numbers* which are almost infinite numbers (see Sect. 2 for some details and references).

Specifically, this paper is structured as follows. Section 2 gives, in a quick way, some basic material, information and references on unimaginable numbers and the arithmetic of infinity obtained by adopting the *grossone*-based system. Section 3 is the core of the paper and describes the experiment conducted in 2023 at the *IIS "Palladio"* in Treviso, North Italy. It had as its object the study of the didactic approach of high school students in the face of unimaginable numbers and the arithmetic of infinity.

2 Mathematical Tools and References

A natural number is called *unimaginable* if it is greater than 1 *googol*, i.e. 10^{100}. The ordinary exponential or scientific notation is far to be able to write unimaginable numbers, so, in the XX Century many special notational methods were developed. Among them we recall *Knuth's up-arrow notation, hyperoperation notation, Conway chained arrow notation, Moser-Steinhaus notation*, etc. For details, examples, basic definitions, etc., we refer the reader to [6,12,20] and the references therein.

The name *infinity computing* is referred to a new computational system proposed by Y. Sergeyev about 20 years ago and able to perform computations with infinite and infinitesimal numbers very easily and handily. This is a great strength of Sergeyev's new system, which allows an immediate approach even towards high school students. It is commonly called the *grossone-based computational (or numerical) system* because it is constructed on the fundamental unit ①, precisely called *grossone*, as well as on the ordinary unit 1. The former allows to write infinite and infinitesimal numbers in the same way as 1 allows to write finite ordinary numbers (see [26,27,29] for introductory surveys on the new system). There are many applications of Sergeyev's new paradigm in several fields of mathematics, physics and applied sciences. For instance [1,4,29] contain applications to differential equations and optimization, [3,7,8,13,14,28] applications to summations and fractals, [15,16] employ the new system combined with cellular automata, [9,21,29,32,33] contain investigations on its mathematical foundations and some discussions about new views of classical paradoxes, applications to logic, etc. In [11] there are also some hints to apply infinity computing to the Carboncettus sequence which originates from Fibonacci numbers (see [10]).

Furthermore, in the previous edition of this conference, NUMTA 2019, a special session of new computational tools and math education has begun to catch on (see for example the paper [30,31] and others), and inside it also some researches on the possible employ of the grossone system in high schools (see [2, 18]). More recently the book [25] and the papers [17,22–24] have been published, and [19] in this same volume "continues" the article [18] but involving also unimaginable numbers. In the next subsection the reader can find more details on them.

3 Activities and Tests in Treviso

3.1 Description of the Experimentation

The involved classes in the experimentations at the IIS "Palladio" in Treviso were two fourth classes, one with 25 students (10 male and 15 female) and another with 22 (14 male and 8 female). The total number of students is 47 (24 male and 23 female), and the age is between 17 and 18 y.o. This research, together with the twin one in Crotone [19], can be considered a second step of [2], which had been organized in 2019 on the basis of [5].

In particular, we proposed to both classes a short cycle of lessons, 7 or 8 h divided in 4–6 days. Just about 1 h was about the grossone-based system, and the remaining ones on unimaginable numbers (this means that the grossone system can be used almost at an intuitive level, see the conclusions in Subsect. 3.2 and cf. [2,18]). During the lectures on unimaginable numbers, tetrations, pentations, etc., many examples and exercises had been proposed to students. A final test with several questions was also administered a week after the conclusion of the cycle of lessons. Each question had 3–5 predefined multi-choice answers, only one of them correct. A selection of 8 questions with the number and percentages of answers by the students are reported in Tables 1, 2, 3, 4, 5, 6, 7 and 8. below. For the reader's convenience the (unique) correct answer will be specified each time.

Using progressive numbering, and not the original one, for the 8 questions that we report as examples, Question 1 asked to find the correct claim about $①+1$ among three different possibilities shown in the first column of Table 1. In the second and fourth column of Table 1, M and F denote the number of male or female students, respectively, who have opted for the corresponding choice. In columns three and five we find the percentage relative to the column M and F, respectively. Column six, denoted by "Tot.", reports the total number of students who opted for this choice (i.e., "Tot. = M+F", roughly speaking), and column seven, the last one in Table 1, reports the percentage relative to the column "Tot.".

Table 1. Question 1, about $①+1$, had three possible suggested answers: (b) is the correct one

Possible choices	M	M%	F	F%	Tot.	%
(a) It is not a number	2	8.33	5	21.74	7	14.89
(b) It is greater than any natural number	17	70.83	17	73.91	34	72.34
(c) It is equal to $①$	5	20.83	1	4.35	6	12.77

Question 2 (for us) asks about the result or meaning of $①+①$, giving three possible choices shown in the first column of Table 2. The subsequent columns 2–7 play the same role as in Table 1.

Table 2. Question 2 about ①+①: (a) is the correct answer

Possible choices	M	M%	F	F%	Tot.	%
(a) It is equal to 2①	11	45.83	15	65.22	26	55.32
(b) It is equal to ①	12	50	5	21.74	17	36.17
(c) It is not a number	1	4.17	3	14.04	4	8.51

Table 3. Question 3 about ①/2: (b) is the correct answer

Possible choices	M	M%	F	F%	Tot.	%
(a) It is equal to ①	2	8.33	5	21.74	7	14.89
(b) It is a positive infinite number	18	75	18	78.26	36	76.60
(c) It is equal to $10^{2\,000\,000\,000\,000}$	4	16.67	0	0	4	8.51

Question 3 asks about the writing ①/2 giving three possible choices as in the first column of Table 3. Question 4 instead asked to find the correct claim among four possible choices as in the first column of Table 4.

Question 5 asked to find the correct claim among four possible choices listed in the first column of Table 5. Question 6 asked for the correct value of $2 \uparrow\uparrow 2$ among four possible choices listed in column 1 of Table 6. Then Question 7 asked to find the correct value of the pentation $2 \uparrow\uparrow\uparrow 2$ among five possible options listed in the first column of Table 7. Finally, Question 8 asks about two Knuth's powers, $3 \uparrow 3$ and $2 \uparrow\uparrow 2$, giving five different options as in column 1 of Table 8. Columns 2–7 in Tables 3, 4, 5, 6, 7 and 8 play obviously the same role as in Table 1.

3.2 Data Analysis and Conclusions

A first interesting and quite singular result emerges from Table 1: among male and female students, we find something like "an X configuration" in the answers. In fact, note that the correct answer (b) was given by an equal number of male and female students (i.e. 17), while the numbers relating to answers (a) and (c) are exactly exchanged in the form of X if we consider the integers 1 and

Table 4. Question 4: (d) is the correct choice

Possible choices	M	M%	F	F%	Tot.	%
(a) ① is less than $+\infty$	14	58.33	15	65.22	29	61.70
(b) ① is greater than $+\infty$	2	8.33	4	17.39	6	12.77
(c) ① is equal to $+\infty$	4	16.67	1	4.35	5	10.64
(d) ① and $+\infty$ are not comparable	4	16.67	3	13.04	7	14.89

Table 5. Question 5: (b) is the correct option

Possible choices	M	M%	F	F%	Tot.	%
(a) A googol is greater than ①	2	8.33	5	21.74	7	14.89
(b) A googol is less than ①	15	62.5	16	69.57	31	65.96
(c) A googol is equal to ①	3	12.5	1	4.35	4	8.51
(d) Googol and grossone are not comparable	4	16.67	1	4.35	5	10.64

Table 6. Question 6 about $2 \uparrow\uparrow 2$: (b) is the correct option

Possible choices	M	M%	F	F%	Tot.	%
(a) It is equal to 2^4	4	16.67	3	13.04	7	14.89
(b) It is equal to $2 \cdot 2$	4	16.67	3	13.04	7	14.89
(c) It is equal to 2^{2^2}	14	58.33	16	69.57	30	63.83
(d) None of the above	2	8.33	1	4.35	3	6.38

2 the closest integer approximations of the average percentage value 6.34%.[1] This simple observation could be the starting point for a series of more in-depth researches on the differences between male and female students in their conception, vision, intuition, previous experiences, stimuli and interest in the study of infinity in mathematics in the broadest sense of the term.

For Question 2 we expected many more answers (a). Although the majority of students, and especially female students, answered correctly, answer (b) still proved to be attractive. The percentage of correct answers to Question 3 rose to 76.60% (from 55.32% of Question 2): *a priori* we would have expected the

Table 7. Question 7 about the pentation $2 \uparrow\uparrow\uparrow 2$: (e) is the correct answer

Possible choices	M	M%	F	F%	Tot.	%
(a) It is equal to 0	8	33.33	0	0	8	17.02
(b) It is equal to $2^{2\,000\,000\,000\,000\,000\,002}$	0	0	0	0	0	0
(c) It is equal to 2002	0	0	0	0	0	0
(d) It is equal to $2^{\left(2^2\right)^{2\left(2^2\right)^{2\left(2^2\right)}}}$	13	54.17	21	91.30	34	72.34
(e) None of the above	3	12.5	2	8.70	5	10.64

---

[1] In other words, consider the three average percentage values taken in the shape of an X: we find 72.39% (average value between 70.83% and 73.91%), 21.28 (average value between 20.83% and 21.74%), 6.34% (average value between 8.33% and 4.35%). Calculating the closest integer approximations on the basis of 24 male and 23 female students, we obtain precisely the values that appear in columns M and F of Table 1.

Table 8. Question 8 on two Knuth's powers, $3\uparrow 3$ and $2\uparrow\uparrow 2$. Obviously (c) is here the correct choice

Possible choices	M	M%	F	F%	Tot.	%
(a) $3\uparrow 3$ is less than $2\uparrow\uparrow 2$	10	41.67	16	69.57	26	55.32
(b) $3\uparrow 3$ and $2\uparrow\uparrow 2$ are not comparable	0	0	0	0	0	0
(c) $3\uparrow 3$ is greater than or equal to $2\uparrow\uparrow 2$	14	58.33	7	30.43	21	44.68
(d) $3\uparrow 3$ and $2\uparrow\uparrow 2$ are equal	0	0	0	0	0	0
(e) None of the above	0	0	0	0	0	0

opposite! It would be very interesting to conduct future studies with this pair of questions and with larger samples of students.

In Question 4, the non-comparable answer probably requires more reflection, and is objectively more difficult and less intuitive. In Question 5, the percentage of correct answers is excellent, close to 70% for girls. It is interesting to note how the answer (d) on non-comparability had almost the same percentages of choices as for Question 4. Indeed, exactly 4 boys chose (d) for Question 4 and for Question 5: are they the same? Are they less reflexive than the girls who have gone from 3 to 1 in parallel? On points like this there would be much to study and debate; an extra fact that we report is that male students seem less inclined to do lengthy calculations and reflections, and were on average quicker in completing the test.

In Question 6 only about 14.9% of students answered correctly, and the percentage of male students is higher in this case. The preference of option (c) is not so clear, it is probably a trick question for them. Turning instead to the pentation $2\uparrow\uparrow\uparrow 2$ considered in Question 7, it seems really difficult to explain the answer (a) chosen by 8 boys (and 0 girls). Maybe something like the so-called *anchoring bias* for (less thoughtful and faster) male students? It should be noted that almost all of the girls chose the complicated expression appearing in (d) and it is a very interesting phenomenon. "If (d) is that complicated, then it must be the right one...", or what is technically called the *seductive detail bias* could be at play.

The answers given to Question 8 show that Knuth's arrow notation, even only for tetrations, needs a medium-long assimilation time by the students. For Question 8 the percentage of correct answers by male students is much higher (about twice).

We conclude the paper with some final remarks and considerations. During all the class-activity, also after the body of the experimentation, the students have shown great facility in the correct use of the gross-based system (cf. also [2,18]). It must be taken into account that almost all the time of the lessons before the test was devoted to unimaginable numbers, tetrations, etc., which require, as confirmed by the test, much more time to be understood or used in simple contexts. Nonetheless we are convinced of the high educational value they can provide as a sort of "mind gym" at various levels.

References

1. Amodio, P., Iavernaro, F., Mazzia, F., Mukhametzhanov, M.S., Sergeyev, Y.D.: A generalized Taylor method of order three for the solution of initial value problems in standard and infinity floating-point arithmetic. Math. Comput. Simul. **141**, 24–39 (2017). https://doi.org/10.1016/j.matcom.2016.03.007
2. Antoniotti, L., Caldarola, F., d'Atri, G., Pellegrini, M.: New approaches to basic calculus: an experimentation via numerical computation. In: Sergeyev, Y.D., Kvasov, D.E. (eds.) NUMTA 2019. LNCS, vol. 11973, pp. 329–342. Springer, Cham (2020). https://doi.org/10.1007/978-3-030-39081-5_29
3. Antoniotti, L., Caldarola, F., Maiolo, M.: Infinite numerical computing applied to Peano's, Hilbert's, and Moore's curves. Mediterr. J. Math. **17**, 99 (2020). https://doi.org/10.1007/s00009-020-01531-5
4. Astorino, A., Fuduli, A.: Spherical separation with infinitely far center. Soft. Comput. **24**, 17751–17759 (2020)
5. Bertacchini, F., Bilotta, E., Caldarola, F., Pantano, P.: The role of computer simulations in learning analytic mechanics towards chaos theory: a course experimentation. Int. J. Math. Educ. Sci. Technol. (2018). https://doi.org/10.1080/0020739X.2018.1478134
6. Blakley, G.R., Borosh, I.: Knuth's iterated powers. Adv. Math. **34**, 109–136 (1979). https://doi.org/10.1016/0001-8708(79)90052-5
7. Caldarola, F.: The exact measures of the Sierpiński d-dimensional tetrahedron in connection with a Diophantine nonlinear system. Commun. Nonlinear Sci. Numer. Simul. **63**, 228–238 (2018). https://doi.org/10.1016/j.cnsns.2018.02.026
8. Caldarola, F.: The Sierpiński curve viewed by numerical computations with infinities and infinitesimals. Appl. Math. Comput. **318**, 321–328 (2018). https://doi.org/10.1016/j.amc.2017.06.024
9. Caldarola, F., Cortese, D., d'Atri, G., Maiolo, M.: Paradoxes of the infinite and ontological dilemmas between ancient philosophy and modern mathematical solutions. In: Sergeyev, Y.D., Kvasov, D.E. (eds.) NUMTA 2019. LNCS, vol. 11973, pp. 358–372. Springer, Cham (2020). https://doi.org/10.1007/978-3-030-39081-5_31
10. Caldarola, F., d'Atri, G., Maiolo, M., Pirillo, G.: New algebraic and geometric constructs arising from Fibonacci numbers. In honor of Masami Ito. Soft Comput. **24(23)**, 17497–17508 (2020). https://doi.org/10.1007/s00500-020-05256-1
11. Caldarola, F., d'Atri, G., Maiolo, M., Pirillo, G.: The sequence of Carboncettus octagons. In: Sergeyev, Y.D., Kvasov, D.E. (eds.) NUMTA 2019. LNCS, vol. 11973, pp. 373–380. Springer, Cham (2020). https://doi.org/10.1007/978-3-030-39081-5_32
12. Caldarola, F., d'Atri, G., Mercuri, P., Talamanca, V.: On the arithmetic of Knuth's powers and some computational results about their density. In: Sergeyev, Y.D., Kvasov, D.E. (eds.) NUMTA 2019. LNCS, vol. 11973, pp. 381–388. Springer, Cham (2020). https://doi.org/10.1007/978-3-030-39081-5_33
13. Caldarola, F., Maiolo, M.: On the topological convergence of multi-rule sequences of sets and fractal patterns. Soft. Comput. **24**, 17737–17749 (2020). https://doi.org/10.1007/s00500-020-05358-w
14. Caldarola, F., Maiolo, M., Solferino, V.: A new approach to the Z-transform through infinite computation. Commun. Nonlinear Sci. Numer. Simul. **82**, 105019 (2020). https://doi.org/10.1016/j.cnsns.2019.105019

15. Cococcioni, M., Cudazzo, A., Pappalardo, M., Sergeyev, Y.D.: Solving the lexicographic multi-objective mixed-integer linear programming problem using branch-and-bound and Grossone methodology. Commun. Nonlinear Sci. Numer. Simul. **84**, 105177 (2020). https://doi.org/10.1016/j.cnsns.2020.105177
16. D'Alotto, L.: Infinite games on finite graphs using Grossone. Soft. Comput. **24**, 17509–17515 (2020)
17. Iannone, P., Rizza, D., Thoma, A.: Investigating secondary school students' epistemologies through a class activity concerning infinity. In: Bergqvist, E. et al. (eds.) Proceedings of the 42nd Conference on International Group for the Psychology of Mathematics Education, vol. 3, pp. 131–138. PME, Umeå, Sweden (2018)
18. Ingarozza, F., Adamo, M.T., Martino, M., Piscitelli, A.: A Grossone-based numerical model for computations with infinity: a case study in an Italian high school. In: Sergeyev, Y.D., Kvasov, D.E. (eds.) NUMTA 2019. LNCS, vol. 11973, pp. 451–462. Springer, Cham (2020). https://doi.org/10.1007/978-3-030-39081-5_39
19. Ingarozza, F., d'Atri, G., Iembo, R.: Unimaginable numbers: a case study as a starting point for an educational experimentation. In: Sergeyev, Y.D., Kvasov, D.E., Astorino, A. (eds.) NUMTA 2023. LNCS, vol. 14478, pp. 113–126. Springer, Cham (2025). https://doi.org/10.1007/978-3-031-81247-7_9
20. Leonardis, A., d'Atri, G., Caldarola, F.: Beyond Knuth's notation for unimaginable numbers within computational number theory. Int. Electron. J. Algebra **31**, 55–73 (2022). https://doi.org/10.24330/ieja.1058413
21. Lolli, G.: Metamathematical investigations on the theory of Grossone. Appl. Math. Comput. **255**, 3–14 (2015)
22. Mazzia, F.: A computational point of view on teaching derivatives. Inform. Educ. **37**, 79–86 (2022)
23. Nasr, L.: The effect of arithmetic of infinity methodology on students' beliefs of infinity. Mediterr. J. Res. Math. Educ. **19**, 5–19 (2022)
24. Nasr, L.: Students' resolutions of some paradoxes of infinity in the lens of the Grossone methodology. Inform. Educ. **38**, 83–91 (2023)
25. Rizza, D.: Primi passi nell'aritmetica dell'infinito. Bonomo (2023). in Italian
26. Sergeyev, Y.D.: Arithmetic of Infinity. Edizioni Orizzonti Meridionali, Cosenza (2003, 2nd edn 2013)
27. Sergeyev, Y.D.: Lagrange Lecture: Methodology of numerical computations with infinities and infinitesimals. Rendiconti del Seminario Matematico dell'Università e del Politecnico di Torino **68**, 95–113 (2010)
28. Sergeyev, Y.D.: Using blinking fractals for mathematical modelling of processes of growth in biological systems. Informatica **22**, 559–576 (2011)
29. Sergeyev, Y.D.: Numerical infinities and infinitesimals: methodology, applications, and repercussions on two Hilbert problems. EMS Surv. Math. Sci. **4**, 219–320 (2017)
30. Serpe, A.: A Computational Approach with MATLAB Software for Nonlinear Equation Roots Finding in High School Maths. In: Sergeyev, Y.D., Kvasov, D.E. (eds.) NUMTA 2019. LNCS, vol. 11973, pp. 463–477. Springer, Cham (2020). https://doi.org/10.1007/978-3-030-39081-5_40
31. Serpe, A., Frassia, M.G.: Task mathematical modelling design in a dynamic geometry environment: Archimedean spiral's algorithm. In: Sergeyev, Y.D., Kvasov, D.E. (eds.) NUMTA 2019. LNCS, vol. 11973, pp. 478–491. Springer, Cham (2020). https://doi.org/10.1007/978-3-030-39081-5_41

32. Tohme, F., Caterina, G., Gangle, R.: A constructive sequence algebra for the calculus of indications. Soft. Comput. **24**, 17621–17629 (2020). https://doi.org/10.1007/s00500-020-05121-1
33. Tohme, F., Caterina, G., Gangle, R.: Observability in the univalent universe. Mediterr. J. Math. **19**, 1–19 (2022). https://doi.org/10.1007/s00009-022-02121-3

The Cantor-Vitali Function and Infinity Computing

Luigi Antoniotti[1], Corrado Mariano Marotta[2,3](✉), Andrea Melicchio[3], and Maria Anastasia Papaleo[3]

[1] IIS "Palladio", Treviso, Italy
[2] IPSEOA "San Francesco", Paola, CS, Italy
corradomariano.marotta@unical.it
[3] University of Calabria, 87036 Arcavacata di Rende, CS, Italy

Abstract. In this work we consider the Cantor-Vitali function $c : [0, 1] \to [0, 1]$, constructed as limit of a sequence of functions $\{f_n\}_{n \in \mathbb{N}_0}$. In particular, we give formulas for the length of the graph of the approximating functions and will discuss them, together with the length of the graph of c, also by using infinity computing.

Keywords: Cantor-Vitali function · Cantor set · Fractals · Grossone · Infinity computing

1 Introduction

The *Cantor-Vitali function* $c : [0, 1] \to [0, 1]$ is a uniformly continuous and surjective function, defined on the closed interval $[0, 1]$ of the real line and having the same image. It is also called the *Cantor ternary function* or *Lebesgue's singular function*, and it has amazing peculiarities: for instance, it is constant on all countable complementary intervals of the Cantor set and yet it is an increasing function on the whole domain despite having derivative zero on all these intervals (see for example [6,35]).

In the present work we use a definition of c as limit of a convergent sequence of functions $\{f_n(x)\}_{n \in \mathbb{N}_0}$, where $\mathbb{N}_0$ is the set of non-negative integers (see Sect. 2). The area under the f_n's is much less interesting than the length l_n of their graphs for which we give a closed formula depending on n (see (6)).

In Sect. 3 we use the *grossone*-based computational system introduced by Y.D. Sergeyev in the early 2000's: we refer the reader to [30,33] for detailed introductory surveys on the subject which show how to work numerically with infinite and infinitesimals numbers in a very easy and handy way, or to the book [29] written in a popular way.

With traditional mathematics we can only say that the limit of the lengths l_n is 2 (see (7)). Instead, by using the new system, we can consider a single sequence or a family of chained ones, obtaining, in this way, a whole range of infinitely many distinct values which make the result more refined and accurate.

© The Author(s), under exclusive license to Springer Nature Switzerland AG 2025
Y. D. Sergeyev et al. (Eds.): NUMTA 2023, LNCS 14478, pp. 232–239, 2025.
https://doi.org/10.1007/978-3-031-81247-7_18

From geometrical considerations we deduce that all these values (obtained after any infinite number of steps) are less than 2 and differ from it by infinitesimal quantity expressed through the new computational method in a precise and sharp way. Easy examples are explicitly given in (11) and (12).

2 The Approximating Functions f_n and Some Length Formulas

The functions

$$f_n : [0,1] \longrightarrow [0,1]$$

are recursively defined for all integer $n \geq 0$ as follows:

$$f_0 := \mathrm{id}_{[0,1]},$$

is the identity on the closed interval $[0,1]$, i.e. $f_0(x) = x$ for all $x \in [0,1]$, and

$$f_{n+1}(x) := \begin{cases} \frac{1}{2} f_n(3x), & \text{if } x \in [0, 1/3], \\ \frac{1}{2}, & \text{if } x \in [1/3, 2/3], \\ \frac{1}{2} + \frac{1}{2} f_n(3x - 2), & \text{if } x \in [2/3, 1]. \end{cases} \qquad (1)$$

Note that the graphic of f_1 is a polygonal through the 4 points $(0,0)$, $(1/3, 1/2)$, $(2/3, 1/2)$, $(2/3, 1)$. Similarly the graph of f_2 is a polygonal through the 8 points

$$(0,0), \ \left(\frac{1}{9}, \frac{1}{4}\right), \ \left(\frac{2}{9}, \frac{1}{4}\right), \ \left(\frac{1}{3}, \frac{1}{2}\right), \ \left(\frac{2}{3}, \frac{1}{2}\right), \ \left(\frac{7}{9}, \frac{3}{4}\right), \ \left(\frac{8}{9}, \frac{3}{4}\right), \ (1,1),$$

as Fig. 1c shows. It is then possible to determine the graphic of f_n as a polygonal through 2^{n+1} points, giving them in a recursive way as for formula (1).

We can also say that the graph of f_1 is made up by 2 oblique line segments equal to the diagonal of the rectangle $\left[0, \frac{1}{3}\right] \times \left[0, \frac{1}{2}\right]$ except for translations parallel to the axes, and 1 horizontal segment line. Then, the graph of f_2 is constituted by 2^2 oblique segment lines equal to the diagonal of $\left[0, \frac{1}{3^2}\right] \times \left[0, \frac{1}{2^2}\right]$ except for translations parallel to the axes, and $1 + 2 = 2^2 - 1$ horizontal line segments. In general we can write the following.

Remark 1. For all $n \in \mathbb{N}_0$, the graph of f_n is made up by 2^n oblique line segments equal to the diagonal of the rectangle

$$\left[0, \frac{1}{3^n}\right] \times \left[0, \frac{1}{2^n}\right]$$

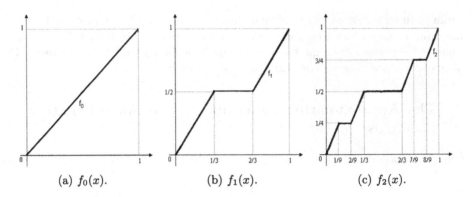

Fig. 1. The first three approximations of the Cantor-Vitali function

except for translations parallel to the axes, and $2^n - 1$ horizontal line segments.[1] Hence, the graph of f_n consists of a total of $2^{n+1} - 1$ line segments through 2^n points, whose ends are $(0, 1)$ and $(1, 1)$.

For the area subtended by the functions f_n the result is trivial. If we define

$$a_n := \int_0^1 f_n(x)\, dx$$

for all $n \in \mathbb{N}_0$, we have

$$a_0 = \frac{1}{2} \quad \text{and} \quad a_{n+1} = \frac{1}{3} + 2 \cdot \frac{1}{3} \cdot \frac{1}{2} \cdot a_n \quad \forall n \in \mathbb{N}_0, \tag{2}$$

from which we get

$$a_n = \frac{1}{2} \quad \forall n \in \mathbb{N}_0.$$

Note that at the same result for a_n we can arrive through considerations of symmetry of f_n with respect to the point $(1/2, 1/2)$.

Now let l_n be the length of the graph of the function f_n, where $n \in \mathbb{N}_0$. From Fig. 1(a-c) it is immediate that

$$l_0 = \sqrt{2}, \quad l_1 = \frac{1 + \sqrt{13}}{3}, \quad l_2 = \frac{5 + \sqrt{97}}{9}. \tag{3}$$

To find a formula for l_n we cannot use the same method seen in (2) for a_n. This because the area of a figure changes linearly with respect to both its width (say x-size) and its height (say y-size), but the length of a curve does not. Obviously, the length of a curve does not vary linearly even with respect to the

---

[1] We can obviously specify the number of segments length $1/3$, $1/9$, etc., but this is not relevant in this paper and it is not relevant for the subsequent computation of l_n.

distance between its two extremes. A simple example is obtained by comparing the two ratios

$$\frac{l_1}{\|(1,1)\|} = \frac{1+\sqrt{13}}{3\sqrt{2}} \approx 1.085539$$

and

$$\frac{\lambda}{\left\|\left(\frac{1}{3},\frac{1}{2}\right)\right\|} = \frac{\frac{1}{9} + 2\sqrt{\left(\frac{1}{9}\right)^2 + \left(\frac{1}{4}\right)^2}}{\sqrt{\left(\frac{1}{3}\right)^2 + \left(\frac{1}{2}\right)^2}} = \frac{2+\sqrt{97}}{3\sqrt{13}} \approx 1.095427,$$

where $\|(x,y)\| = \sqrt{x^2+y^2}$ is the Euclidean norm of a vector $(x,y) \in \mathbb{R}^2$ and λ denotes the length of the graph of f_2 moving from $(0,0)$ to $(1/3,1/2)$.

To find a formula for l_n, therefore, we use another strategy. Recalling Remark 1, we know that the graph of f_n is made up by 2^n oblique line segments, i.e. 2^n diagonals of length

$$\sqrt{\left(\frac{1}{3^n}\right)^2 + \left(\frac{1}{2^n}\right)^2}, \qquad (4)$$

and $2^n - 1$ horizontal line segments for a total length equal to

$$1 - 2^n \cdot \frac{1}{3^n} \qquad (5)$$

(just remove from the length of the unit segment $[0,1]$ the projections on the x-axis of the 2^n oblique segments). Therefore, from (4) and (5) we conclude that

$$l_n = 1 - \left(\frac{2}{3}\right)^n + \sqrt{1 + \left(\frac{2}{3}\right)^{2n}}, \qquad (6)$$

for all integers $n \geq 0$. In fact, we note that, for $n = 0, 1, 2$, we recover respectively the values in (3).

3 Highlights Using the Grossone System

Using standard analysis, all we can say on the behavior of the sequence (6) when n approaches infinity is that

$$\lim_{n\to\infty} l_n = \lim_{n\to\infty} 1 - \left(\frac{2}{3}\right)^n + \sqrt{1 + \left(\frac{2}{3}\right)^{2n}} = 2, \qquad (7)$$

and this is consistent with the observation that taking x- and y-projections of the $2^{n+1} - 1$ line segments which constitute the graph of f_n, they go to cover the two segments

$$[0,1] \times \{0\} \qquad \text{and} \qquad \{0\} \times [0,1]$$

without overlays, and the slope of the oblique segments approaches $+\infty$.

In the remainder of this section, as already announced in the Introduction, we will use the grossone-based system introduced by Sergeyev in the early 2000s to say something different and more precise than the limit in (7).

The grossone-based numeral system, roughly speaking, is founded on two different fundamental units: the ordinary unit 1 which give rise to natural numbers, integers, rationals, etc. and a correspondent "infinite unit" ① called *grossone*, which generate a whole range of infinite numbers like

$$① , \quad 2① , \quad -① , \quad -5① , \quad \frac{3}{7}① , \quad -\frac{9}{7}① , \tag{8}$$

and also numbers with a finite and an infinite part like

$$① + 6, \quad 3① - 1, \quad -① + \frac{4}{3}, \quad -5①^2 + 7① - \sqrt{6}, \quad \frac{8}{7}①^4 - \frac{5}{7}① + 63, \tag{9}$$

etc. Introducing ① implies to introduce also infinitesimal numbers, i.e. inverses of infinite numbers, like

$$\frac{1}{①}, \quad ①^{-3/2}, \quad 5①^{-6} + 7①^{-5/2}, \quad ①^{-5/2} - \frac{① + 2}{①^3 - 4① + 4}, \tag{10}$$

and, obviously, additions, multiplications, divisions between elements as in (9) and (10).

Much more details on the grossone-based system can be found in Sergeyev's papers [30,33] or the popular book [29]. In recent years many applications of the grossone system have been found, for example to fractals (see [4,7,8,32]) and *blinking fractals* (see [12,31]), summations, ordering, probability, game theory (see [13–16,27,34] and the references therein), optimization, differential equations, Infinity Computer (see [1,5,17,19]), and many other fields. Links with logic, mathematics foundations, Fibonacci numbers and unimaginable ones can be found in [9–11,21–23]), and very interesting are also a series of recent didactic studies and experimentations in schools about the grossone systems (see [2,3,18,20,24–26,28]).

Applying grossone to Cantor-Vitali function allows us to make mare precise computations. For the areas a_n we have no changes because it is constant for all n. Instead, for the length l_n of the graphic of the approximation function $f_n(x)$, we get a different result depending on the infinite level n. For instance, if $n = ①$ we get from (6)

$$l_① = 1 - \left(\frac{2}{3}\right)^① + \sqrt{1 + \left(\frac{2}{3}\right)^{2①}}, \tag{11}$$

which is strictly less than 2, but it differs from it by an infinitesimal quantity. We can also decrease this infinitesimal difference by considering chained sequences (see [29,30,33]). For example, in virtue of the monotonicity of (6), clear from the geometric discussion made above, we get that

$$l_{5①/2+6} = 1 - \left(\frac{2}{3}\right)^{5①/2+6} + \sqrt{1 + \left(\frac{2}{3}\right)^{5①+12}} \tag{12}$$

is strictly greater than (11) and strictly less than 2. Now it should be clear how it is possible to give precise numerical values not only to infinite quantities, but also to infinitesimal ones. Making comparisons between them is then straightforward through the new system.

Acknowledgments. This research was partially supported by Seeweb s.r.l., Cloud Computing provider based in Frosinone, Italy, and part of DHH Group.

References

1. Amodio, P., Iavernaro, F., Mazzia, F., Mukhametzhanov, M.S., Sergeyev, Y.D.: A generalized Taylor method of order three for the solution of initial value problems in standard and infinity floating-point arithmetic. Math. Comput. Simul. **141**, 24–39 (2017). https://doi.org/10.1016/j.matcom.2016.03.007
2. Antoniotti, L., Astorino, A., Caldarola, F.: Unimaginable numbers and infinity computing at school: an experimentation in Northern Italy. In: Sergeyev, Y.D., Kvasov, D.E., Astorino, A. (eds.) NUMTA 2023. LNCS, vol. 14478, pp. 223–231. Springer, Cham (2025). https://doi.org/10.1007/978-3-031-81247-7_17
3. Antoniotti, L., Caldarola, F., d'Atri, G., Pellegrini, M.: New approaches to basic calculus: an experimentation via numerical computation. In: Sergeyev, Y.D., Kvasov, D.E. (eds.) NUMTA 2019. LNCS, vol. 11973, pp. 329–342. Springer, Cham (2020). https://doi.org/10.1007/978-3-030-39081-5_29
4. Antoniotti, L., Caldarola, F., Maiolo, M.: Infinite numerical computing applied to Peano's, Hilbert's, and Moore's curves. Mediterr. J. Math. **17**, 99 (2020). https://doi.org/10.1007/s00009-020-01531-5
5. Astorino, A., Fuduli, A.: Spherical separation with infinitely far center. Soft. Comput. **24**(23), 17751–17759 (2020). https://doi.org/10.1007/s00500-020-05352-2
6. Bass, R.F.: Real analysis for graduate students, 2nd edn. Createspace Independent Publishing (2013)
7. Caldarola, F.: The exact measures of the Sierpiński d-dimensional tetrahedron in connection with a Diophantine nonlinear system. Commun. Nonlinear Sci. Numer. Simul. **63**, 228–238 (2018). https://doi.org/10.1016/j.cnsns.2018.02.026
8. Caldarola, F.: The Sierpiński curve viewed by numerical computations with infinities and infinitesimals. Appl. Math. Comput. **318**, 321–328 (2018). https://doi.org/10.1016/j.amc.2017.06.024
9. Caldarola, F., Cortese, D., d'Atri, G., Maiolo, M.: Paradoxes of the infinite and ontological dilemmas between ancient philosophy and modern mathematical solutions. In: Sergeyev, Y.D., Kvasov, D.E. (eds.) NUMTA 2019. LNCS, vol. 11973, pp. 358–372. Springer, Cham (2020). https://doi.org/10.1007/978-3-030-39081-5_31
10. Caldarola, F., d'Atri, G., Maiolo, M., Pirillo, G.: New algebraic and geometric constructs arising from Fibonacci numbers. In honor of Masami Ito. Soft Comput. **24(23)**, 17497–17508 (2020). https://doi.org/10.1007/s00500-020-05256-1
11. Caldarola, F., d'Atri, G., Maiolo, M., Pirillo, G.: The sequence of carboncettus octagons. In: Sergeyev, Y.D., Kvasov, D.E. (eds.) NUMTA 2019. LNCS, vol. 11973, pp. 373–380. Springer, Cham (2020). https://doi.org/10.1007/978-3-030-39081-5_32
12. Caldarola, F., Maiolo, M.: On the topological convergence of multi-rule sequences of sets and fractal patterns. Soft. Comput. **24**, 17737–17749 (2020). https://doi.org/10.1007/s00500-020-05358-w

13. Caldarola, F., Maiolo, M., Solferino, V.: A new approach to the Z-transform through infinite computation. Commun. Nonlinear Sci. Numer. Simul. **82**, 105019 (2020). https://doi.org/10.1016/j.cnsns.2019.105019
14. Calude, C.S., Dumitrescu, M.: Infinitesimal probabilities based on grossone. SN Comput. Sci. **1**, 36 (2020). https://doi.org/10.1007/s42979-019-0042-8
15. Cococcioni, M., Cudazzo, A., Pappalardo, M., Sergeyev, Y.D.: Solving the lexicographic multi-objective mixed-integer linear programming problem using branch-and-bound and grossone methodology. Commun. Nonlinear Sci. Numer. Simul. **84**, 105177 (2020). https://doi.org/10.1016/j.cnsns.2020.105177
16. D'Alotto, L.: Infinite games on finite graphs using grossone. Soft. Comput. **24**, 17509–17515 (2020). https://doi.org/10.1007/s00500-020-05167-1
17. Falcone, A., Garro, A., Mukhametzhanov, M.S., Sergeyev, Y.D.: Simulation of hybrid systems under zeno behavior using numerical infinitesimals. Commun. Nonlinear Sci. Numer. Simul. **111**, 106443 (2022). https://doi.org/10.1016/j.cnsns.2022.106443
18. Iannone, P., Rizza, D., Thoma, A.: Investigating secondary school students' epistemologies through a class activity concerning infinity. In: Bergqvist, E., Österholm, M., Granberg, C., Sumpter, L. (eds.) Proceedings of the 42nd Conference of the International Group for the Psychology of Mathematics Education, vol. 3, pp. 131–138. PME, Umeå, Sweden (2018)
19. Iavernaro, F., Mazzia, F., Mukhametzhanov, M.S., Sergeyev, Y.D.: Conjugate-symplecticity properties of Euler'Maclaurin methods and their implementation on the Infinity Computer. Appl. Numer. Math. **155**, 58–72 (2020). https://doi.org/10.1016/j.apnum.2019.06.011
20. Ingarozza, F., Adamo, M.T., Martino, M., Piscitelli, A.: A grossone-based numerical model for computations with infinity: a case study in an Italian high school. In: Sergeyev, Y.D., Kvasov, D.E. (eds.) NUMTA 2019. LNCS, vol. 11973, pp. 451–462. Springer, Cham (2020). https://doi.org/10.1007/978-3-030-39081-5_39
21. Leonardis, A., d'Atri, G., Caldarola, F.: Beyond Knuth's notation for unimaginable numbers within computational number theory. Int. Electron. J. Algebra **31**, 55–73 (2022). https://doi.org/10.24330/ieja.1058413
22. Lolli, G.: Metamathematical investigations on the theory of grossone. Appl. Math. Comput. **255**, 3–14 (2015). https://doi.org/10.1016/j.amc.2014.03.140
23. Margenstern, M.: Fibonacci words, hyperbolic tilings and grossone. Commun. Nonlinear Sci. Numer. Simul. **21**, 3–11 (2015). https://doi.org/10.1016/j.cnsns.2014.07.032
24. Mazzia, F.: A computational point of view on teaching derivatives. Inform. Educ. **37**, 79–86 (2022). https://doi.org/10.32517/0234-0453-2022-37-1-79-86
25. Nasr, L.: The effect of arithmetic of infinity methodology on students' beliefs of infinity. Mediterr. J. Res. Math. Educ. **19**, 5–19 (2022)
26. Nasr, L.: Students' resolutions of some paradoxes of infinity in the lens of the grossone methodology. Inform. Educ. **38**, 83–91 (2023)
27. Pepelyshev, A., Zhigljavsky, A.: Discrete uniform and binomial distributions with infinite support. Soft. Comput. **24**, 17517–17524 (2020). https://doi.org/10.1007/s00500-020-05190-2
28. Rizza, D.: Primi passi nell'Aritmetica dell'Infinito. Bonomo Editore (2023)
29. Sergeyev, Y.D.: Arithmetic of Infinity. Edizioni Orizzonti Meridionali, Cosenza (2003, 2nd ed 2013)
30. Sergeyev, Y.D.: Lagrange Lecture: Methodology of numerical computations with infinities and infinitesimals. Rendiconti del Seminario Matematico dell'Università e del Politecnico di Torino **68**, 95–113 (2010)

31. Sergeyev, Y.D.: Using blinking fractals for mathematical modelling of processes of growth in biological systems. Informatica **22**, 559–576 (2011)
32. Sergeyev, Y.D.: The exact (up to infinitesimals) infinite perimeter of the Koch snowflake and its finite area. Commun. Nonlinear Sci. Numer. Simul. **31**, 21–29 (2016). https://doi.org/10.1016/j.cnsns.2015.07.004
33. Sergeyev, Y.D.: Numerical infinities and infinitesimals: methodology, applications, and repercussions on two Hilbert problems. EMS Surv. Math. Sci. **4**, 219–320 (2017). https://doi.org/10.4171/EMSS/4-2-3
34. Sergeyev, Y., De Leone, R. (eds.): Numerical Infinities and Infinitesimals in Optimization. Springer, Cham (2022)
35. Vestrup, E.M.: The theory of measures and integration. Wiley (2003)

New Probabilistic Methods for Generating Risk Maps

Arrigo Bertacchini[1], Pierpaolo Antonio Fusaro[2(✉)], and Massimo Zupi[3]

[1] Department of Architecture and Industrial Design, University of Campania "Luigi Vanvitelli", Aversa, Italy
arrigo.bertacchini@unicampania.it
[2] Department of Physics, University of Calabria, Rende, Italy
pierpaolo.fusaro@unical.it
[3] Department of DIAm, University of Calabria, Rende, Italy
massimo.zupi@unical.it

Abstract. The year 2022 in Italy registered an increase of 170% in forested and non-forested areas devastated by fire (in 2021, fires affected 159,437 hectares). In detail, the highest number of calls to the fire fighters was recorded in Sicily and Calabria, with 301 and 223 requests respectively. The Calabria Region, with 35,480 hectares of forest area destroyed in the year 2021, is a strategic study area for conducting an in-depth analysis of how fires are spread and distributed. In light of this, the present work proposes to estimate the fire risk in the region using a non-parametric statistical technique called Smooth Kernel Distribution (SKD). This technique allows us to estimate the probability density function of a continuous random variable based on a sample dataset of the temporal sequence and geospatial location of fires in the region over the last decade. The main potential of the adopted method is that the SKD allows the creation of risk maps based on historical data, identifying high-risk areas and ensuring the development of targeted preventive actions. The preliminary results obtained in the study area show that the SKD is a powerful tool for analysing fire risk. The identified risk maps show the areas in the region with the highest fire risk, enabling the competent authorities to take preventive measures to avoid further damage. Furthermore, the research conducted has shown that the use of the SKD in combination with other statistical techniques, such as multivariate analysis, can provide a more comprehensive understanding of fire risk in the region (intensity of the fire, vegetation maps, terrain orography).

Keywords: Smooth Kernel Distribution · Fire · Risk map

1 Introduction

Fires pose a significant threat to public safety worldwide, and are responsible for causing numerous deaths, property damage, and environmental impacts [12, 13]. In addition to affecting entire natural ecosystems in herbaceous-dominated regions with human activities, fires are a major contributor to greenhouse gas emissions and are a key driver of global climate change [14]. Changes in global climate and weather patterns can alter vegetation growth and aridity, influencing fire activity [5,15].

EFFIS, the European Forest Fire Information System, monitored fires in 45 countries worldwide in 2022. These countries experienced 16,941 fires, which burned 1,624,381 hectares, representing a 4% increase in the area burned and a 48% increase in the total number of fires between 2021 and 2022. In Europe, EFFIS estimated that fires in 2022 caused a total burnt area of 837,212 hectares, which is an 86% increase compared to 2021. In Italy, the year 2022 witnessed a 170% increase in wooded and non-wooded areas devastated by fires (in 2021 fires affected 159,437 hectares, which was 154.8% more than in 2020) [6,10,16]. These figures reveal that the fire risk is particularly high both in Italy and in the rest of Europe and the world.

Understanding the spatial distribution of fire ignition and the environmental and human factors that influence it is crucial for decision-making in fire management [8,17,21].

Fire risk mapping represents a suitable tool for studying the problem and mitigating the negative impacts of fire [1]. Typically, the spatial distribution of fires is modeled continuously, but the precision of the recorded x and y coordinates of ignition points may be lacking [8,11]. In this study, kernel density estimation was used to estimate the probability density of the fire distribution in the study area and create a map that identifies areas with a high risk of fire. The potential of this methodology is to assist competent authorities in taking targeted preventive measures to avoid the risks that fires can cause in the future.

2 Material

2.1 Study Area

The region of Calabria is the southernmost region of Italy, located at a latitude of 37°48' - 40°0' N and a latitude of 15°24' - 17°12' S. Calabria has a population of about 1.9 million inhabitants and is divided into five provinces Cosenza, Crotone, Catanzaro, Vibo Valentia and Reggio Calabria. It covers an area of 15,081 km², with altitudes varying from 2200 m to sea level, and is mainly characterised by mountains, 39.4%, and hills, 36.1%. The remaining part of the territory consists of plains, 6.9%, and other forms of land cover 17.6%. The vegetation is wide and varied, including woods, maquis, pine forests, grasslands and wetlands. In particular, woods make up 24% of the land, while Mediterranean scrub and pine forests cover 16 and 17%. Precipitation is much more frequent in autumn and winter, with average values varying between 700 and 1200 mm. Temperatures vary greatly depending on the geographical area, but on average, summer highs reach 30°, while in winter they can drop as low as 0°. The average wind speed is moderate to high in both winter and summer. Only in 2021 in Calabria, 35,480 hectares of wooded and non-wooded areas were destroyed by fire, with a total number of fires of 191 [6]. For the regional plan for the prevention and active fight against fires, the Region of Calabria draws up the "Map of potential forest fire risk" based on the "Map of probable ignition zones". This map is drawn up by analysing the points of origin and the areas affected by fire, which are taken care of by the Regional Corps of Forest Rangers [23] (Fig. 1).

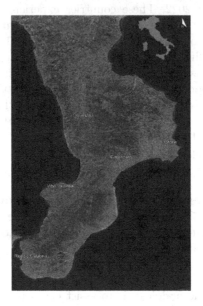

Fig. 1. Study area (Calabria region, southern Italy) and distribution of ignition point from 2012–2021.

2.2 Ignition Point Database

NASA's FIRMS, Fire Information for Resource Management System, is a powerful tool used to monitor forest fires worldwide and provides real-time information on active fires and has a historical archive of past fires. Among the sources used by FIRMS are MODIS satellite sensors, Moderate Resolution Imaging Spectroradiometer, which has a spatial resolution of 250 m, 500 m, and 1 km, with a detection wavelength ranging from 0.4 to 14 4 micrometres, and the Active Fire product of the Visible Infrared Imaging Radiometer Suite (VIIRS/S-NPP) with a spatial resolution of 375 m and 750 m, and a spectral bandwith of 16 bands between 0.4 and 12.5 micrometres reprocessed by the Land Science Investigator Processing System, in addition to the ground observation networks [7,18]. The study used highly accurate data from the FIRMS VIIRS/S-NPP instrumentation. These data were processed in the cross-platform symbolic and numerical calculation environment Wolfram Mathematica.

2.3 Kernel Density Estimates

The Smooth Kernel Distribution is a non-parametric estimation technique utilized for the analysis in various academic disciplines. It enables the estimation of a continuous surface with its density function [12]. It can be applied to estimate the density of fire points, the temporal spread of Sars-CoV-2, space utilization by animals, and the peak frequency of neural pulses [2,3,11,19,20].

The kernel density function is defined as [19]:

$$f(x) = \frac{1}{nh^2} \sum_{i=1}^{n} K\{\frac{s-s_i}{h}\} \quad (1)$$

where n is the total number of observed events, K is the kernel function, h is the bandwidth that produces a density estimate, s is the position at which the density estimate is calculated and s_i is the position of each observed event. The kernel function is used to 'smooth' the distribution of the data, providing a continuous estimate of the distribution underneath. Different kernel functions exist, in this study the Gaussian kernel function was used [4].

The density estimate is affected by the bandwidth parameter, which determines the degree of smoothing applied to the data, and subsequently, the accuracy of the fit. A narrower bandwidth produces a more detailed density estimate, while a wider bandwidth produces a less detailed estimate. [9] Silverman's rule was used as the band length [22].

The Smooth Kernel Distribution is a significant statistical data analysis tool, with applications in constructing continuous histograms and creating density plots.

3 Methods

The methodological steps that were followed to collect the data, analyse the results and visualise them are described below. Data on the number of fires from 1 January 2012 until 31 December 2021 are considered.

The parameters considered for analysing the fires were:

- *Lat* - Fire Latitude Coordinates;
- *Long* - Fire Longitude Coordinates;

To process and analyse the data we have used the Mathematica scientific calculation system that contains a series of useful methods and tools [24].

Here is an overview of the functions used in Mathematica:

- *GeoSmoothHistogram* - is a function to create geographical histograms by combining the geographical distribution of the data with the distribution histogram, providing a visual representation of the probability density of the data on a map [Fig. 2];

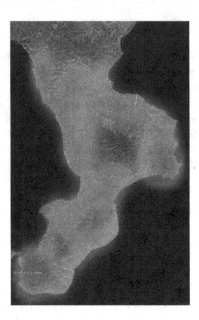

Fig. 2. GeoSmoothHistogram.

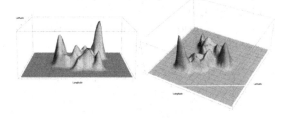

Fig. 3. Smooth Kernel Distribution - Plot 3D.

- *SmoothKernelDistribution* - is a non-parametric estimation function used to estimate the probability density of an unknown continuous variable, based on a sample of data. This function visualised in 3D returns a continuous density surface with peaks in the areas with the highest concentration of fires in the area under consideration, estimating the probable presence of ignition points even in areas that were not actually affected [Fig. 3];
- *ContourPlot* - is a graphical representation of a function of two variables that uses contour lines to show the variations of the function in the plane. This function represents the set of curves that connect the points at which the function has the same value, thus creating curved lines that follow the contour lines of the function, making it easy to identify the function's maximum and minimum zones;
- *GeoGraphics* - this is a function used to display geographical data on a map of the area being studied; [Fig. 4].

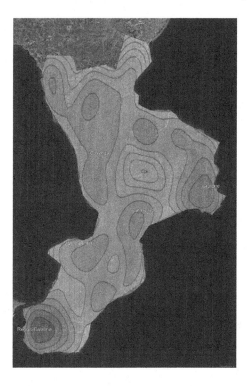

Fig. 4. Fire risk map.

4 Conclusions

Fire risk is a complex and challenging phenomenon to predict, requiring a considerable amount of information. Currently, fire risk maps do not meet the requirements necessary to implement effective preventive measures, as they are based on outdated methodologies, such as the analysis of fire-affected areas. The methodology of this study considers not only the areas effectively affected by fire, but also a spatial distribution that estimates the probability density of the distribution. Therefore, the fire risk map generated with the Smooth Kernel Distribution could be more accurate and detailed than one based solely on burnt area. Moreover, it could estimate the hazard risk in areas that have not yet been affected by fire. This study provides valuable insights into fire risk assessment. The spatial distribution of real fires made it possible to identify high-risk areas through the probability distribution of the real data. The inclusion of additional information, such as vegetation, soil type, weather conditions, infrastructure and human activities, could further improve the accuracy of the map and increase the effectiveness of policies. In future research, it is expected that this methodology will be implemented using spatial information within predictive models to predict fire behaviour and progression based on historical and real time data. In addition, the use of ignition data from NASA's FIRMS system provides a real time representation of fire danger. This data is used to dynamically update the map, facilitating accurate and timely risk assessment, unlike conventional fire hazard maps that are prepared on an annual basis. The dynamic maps provide continuous and current monitoring of prevailing conditions, enabling competent authorities and emergency responders to take more immediate and effective prevention and response measures.

Foundings

This research was granted by Next Generation UE - PNRR "Tech4You Project" funds assigned to University of Calabria (PP1.4.3 - Probabilistic space-time models for forest fire spreading, Scientific Coordinator: Professor Pietro Pantano).

References

1. Abedi Gheshlaghi, H.: Using GIS to develop a model for forest fire risk mapping. J. Indian Soc. Remote Sens. **47**(7), 1173–1185 (2019)
2. Benhamou, S., Cornélis, D.: Incorporating movement behavior and barriers to improve kernel home range space use estimates. J. Wildl. Manag. **74**(6), 1353–1360 (2010)
3. Bertacchini, F., Bilotta, E., Pantano, P.S.: On the temporal spreading of the SARS-COV-2. PLoS ONE **15**(10), e0240777 (2020)
4. Botev, Z.I., Grotowski, J.F., Kroese, D.P.: Kernel density estimation via diffusion (2010)
5. Bradstock, R.A.: A biogeographic model of fire regimes in Australia: current and future implications. Glob. Ecol. Biogeogr. **19**(2), 145–158 (2010)
6. Fontana, E., Morabito, A., Nicoletti, A.: L'Italia in fumo, Gli incendi del patrimonio naturale, i fattori di rischio e le proposte di Legambiente. Technical report, Legambiente (2022)
7. Giglio, L., Schroeder, W., Hall, J.V., Justice, C.O.: Modis collection 6 active fire product user's guide revision a. Department of Geographical Sciences. University of Maryland, vol. 9 (2015)
8. Hao, W.M., Liu, M.H.: Spatial and temporal distribution of tropical biomass burning. Global Biogeochem. Cycles **8**(4), 495–503 (1994)
9. Jones, M., Kappenman, R.: On a class of kernel density estimate bandwidth selectors. Scand. J. Stat. 337–349 (1992)
10. Kok, E., S.C.: Country report for the Netherlands, in san-miguel-ayanz et al. (eds), forest fires in europe, middle east and north africa 2021. Technical report, Publications Office of the European Union (2022)
11. Koutsias, N., Kalabokidis, K.D., Allgöwer, B.: Fire occurrence patterns at landscape level: beyond positional accuracy of ignition points with kernel density estimation methods. Nat. Resour. Model. **17**(4), 359–375 (2004)
12. McCaffrey, S.: Thinking of wildfire as a natural hazard. Soc. Nat. Resour. **17**(6), 509–516 (2004)
13. Moritz, M.A., et al.: Learning to coexist with wildfire. Nature **515**(7525), 58–66 (2014)
14. Oom, D., Pereira, J.M.: Exploratory spatial data analysis of global modis active fire data. Int. J. Appl. Earth Obs. Geoinf. **21**, 326–340 (2013)

15. Richardson, D., et al.: Global increase in wildfire potential from compound fire weather and drought. NPJ Clim. Atmos. Sci. **5**(1), 23 (2022)
16. San-Miguel-Ayanz, J., et al.: Forest fires in europe, middle east and north africa 2021. Technical report, Publications Office of the European Union (2022)
17. Schneider, P., Roberts, D., Kyriakidis, P.: A vari-based relative greenness from modis data for computing the fire potential index. Remote Sens. Environ. **112**(3), 1151–1167 (2008)
18. Schroeder, W., Giglio, L.: Visible infrared imaging radiometer suite (VIIRS) 375 m\& 750 m active fire detection data sets based on nasa viirs land science investigator processing system {(SIPS)} reprocessed data-version 1 product user's guide version 1.2 (2017). https://lpdaac.usgs.gov/documents/132
19. Shimazaki, H., Shinomoto, S.: Kernel bandwidth optimization in spike rate estimation. J. Comput. Neurosci. **29**, 171–182 (2010)
20. Shuo, Z., Jingyu, Z., Zhengxiang, Z., Jianjun, Z.: Identifying the density of grassland fire points with kernel density estimation based on spatial distribution characteristics. Open Geosci. **13**(1), 796–806 (2021). https://doi.org/10.1515/geo-2020-0265
21. y Silva, F.R., Martínez, J.R.M., González-Cabán, A.: A methodology for determining operational priorities for prevention and suppression of wildland fires. Int. J. Wildland Fire **23**(4), 544–554 (2014)
22. Silverman, B.W.: Density estimation for statistics and data analysis, vol. 26. CRC Press (1986)
23. Piano regionale per la prevenzione e lotta attiva agli incendi boschivi anno 2022. Technical report, Regione Calabria (2022)
24. Wolfram, S.: An elementary introduction to the wolfram langauge (2015)

How to Deal with Different Densities of Urban Spatial Data? A Comparison of Clustering Approaches to Detect City Hotspots

Eugenio Cesario[1], Paolo Lindia[1(✉)], and Andrea Vinci[2]

[1] University of Calabria, 87036 Rende, CS, Italy
eugenio.cesario@unical.it,paolo.lindia@dimes.unical.it
[2] ICAR-CNR, 87036 Rende, CS, Italy
andrea.vinci@icar.cnr.it

Abstract. In the field of urban data analysis, the detection of city hotspots is becoming a fundamental activity aimed at showing functions and roles played by each city area and providing valuable support for policymakers, scientists, and planners. However, since metropolitan cities are heavily characterized by variable densities, multi-density clustering algorithms might be more reliable than classic approaches to discover proper hotspots from urban data. This paper presents a study on hotspots detection in urban environments, by comparing two approaches, i.e., single-threshold and multi-density threshold ones, for clustering urban data. The experimental evaluation, carried out on a synthetic state-of-the-art multi-density dataset, shows that a multi-density approach achieves higher clustering quality than classic techniques.

Keywords: Urban data mining · Multi-density clustering · Smart City

1 Introduction

Nowadays, massive amounts of geo-referenced urban data are daily collected in metropolitan areas [4,15]. The availability of such a large amount of data can be effectively exploited to discover knowledge models, which can support city managers to takle the main challenges and issues our cities deal with every day. In particular, the identification of *city hotspots*, i.e. areas where urban events (crimes, electric peaks, traffic spikes, viral infections, pollution peaks, etc.) occur more frequently than in other parts of the city [1,3,11,13], is particularly helpful in framing urban territories across a variety of scales. In fact, the shape, area, and borders of each hotspot provide high-level spatial knowledge summaries to functions and roles played by each city area, which is a valuable knowledge for decision-makers, researchers, and planners [7].

Among the techniques proposed in literature, density-based clustering is the most appropriate methodology to discover urban hotspots, whose approaches may be divided into two major groups. The first group consists of algorithms (such as DB-Scan) that, as a result of the use of global parameters, compute a single minimal threshold value to identify dense (and not dense) regions [2,6,8]. On the other side, the second group includes algorithms that compute multiple minimum threshold values. These algorithms try to differentiate between various density zones that may or may not be nested and are often non-convex in shape. They typically identify numerous pattern distributions of varied densities. From an applicative viewpoint, the application of multi-density clustering algorithms to urban data can produce better results than using traditional state-of-the-art techniques. In fact, events occurring in metropolitan areas may be very different in terms of density, due to the huge variations in population, traffic, and event density from one location to the next. This has been examined in various recent studies [7,9,12], which assert that there are significant differences between the regions of cities in terms of inter-city and intra-city densities. For such a reason, the application of multi-density clustering algorithms to urban data can be more effective than using traditional approaches.

This paper presents a study on hotspots detection in urban environments, by comparing two approaches, i.e., single-threshold and multi-density threshold ones, for clustering urban data. In order to assess the clustering quality and the ability of the algorithms to detect accurate hotspots, an experimental assessment on a state-of-the-art multi-density dataset has been conducted. To do this, we use a labeled dataset comprising target clusters, which are taken into account as ground truth clusters during the evaluation process. By utilizing such ground truth information, the algorithms may be assessed both subjectively and quantitatively.

The remainder of the paper is structured as follows. Section 2 introduces the density-based clustering algorithms exploited in this paper to analyze multi-density data. Section 3 provides the comparative experimental evaluation of the two approaches considered in this paper. Finally, Sect. 4 concludes the paper and plans future research works.

2 Description of the Algorithms

The two density-based clustering techniques (DBSCAN [8] and CHD [7]) we selected from the literature, are briefly explained in this section.

The well-known algorithm for finding general spatial clusters is DBSCAN [8] (density-based spatial clustering of applications with noise), a density-based clustering algorithm based on the definitions of *core-points* and *density reachability*. Also, DBSCAN can differentiate noise points within a dataset and can identify clusters of various shapes without prior knowledge about the expected number of clusters. A point is considered to be a *core point* if it is surrounded by at least a certain number of other points ($minPts$), within a certain radius ($epsilon$). A point p is *density-reachable* from a core point if it is either immediately reachable from the core point, or if a path connecting the core point and the specified point contains directly reachable points. Lastly, two points, p and q, are considered *density-connected* if there exists a *core point* cp, such that both p and q

are *density-reachable* from *cp*. By combining data points that are connected to one another, DBSCAN creates clusters. Outliers, or noise points, are defined as points that are not reached from any core point. It is important to highlight that based on the above definitions, DBSCAN is capable of identifying clusters that exhibit a predetermined density d, determined directly by the ϵ and *minPts* parameters. Consequently, DBSCAN may not be able to detect clusters that possess varying densities, as it relies on a fixed density threshold.

A multi-density clustering technique named City Hotspot Detector (CHD) was proposed in [7], specifically designed to analyze urban spatial data. The algorithm is made up of the following phases. First, given a fixed *min_pts*, the reachability distance for each point is calculated and used as an estimator of the density of each data point. This reachability distance serves as an estimate of the density for each point. The points are then ordered by the estimated density, and the density variation between each consecutive pair of points in the ordered list is computed. A rolling mean operator on windows of size s is used to smooth the resultant density variation list. On the basis of the smoothed density variations, the points are subsequently divided into several density level sets, and a different *epsilon* value is estimated for each level set. As an end step, the DBSCAN algorithm is used to analyze each density level set. Each DBSCAN instance specifically accepts as input a certain *epsilon* value computed for the examined density level set. The CHD algorithm's final output is the set of clusters found for each partition [5].

3 Experimental Evaluation and Results

In order to assess the ability of the clustering algorithms (*CHD* and *DBSCAN*, described in Sect. 2) to analyze datasets characterized by areas with different densities, here we present a comparative analysis of the two approaches.

Data Description. Zhan Compound, a cluster-labeled dataset that can be found in the literature [14] and whose data instances and target clusters are shown in Fig. 1, is used as the basis for the comparison analysis. The dataset consists of 399 instances that are separated into six target groups using X and Y variables (Fig. 1). The clusters have a variety of sizes and densities, as well as irregular multi-geometric forms.

Table 1 reports some details about the composition of each cluster: number of points, area and density. The table shows that the highest density cluster has a density of 6.19 (cluster n. 6), while the lowest density cluster has a density of 0.21 (cluster n. 1).

Results and Performance Evaluations. We compare the results of the cluster analysis to the ground truth labels that are supplied by the datasets, in order to evaluate the performance of the selected clustering methods over the datasets previously mentioned. By comparing the identified clusters with the provided target clusters, the efficacy of the clustering algorithms may be assessed. The following set of external metrics [10] have been adopted in order to achieve

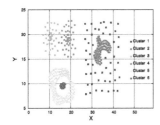

Fig. 1. The *Comound* dataset.

Table 1. Target clusters descriptions

Cluster	# points	Area	Density
Cluster 1	50	236.60	0.21
Cluster 2	92	19.93	4.76
Cluster 3	38	35.29	1.08
Cluster 4	45	37.05	1.21
Cluster 5	158	66.95	2.36
Cluster 6	16	2.58	6.19

this: Gamma, Rand, Adjusted Rand, Homogeneity and Fowlkes. All metrics can assume values between 0 and 1, with high values indicating a good match between the found and target clusters, and low values indicating the presence of more incorrectly assigned items.

It is crucial to remember that the choice of input parameters directly influences the level of quality of the outcomes for any clustering technique. As a result, the input parameters must be carefully selected in relation to the examined dataset in order to properly compare the clustering methods. We used a parameter sweeping process, which involves running many instances of each algorithm while experimenting with various parameter values, to identify the most suitable input parameter setting. The parameter values for each algorithm that produce the best average performance-calculated as the average of the metrics described above-are chosen. For $\omega* = 2.50$, $min\_pts = 64$, and $s = 1$, we find that CHD produces the best results, whereas DBSCAN performs best for $\epsilon* = 1.53$ and $min\_pts = 64$.

The metrics mentioned above can be utilized to compare clustering algorithm performance outcomes based on accurate and measurable standards.

Figure 2a compares the quantitative performance of the algorithms under consideration by using the run with the optimal set of input parameters to display the values of the indices. The figure shows that the *CHD* algorithm achieves better results than *DBSCAN* for all indexes. This result shows how the multi-density approach adopted by *CHD* allows to detect higher quality cluster than *DBSCAN* in multi-density contexts.

Finally, Figs. 2b and 2c show the clustering models found by the two algorithms on a qualitative level. While both algorithms achieve good custers separability, as can be seen in the figure, *DBSCAN* was unable to identify the huge low-density cluster (cluster 1 in Fig. 2b), classifying it as noise. Instead, *CHD* detects that cluster.

Table 2 reports some details about the composition of clusters for each algorithm (*CHD* and *DBSCAN*). In particular table shows the number of points, the area and the density. What is shown in the table confirms what we had seen above in Fig. 2, both algorithms achieve good clusters separability, but DBSCAN fails to detect clusters with low density. This confirms what was reported in Sect. 2, DBSCAN is able to identify clusters that have a specific density, which is set

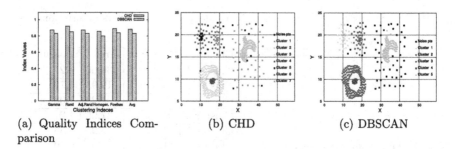

(a) Quality Indices Comparison (b) CHD (c) DBSCAN

Fig. 2. (a) Clustering quality indices vs. different input parameter values. (b) (c) The clusters detected by the two algorithms

by the *epsilon* and *minPts* parameters. As a result, since DBSCAN relies on a single density threshold, it's not able to detect clusters that have different densities.

Table 2. CHD and DBSCAN clusters exploratory metrics

Cluster	CHD			DBSCAN		
	# points	Area	Density	# points	Area	Density
Cluster 1	17	4.66	3.65	34	28.72	1.18
Cluster 2	91	19.74	4.61	42	30.79	1.36
Cluster 3	157	66.84	2.35	158	66.95	2.36
Cluster 4	16	1.92	8.33	16	1.92	8.33
Cluster 5	10	1.85	5.41	94	24.27	3.87
Cluster 6	46	63.30	0.72	-	-	-
Cluster 7	41	116.09	0.35	-	-	-
Noise Points	21	-	-	55	-	-

4 Conclusions

This paper presented a study on hotspots detection in urban environments. Specifically, a comparative analysis between single-threshold and multi-density threshold algorithms for clustering spatial data is shown, in order to sketch the main differences of these two approaches when they are applied to identify hotspots in urban environments. The experimental evaluation, carried out on a synthetic state-of-the-art multi-density dataset, has shown that multi-density clustering approaches outperform classic techniques to retrieve proper hotspots. In future work, other research issues will be investigated. For example, we will further explore the application of these algorithms on specific real-world domains, such as crime and pandemic scenarios. In addition, we will work on a parallel implementation of the approach, to overtake some scalability issues that we have experienced during our tests.

References

1. Altomare, A., Cesario, E., Vinci, A.: Data analytics for energy-efficient clouds: design, implementation and evaluation. Int. J. Parallel Emergent Distrib. Syst. **34**(6), 690–705 (2019)
2. Ankerst, M., Breunig, M.M., Kriegel, H.P., Sander, J.: Optics: ordering points to identify the clustering structure. In: ACM Sigmod Record, vol. 28, pp. 49–60. ACM (1999)
3. Canino, M.P., Cesario, E., Vinci, A., Zarin, S.: Epidemic forecasting based on mobility patterns: an approach and experimental evaluation on COVID-19 data. Soc. Netw. Anal. Min. **12**(1), 116 (2022)
4. Cesario, E.: Big data analysis for smart city applications. In: Sakr, S., Zomaya, A.Y. (eds.) Encyclopedia of Big Data Technologies. Springer, Cham (2019)
5. Cesario, E., Lindia, P., Vinci, A.: Detecting multi-density urban hotspots in a smart city: approaches, challenges and applications. Big Data Cogn. Comput. **7**(1) (2023). https://doi.org/10.3390/bdcc7010029. https://www.mdpi.com/2504-2289/7/1/29
6. Cesario, E., Talia, D.: Distributed data mining patterns and services: an architecture and experiments. Concurr. Comput. Pract. Exp. **24**(15), 1751–1774 (2012)
7. Cesario, E., Uchubilo, P.I., Vinci, A., Zhu, X.: Multi-density urban hotspots detection in smart cities: a data-driven approach and experiments. Pervasive Mob. Comput. **86**, 101687 (2022)
8. Ester, M., Kriegel, H.P., Sander, J., Xu, X., et al.: A density-based algorithm for discovering clusters in large spatial databases with noise. In: KDD, vol. 96, pp. 226–231 (1996)
9. Nelson, G.D.: What micro-mapping a city's density reveals (2021). https://www.citylab.com/perspective/2019/07/urban-density-map-city-population-data-geography/591760/
10. Jain, A.K., Dubes, R.C.: Algorithms for clustering data. Prentice-Hall Inc. (1988)
11. Mastroianni, C., Cesario, E., Giordano, A.: Efficient and scalable execution of smart city parallel applications. Concurr. Comput. Pract. Exp. **30**(20) (2018)
12. Tayebi, M., Ester, M., Glasser, U., Brantingham, P.: Crimetracer: activity space based crime location prediction. In: 2014 IEEE/ACM International Conference on Advances in Social Networks Analysis and Mining (ASONAM), pp. 472–480 (2014)
13. Yuan, J., et al.: T-drive: driving directions based on taxi trajectories. In: GIS 2010 (2010)
14. Zahn, C.T.: Graph-theoretical methods for detecting and describing gestalt clusters. IEEE Trans. Comput. **100**(1), 68–86 (1971)
15. Zheng, Y., Capra, L., Wolfson, O., Yang, H.: Urban computing: concepts, methodologies, and applications. ACM Trans. Intell. Syst. Technol. **5**(3), 38:1–38:55 (2014)

The Impact of Vectorization on the Efficiency of a Parallel PIC Code for Numerical Simulation of Plasma Dynamics in Open Trap

Igor Chernykh[(✉)], Igor Kulikov, Vitaly Vshivkov, Anna Efimova, Dmitry Weins, Ivan Chernoshtanov, and Marina Boronina

Institute of Computational Mathematics and Mathematical Geophysics, Siberian Branch of the RAS, 6 Ac. Lavrentieva ave., Novosibirsk, Russia
{chernykh,kulikov,vsh,efimova,boronona}@ssd.sscc.ru, vins@sscc.ru,
cherivn@ngs.ru

Abstract. Last decades, we have seen a growing interest in thermonuclear fusion. There are a lot of installation prototypes created for controlled thermonuclear fusion. Our group developing software for numerical simulation of plasma physics processes in an open magnetic trap, which can be used for controlled thermonuclear fusion. This code is Fortran-based code. Parallel implementation using MPI for parallel computations. This parallel implementation is optimized for using CPUs with advanced vectorization instructions such as AVX2/AVX512. In this paper, we will show some optimization techniques for maximizing the performance of our parallel Particle-in-Cell (PIC) code using data alignment. This feature helps the compiler to optimize data and cycles in code during the compilation and building of the program. Also, we will show the performance comparison between optimized and nonoptimized programs and the effect of vector instructions utilization on the total code performance.

Keywords: high performance computing · data structure optimization · massive parallel system · performance optimization

1 Introduction

The Particle-in-Cell method was created in the 1960s by Hockney and Eastwood [1], and at this moment this is one of the most widely used methods for solving problems of collisionless particle dynamics [2,3]. This method has high physical clarity, ease of implementation, and cost-effectiveness in comparison with finite-difference methods for calculating the distribution function of particles. One of the main features of the PIC method is the dependence of its accuracy on the number of particles in the cell. Therefore, to obtain a more accurate solution, a larger number of particles is required [4] with more processing time. At the same time, the difficulties of parallelization of the particle-in-cell method are associated with the movement of particles between grid cells, as well as with the transition from the grid to the particle (calculation of the force acting on the ion) and vice versa (calculation of the charge density and average ion velocities). At the moment, there is no single approach to optimizing the particle method. Abundance of existing codes such as OSIRIS [5], VORPAL [6], WARP [7], Smilei [8], including GPU codes PIConGPU [9], PICADOR [10], as well as different vectorization approaches for PIC codes [11,12] and their focus on specific tasks, taking into account their characteristics underlines the relevance of our project.

2 Mathematical Model

In this section, we will give a brief description of the mathematical model and numerical method. The problem of plasma dynamics in an open trap is based on the cylindrical trap sizes $[R_{max} \times L_{max}]$ with the magnetic field H. The fully ionized hydrogen plasma with zero temperature and density n_0 at the initial time $t = 0$ is inside the trap. The injection of the particles into the trap starts at the time $t = 0$ from its center. The injection of the ions and electrons at the injection point is typical for this kind of problem. The self-consistent non-linear interaction of the injected beam with the plasma and the magnetic field of the trap is the subject of research.

We assume the axial symmetry of the problem, and the plasma quasi-neutrality $(n_i = n_e = n)$. The massless electron component can be obtained from the following equations:

- time: $t_0 = 1/\omega$, where $\omega = \omega_{ci} = eH_0/cm_i$ is the ion gyrofrequency;
- length: $L_0 = c/\omega_{pi}$, where $\omega_{pi} = \sqrt{4\pi n_0 e^2/m_i}$ is the ion plasma frequency;
- velocity: the Alfven velocity $V_A = H_0/\sqrt{4\pi m_i n_0}$.

The motion equations:

$$\frac{d\boldsymbol{r}_i}{dt} = \boldsymbol{v}_i, \quad (1)$$

$$\frac{d\boldsymbol{v}_i}{dt} = \boldsymbol{F}_i, \quad (2)$$

where $\boldsymbol{v}_i$ are the ion velocities, and $\boldsymbol{r}_i$ are the ion coordinates.

$\boldsymbol{F}_i = \boldsymbol{E} + [\boldsymbol{v}_i, \boldsymbol{H}] - \kappa \boldsymbol{j}/n$ is the combination of the electromagnetic force with the force of the friction between the ions and the electrons.

The velocities of the electrons $\boldsymbol{V}_e$ can be obtained from the equation:

$$\boldsymbol{j} = (\boldsymbol{V}_i - \boldsymbol{V}_e)n. \quad (3)$$

The distribution function of the ions $f_i(t, r, v)$ defines their density n_i and mean velocity V_i:

$$n(r) = \int f_i(t, r, v) dv, \qquad (4)$$

$$V_i(r) = \frac{1}{n(r)} \int f_i(t, r, v) v \, dv. \qquad (5)$$

The electric field E is defined by the following equation:

$$E + [V_e, H] + \frac{\nabla p_e}{2n} - \kappa \frac{j}{n} = 0. \qquad (6)$$

The electric field E and magnetic field H are defined from Maxwell's equations:

$$\frac{\partial H}{\partial t} = -rot E, \qquad (7)$$

$$rot H = j. \qquad (8)$$

The temperature T_e is defined by the following equation:

$$n\left(\frac{\partial T_e}{\partial t} + (V_e \nabla) T_e\right) = (\gamma - 1)\left(2\kappa \frac{j^2}{n} - p_e div V_e\right) \qquad (9)$$

for $\gamma = 5/3$ and the electron pressure $p_e = nT_e$.

The detailed description of the numerical method and some tests with simulation results can be obtained from [13–15].

3 Data Structure and Performance Impact of Vectorization

We used compute nodes from the NKS-1P system of the Siberian Supercomputer Center for our simulation. Each node has two Intel Xeon Scalable 2nd Gen 6248R (24 cores, 3GHz) CPUs with 192GB of DRAM4 memory. These CPUs support AVX-512 [16] instructions. We used Intel oneAPI [17] base and HPC software tools as well as Intel Fortran Compiler from the Intel Parallel Studio 2019XE. The difference between these versions of the Intel Fortran compiler shows more than 30% performance boost for our PIC code. We compiled our code on Intel Parallel Studio 2019XE with and without data structuring for optimal compilers auto-vectorization [18–20]. For the comparison, we compiled our PIC code on Intel oneAPI software tools without manual data aligning. The same data alignment part of the code can't be compiled with Intel oneAPI because of the Intel Fortran Compiler restrictions.

The next listing 1.1 shows the main data structure and manual data alignment technique for our PIC code. This data alignment helps the compiler to optimize data for auto-vectorization.

Listing 1.1. Main data structure and manual data alignement for PIC code

```
type coo
  real*8, allocatable :: r(:)
  real*8, allocatable :: z(:)
  real*8, allocatable :: u(:)
  real*8, allocatable :: v(:)
  real*8, allocatable :: w(:)
  real*8, allocatable :: a(:)
  real*8, allocatable :: it(:)
  dir$ attributes align:64 :: r
  dir$ attributes align:64 :: z
  dir$ attributes align:64 :: u
  dir$ attributes align:64 :: v
  dir$ attributes align:64 :: w
  dir$ attributes align:64 :: a
end type coo
```

For the performance evaluation, we used the Intel Advisor software tool from Intel oneAPI 2023. This tool shows the roofline model for the PIC code. This model shows the total performance as well as the performance of each function. We can compare the performance of the main functions and the total performance of our PIC code with different compilation parameters and optimizations. We used the Fortran compiler option $-xsacsadelake$ for building AVX512 optimized code and $-xsse4.2$ option for building nonAVX512 code. Figure 1 shows roofline analysis for the same version of code with data alignment from listing 1.1 with AVX512 auto-vectorization compiler parameters (bottom picture) and without these optimizations (top picture). The total performance of code without AVX512 compiler options and data alignment is equal to 1.23 GFLOPS. And the total performance of this code with AVX512 is equal to 1.82 GFLOPS. Data alignment with advanced vectorization gave a performance boost of up to 48%. As we said before, we can't compile our PIC code with data alignment using the Intel oneAPI Fortran compiler, because of the language restrictions on this structure type. Figure 2 shows the performance comparison between our PIC code compiled without data alignment and with AVX512 auto-vectorization of Intel Fortran Compiler from Intel Parallel Studio 2019XE (bottom picture) and Intel oneAPI 2021 (top picture). The total performance of the code compiled with the old compiler is equal to 1.24 GFLOPS, and for the Intel, oneAPI compiler is equal to 2.66 GFLOPS.

The newer compiler did better data alignment and optimizations without any manual help. We achieved a 2.1× performance boost of the PIC code compiled with Intel oneAPI tools, and a 1.46× performance boost in comparison with

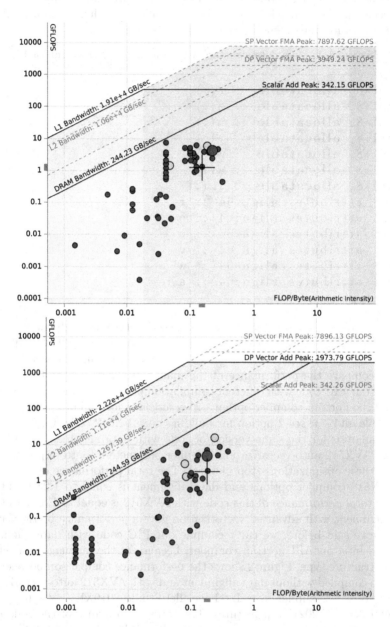

Fig. 1. Roofline analysis results for the PIC code on compute node with two Intel Xeon Scalable 6248R CPUs. The top picture shows the performance and arithmetic intensity for the PIC code without AVX512 compiler auto-vectorization options. The bottom picture shows the same data for AVX512 compiler auto-vectorization.

Vectorization of a Parallel PIC Code 259

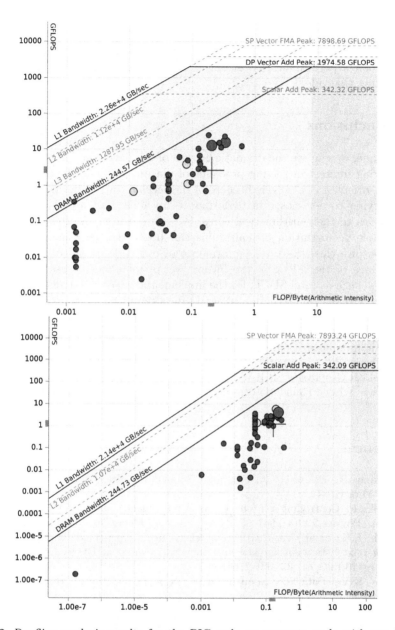

Fig. 2. Roofline analysis results for the PIC code on compute node with two Intel Xeon Scalable 6248R CPUs. The bottom picture shows the performance and arithmetic intensity for the PIC code without data alignment and with AVX512 compiler auto-vectorization options of the Intel oneAPI Fortran compiler. The bottom picture shows the same data for the Intel Parallel Studio 2019XE Fortran compiler.

manually optimized data structure and AVX512 auto-vectorization options of the old Intel compiler.

We also can see from this analysis, that the most "heavy" functions such as particle density calculation or calculation of new coordinates of particles in a cell achieved 4–6× performance boost only by the adding AVX512 compiler option for building the code.

4 Conclusions

Last decade, we can see the dramatic growth of the particle-in-cell codes, due to the growing interest in solving problems of plasma physics. The most of codes are GPU-oriented PIC solvers because of the accelerator's performance. Modern CPUs trying to get closer in performance by adding advanced vectorization instructions to their arithmetical cores. We can see from the roofline analysis, that simple vectorization optimizations can boost the performance up to 6x times. Despite this, the total performance of our code is still under the peak performance of the CPU. In the future, we are planning to use special data templates, such as Intel SDLT, for the implementation of the structure of arrays approach. Usually, the structure of arrays data templates helps to overcome the limitations of the DRAM memory because of the effective CPU cache usage.

Acknowledgements. Computations were performed on the NKS-1P supercomputer at the Siberian Supercomputer Center, Institute of Computational Mathematics and Mathematical Geophysics SB RAS, Novosibirsk, Russia. This work was supported by the Russian Science Foundation (project 19-71-20026) https://rscf.ru/project/19-71-20026/.

References

1. Hockney, R., Eastwood, J.: Computer Simulation Using Particles. McGraw-Hill, New York (1981)
2. Colella, P.: Controlling self-force errors at refinement boundaries for AMR-PIC. J. Comp. Physics **229**(4), 947–957 (2010). https://doi.org/10.1016/j.jcp.2009.07.004
3. Singh, P.K., et al.: Vacuum laser acceleration of super-ponderomotive electrons using relativistic transparency injection. Nat. Commun. **13**, 54 (2022). https://doi.org/10.1038/s41467-021-27691-w
4. Lotov, K.V., et al.: Note on quantitatively correct simulations of the kinetic beam-plasma instability. Phys. Plasmas **22**, 024502 (2015). https://doi.org/10.1063/1.4907223
5. Fonseca, R.A., et al.: OSIRIS: a three-dimensional, fully relativistic particle in cell code for modeling plasma based accelerators. In: Sloot, P.M.A., Hoekstra, A.G., Tan, C.J.K., Dongarra, J.J. (eds.) ICCS 2002. LNCS, vol. 2331, pp. 342–351. Springer, Heidelberg (2002). https://doi.org/10.1007/3-540-47789-6_36
6. Nieter, C., et al.: VORPAL: a versatile plasma simulation code. J. Comp. Phys **196**(2), 448–473 (2004). https://doi.org/10.1016/j.jcp.2003.11.004

7. Vay, J.L., et al.: Novel methods in the particle-in-cell accelerator code-framework warp. Comput. Sci. Discov. **5**, 014019 (2013). https://doi.org/10.1088/1749-4699/5/1/014019
8. Derouillat, J., et al.: Smilei: a collaborative, open-source, multi-purpose particle-in-cell code for plasma simulation. Comput. Phys. Commun. **222**, 351–373 (2018). https://doi.org/10.1016/j.cpc.2017.09.024
9. Gonoskov, A., et al.: Extended particle-in-cell schemes for physics in ultrastrong laser fields: review and developments. Phys. Rev. **92**, 0233305 (2015). https://doi.org/10.1103/PhysRevE.92.023305
10. Bastrakov, S., et al.: Particle-in-cell plasma simulation on heterogeneous cluster systems. J. Comp. Sci. **3**(6), 474–479 (2012). https://doi.org/10.1016/j.jocs.2012.08.012
11. Decyk, V.K., et al.: Particle-in-Cell algorithms for emerging computer architectures. Comput. Phys. Commun. **185**(3), 708–719 (2014). https://doi.org/10.1016/j.cpc.2013.10.013
12. Vincenti, H., et al.: An efficient and portable SIMD algorithm for charge/current deposition in particle-in-cell codes. Comput. Phys. Commun. **210**, 145–154 (2017). https://doi.org/10.1016/j.cpc.2016.08.023
13. Chernykh, I., et al.: High-performance simulation of high-beta plasmas using PIC method. Commun. Comput. Inf. Sci. **1331**, 207–215 (2020). https://doi.org/10.1007/978-3-030-64616-5_18
14. Boris, J.P.: Relativistic plasma simulation—optimization of a hybrid code. In: Proceedings of the Fourth Conference on Numerical Simulations of Plasmas, Washington, DC, USA, pp. 3–67 (1970)
15. Boronina M.A., Chernykh I.G., Genrikh E.A., Vshivkov V.A.: Performance improvement of particle-in-cell method for numerical modelling of open magnetic system, J. Phys. Conf. Ser. **1640**, 012014 (2020). https://doi.org/10.1088/1742-6596/1640/1/012014
16. Advanced Vector Extensions. https://en.wikipedia.org/wiki/Advanced_Vector_Extensions. Accessed 12 Dec 2023
17. Intel One API. https://www.intel.com/content/www/us/en/developer/tools/oneapi/overview.html. Accessed 12 Dec 2023
18. Kulikov, I.M., Chernykh, I.G., Glinskiy, B.M., Protasov, V.A.: An efficient optimization of hll method for the second generation of intel xeon phi processor. Lobachevskii J. Math. **39**(4), 543–551 (2018). https://doi.org/10.1134/S1995080218040091
19. Kulikov, I., et al.: A new approach to the supercomputer simulation of carbon burning sub-grid physics in ia type supernovae explosion. Commun. Comput. Inf. Sci. **1618**, 210–232 (2022). https://doi.org/10.1007/978-3-031-11623-0_15
20. Chernykh, I., Vorobyev, E., Elbakyan, V., Kulikov, I.: The impact of compiler level optimization on the performance of iterative poisson solver for numerical modeling of protostellar disks. Commun. Comput. Inf. Sci. **1510**, 415–426 (2021). https://doi.org/10.1007/978-3-030-92864-3_32

Algorithms for Design with CNC Machines: The Case Study of Wood Furniture

Francesco Demarco[1(✉)], Francesca Bertacchini[2], Eleonora Bilotta[1], Carmelo Scuro[1], and Pietro Pantano[1]

[1] Department of Physics, University of Calabria, Rende, Italy
{francesco.demarco,eleonora.bilotta,carmelo.scuro,
pietro.pantano}@unical.it
[2] Department of Mechanical, Energy and Management Engineering,
University of Calabria, Rende, Italy
francesca.bertacchini@unical.it

Abstract. Thanks to the application of the Industry 4.0 paradigm, contemporary factories consist of flexible production lines that can generate countless product variations without substantial increases in production costs. This study highlighted a scientific and technological gap between the flexible manufacturing system and the design system adopted to make products. In fact, commonly used CAD design technologies are static and do not allow the generation of dynamic and variable designs, causing the need to redesign models in whole or in part in order to realize variations in the generated shapes. In this paper, an algorithm-based generative design methodology oriented to the flexible manufacturing paradigm is proposed. This design approach, based on parametric modeling in Grasshopper, allows countless geometric variations of a product to be automatically generated while returning input CAD files for CNC machines. Specifically, the proposed design approach was tested by making two applications for the wood furniture industry; the output obtained in the case studies consists of a generative and parametric algorithm. The generative system provides a file for advanced manufacturing systems; in the case study, a numerically controlled laser cutter, one of the most popular machines for making flat panels from wood and metal, was chosen. The results obtained showed how the algorithmic design approach is of great importance in order to ensure customized production without substantial cost increases. This is made possible through algorithmic design automation and contemporary manufacturing technologies.

Keywords: Parmaetric Design · CNC · Industry 4.0

1 Introduction

Design for manufacture and assembly (DfMA) has become an efficient design method to control the process of manufacturing and assembling. However, the existing design systems are not well suited for combining the parametric design with DfMA. The primary purpose of DfMA method provides the ease of component prefabrication and assembly tolerance control through data model of design, modeling, fabrication and assembly. DfMA provides the potential to ensure maximum integration of knowledge from design, manufacturing and assembly. It requires customizable prefabricated elements and highly automated production systems. Current practices of designing furniture elements, and in general for wood manufacturing, often lack integrated information delivery between design, fabrication and assembly phases. In general, the integration of these processes is poorly researched in scientific papers [5,7,16], and the main advances are related to industrial research and are not disclosed for that reason. This paper aims to study a methodology to establish a algorithm-based generative design oriented to manufacture and assembly. Te method allows the generation of parametric 3D models accompanied by information needed for manufacturing machines, such as plane representations appropriately grouped and oriented according to the characteristics of the manufacturing process in use (Laser cutting), and information for the assembly of parts. The implementation of the proposed method leads to productivity increase, to production flexibility and, since CNC machine tools are involved, to resource costs reduction. In this paper is presented an approach that aims to position itself in the scientific gap identified in the lack of integration processes between parametric and customizable CAD models and files for CNC machines.

2 Literature Review

In literature, only few researchers have tried to approach this Design for manufacture and assembly automation problem as a whole over the last years. However, it is possible to find research and insights related to specific aspects of the problem itself, for example, it is possible to identify studies that have focused on reducing the geometric deviation between the designed parts and the resulting prototypes. The authors themselves investigated this problem [1,4,8,9] by identifying a possible solution to geometric deviations in high-resolution SLA prints for jewelry. This geometric deviation problem is common to different categories of advanced manufacturing machines, Efendy and French [17] tried to reduce prototype deviation in CNC technologies for wood parts by acting directly on G-codes. In both cases, the results indicate that the possible variations were reduced. In addition, it is possible to identify design-related research for manufacturing at smaller scales, especially for jewelry [2,3,6,10]. Thus, these are applications that employ high-resolution tools but generate prototypes consisting of a single or few components, so little weight is given to assembly in the parametric and generative design [11,13,14]. In larger-scale prototyping, however, assembly design plays a major role in the ideation and production process [12,18,19]. The example of prefabricated bridges [21], although far from the discussion in this paper, is emblematic in that it highlights how the design of assemblies combined with a process geared toward the manufacturability of the object allows for substantial reductions in cost and lead time. A further point explored in depth in the literature are studies on the effects of laser cutting on panels made for the wood furniture industry. Cebrail and Tutuş [15] made comparisons between the effects on the finish of wood panels made by traditional cutting methods and by laser cutters. They identified a cutting power level of the CNC laser machine and two distinct cutting speeds, related to panel thicknes, that returned an acceptable level of surface roughness of woodmaterials. The parameters identified in the work described, were then employed by the authors in the implementation of the algorithmic approach presented in this paper.

3 Methodology

The main purpose of the work presented is to create a system that integrates parametric modeling to manufacturing with numerically controlled machines, in the specific case study described a laser cutter is used. To this end, a design-for-manufacture and assembly approach was implemented in Grasshopper, a parametric modeling plug-in for Rhinoceros software. Parametric modeling (PD) allows an object to be designed by an explicit flow of data in the form of a graph. The dataflow permits modeling a design object as a constrained collection of schemata that are represented in the form of a graph, called Direct Acyclic Graph (DAG). DAGs can be realized by means of pre-programmed components, components produced by third parties, and by the realization of scripts in python or C#, for the realization of specific tasks, which can be combined with the grasshopper graphics language. In programming specific components, it is possible to employ the Rhinocommon library to create, manage and transform geometries, but also third-party libraries that allow temporary files to be written to the system. Specifically, such a system has been used to create files for machines containing the coordinates of the polylines describing the geometries, the speed of movement of the laser cutter head, and the cutting power level. The parameter and rule-based algorithmic design of the PD aids the implementation of the DfMA approach by allowing the unique identification of components, their positioning with respect to the global model, and then the definition of assemblies to achieve what was designed. This approach allows the creation of designs that are not only aesthetically appealing and functional, but also optimized for efficient manufacturing and assembly. Figure 1 shows the flowchart of the methodology implemented in the present work. The implemented method consists of the creation of a parametric digital model of the object to be produced, from the latter the individual components and their relative positions are identified for assembly instructions. A script extracts the polylines that will then be converted into the laser cutter's machining instructions along with the specific parameters of the machine used, such as speed and power. When a variant of the parametric model is fixed, this results in a set of G-codes, one for each component, ready for transfer to the cutting machine. Loading and unloading the machined wood panels remains the only manual operation to be performed. It is clear from the description provided that the methodology presented is generic and can be scaled and implemented for different processing chains. In the present work, a case study was carried out for wood furniture; this choice is mainly due to the laser cutter present in the laboratory, namely a VD 1325 $CO2$ laser. The machine in question is an industrial laser cutter specifically for cutting organic material. The machine used has a large work surface and has a power range that allows the cutting of even thick wood panels. These features make it particularly high-performing in processing for wood furniture.

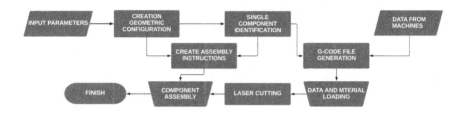

Fig. 1. Flow chart of the proposed method.

4 Results

In the present section authors report two different parametric models developed for the wood furniture sector in Grasshopper, in both case studies the proposed model was implemented. Developing visual scripts combined with components programmed in python related to both the generation of the shapes themselves and the writing of the G-code. Specifically, it is therefore possible to distinguish two main steps common to both case studies: The first is related to the creation of the shapes and the identification of the components, while the second step transforms the components into lists, the geometries are then identified as discrete sets of points and polylines. The first step is strongly related to the designed form and is described in the two Subsects. 4.1 and 4.2. The second step consists of an algorithm in python that exploits "write()" method to create G-code files containing the discrete set of coordinates in the plane and the oriented segments joining them. This data represents the path that the machine's cutting head must take to accomplish the required machining. In the specific case, the machine is a 150-watt $CO2$ laser with a working plane of 2500×1350 mm. In addition to the path information the G-code must provide the power and speed of movement of the cutting head, literature analysis [15] and experimentation have shown that the optimal parameters for achieving a good finish of the cut surfaces are a constant power of 130 W and a speed of 10 mm/s for panels with thickness $>=$ 10 mm or 20 mm/s for panels with less thickness.

4.1 Case Study 1

The first case study involved the implementation of the proposed method for building a parametric kitchen module. The entire code was realized medinate the Grasshopper visual script, with some components made in python for the distribution of components on the XY plane and the component that by means of the "write()" method transforms coordinates and polylines into axis motion for the CNC machine. Figure 2 shows the visual script by means of which the first case study was implemented. As is easily deduced from the image as the complexity of the problem represented increases the complexity of the graph by means of the visual script increases and it becomes particularly difficult for external users to read. The use of groupings by color can help identify the main functions used in the visual code (Fig. 2), with red highlighting the components that define anchor points, "white" highlighting the components that define the parametric shape, yellow highlighting the assembly instructions, and green highlighting the group of components that manage the distribution of plan components and the writing of the G-code. Figure 3 shows the main outputs obtained through the script made for the case study, parametric modeling combined with the DfM approach. As illustrated by Fig. 3.a.b and c the visual script returns an editable digital model by changing the input parameters, with each change corresponding to a change in the components and consequently in the generatable cut file by fixing a given configuration (Fig. 3.d).

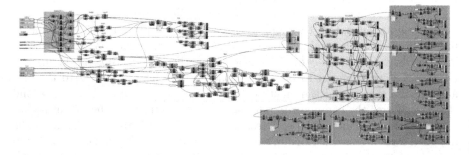

Fig. 2. DAG for the DfMA approach; the visual script generates a digital model, divided into components and G-code files for laser cutter.

4.2 Case Study 2

The second case study involved the creation of a library, the script for generating the editable geometry was developed mainly by implementing a C# script inspired by fractal subdivision. While it has components for assembly and G-code generation common to the first case study, for that reason not discussed in depth in the current section. The script uses the RhinoCommon API specifically

Algorithms for Design with CNC Machines   267

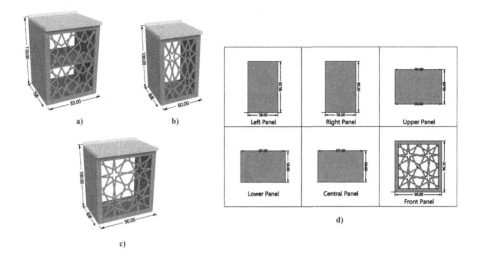

Fig. 3. Results of the first case study: the implemented method returns a parametric digital model a, b, c) and its components d) from which G-codes are extracted.

the Rhino.Geometry namespace. The first step in the process involves creating a starting rectangle that represents the overall dimensions of the library, and its center is identified. Next, a list (List <Rectangle3d>()) is initialized to contain successive subdivisions of the starting rectangle, as well as a list of the coordinates of the centers of each rectangle. After specifying two input parameters–a ratio to define the first subdivision and the desired number of iterations–a "for" loop is started that, using the principle of fractal subdivision, generates a series of rectangles within the main one. At the end of the loop, the resulting curves are extruded to create a solid. The output of the script developed for the second case study allows design variants and files to be generated for CNC machines, some of the variations obtained are shown in Fig. 4.

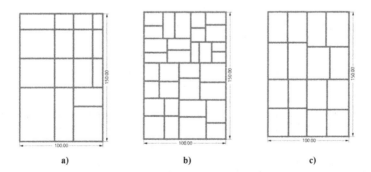

Fig. 4. Examples of parametric models obtained by providing different input parameters such as ratio and number of iterations

5 Conclusion and Future Works

Currently, numerically controlled technologies have greater production flexibility than design methodologies. The implementation of algorithmic and generative methodologies helps to overcome this scientific gap. The objective of this study is to present one of the potential solutions to address this problem, it allows the generation of parametric shapes oriented to manufacturing and assembly. This algorithmic approach makes it possible to go directly from the digital model to prototypes made with advanced manufacturing techniques. Future developments for the present approach involve the implementation of genetic algorithms to optimize the distribution of components to be cut by providing as input dimensions of the base panels used in order to minimize machining waste [20].

Funding. This research was granted by Next Generation UE - PNRR Tech4You Project funds assigned to University of Calabria (PP1.4.2 - Process of chaotic design, Scientifc Coordinator: Professor Pietro Pantano).

References

1. Bertacchini, F., Pantano, P., Bilotta, E.: Shaping the aesthetical landscape by using image statistics measures. Acta Physiol. (Oxf) **224**, 103530 (2022)
2. Bertacchini, F., Bilotta, E., Caldarola, F., Pantano, P.: The role of computer simulations in learning analytic mechanics towards chaos theory: a course experimentation. Int. J. Math. Educ. Sci. Technol. **50**(1), 100–120 (2019)
3. Bertacchini, F., Bilotta, E., Carini, M., Gabriele, L., Pantano, P., Tavernise, A.: Learning in the smart city: a virtual and augmented museum devoted to chaos theory. In: Chiu, D.K.W., Wang, M., Popescu, E., Li, Q., Lau, R. (eds.) ICWL 2012. LNCS, vol. 7697, pp. 261–270. Springer, Heidelberg (2014). https://doi.org/10.1007/978-3-662-43454-3_27
4. Bertacchini, F., et al.: Preliminary study of an innovative method to increase the accuracy in direct 3d-printing of nurbs objects. In: 2021 IEEE International Workshop on Metrology for Industry 4.0 & IoT (MetroInd4. 0&IoT), pp. 94–98. IEEE (2021)
5. Bertacchini, F., Bilotta, E., De Pietro, M., Demarco, F., Pantano, P., Scuro, C.: Modeling and recognition of emotions in manufacturing. Int. J. Interactive Des. Manuf. (IJIDeM) **16**(4), 1357–1370 (2022)
6. Bertacchini, F., Bilotta, E., Demarco, F., Pantano, P., Scuro, C.: Multi-objective optimization and rapid prototyping for jewelry industry: methodologies and case studies. Int. J. Adv. Manuf. Technol. **112**, 2943–2959 (2021)
7. Bertacchini, F., Bilotta, E., Demarco, F., Pantano, P., Scuro, C.: Increase the accuracy in direct 3D-printing of mathematical patterns for smart manufacturing. In: 2022 IEEE International Workshop on Metrology for Industry 4.0 & IoT (MetroInd4. 0&IoT), pp. 387–391. IEEE (2022)
8. Bertacchini, F., Bilotta, E., Gabriele, L., Pantano, P., Servidio, R.: Using lego mindstorms in higher education: cognitive strategies in programming a quadruped robot. In: Workshop proceedings of the 18th international conference on computers in education, ICCE, pp. 366–371 (2010)

9. Bertacchini, F., et al.: An emotional learning environment for subjects with autism spectrum disorder. In: 2013 International Conference on Interactive Collaborative Learning (ICL), pp. 653–659. IEEE (2013)
10. Bertacchini, F., Gabriele, L., Tavernise, A., et al.: Bridging educational technologies and school environment: implementations and findings from research studies. In: Educational Theory, pp. 63–82. Nova Science Publishers, Inc. (2011)
11. Bertacchini, F., Pantano, P.S., Bilotta, E.: Jewels from chaos: a fascinating journey from abstract forms to physical objects. Chaos Interdiscipl. J. Nonlinear Sci. **33**(1) (2023)
12. Bertacchini, F., Scuro, C., Pantano, P., Bilotta, E.: Modelling brain dynamics by Boolean networks. Sci. Rep. **12**(1), 16543 (2022)
13. Bertacchini, F., Scuro, C., Pantano, P., Bilotta, E.: A project based learning approach for improving students' computational thinking skills. Front. Robot. AI **9** (2022)
14. Bilotta, E., Gabriele, L., Servidio, R., Tavernise, A.: Edutainment robotics as learning tool. Trans. Edutainment III, 25–35 (2009)
15. Cebrail, A., Tutuş, A.: The effect of traditional and laser cutting on surface roughness of wood materials used in furniture industry. Wood Industry Eng. **2**(2), 45–50 (2020)
16. Demarco, F., Bertacchini, F., Scuro, C., Bilotta, E., Pantano, P.: The development and application of an optimization tool in industrial design. Int. J. Interact. Des. Manuf. (IJIDeM) **14**(3), 955–970 (2020). https://doi.org/10.1007/s12008-020-00679-4
17. Efendy, E., French, M.: Reducing build variation in arched guitar plates. In: Proulx, T. (eds.) Experimental and Applied Mechanics, Volume 6: Proceedings of the 2010 Annual Conference on Experimental and Applied Mechanics, pp. 607–620. Springer, New York (2011). https://doi.org/10.1007/978-1-4419-9792-0_89
18. Gabriele, L., Tavernise, A., Bertacchini, F.: Active learning in a robotics laboratory with university students. In: Increasing Student Engagement and Retention Using Immersive Interfaces: Virtual Worlds, Gaming, and Simulation, vol. 6, pp. 315–339. Emerald Group Publishing Limited (2012)
19. Herrmann, J.W., et al.: New directions in design for manufacturing. In: International Design Engineering Technical Conferences and Computers and Information in Engineering Conference, vol. 46962, pp. 853–861 (2004)
20. Marco, L., Farinella, G.M.: Computer Vision for Assistive Healthcare. Academic Press (2018)
21. Nguyen, D.C., Shim, C.S.: Design for manufacturing and assembly-oriented parametric modelling of prefabricated bridges. In: Proceedings of IABSE Symposium: Challenges for Existing and Oncoming Structures, Prague, Czech Republic, 25–27 May 2022 (2022)

PyGrossone: A Python Library for the Infinity Computer

Alberto Falcone[(✉)], Alfredo Garro, and Yaroslav D. Sergeyev

Department of Informatics, Modeling, Electronics and Systems Engineering,
University of Calabria, 87036 Rende, (CS), Italy
{alberto.falcone,alfredo.garro,yaro}@dimes.unical.it

Abstract. Computers typically adopt the IEEE 754-1985 binary floating point standard to represent and work with numbers. Although computers can work with finite numbers, numerical computations that involve infinite and infinitesimal quantities are impossible due to architectural limitations. This paper is dedicated to the Infinity Computer, a new type of supercomputer that allows one to perform calculations involving finite, infinite, and infinitesimal numbers. The existent simulators of the Infinity Computer available for the Matlab/Simulink environment are already used in several research domains to solve real-world problems, where accuracy is a crucial aspect. However, the Matlab/Simulink simulators are not well suited for solving problems in the Artificial Intelligence and Machine Learning domains, where the Python programming language is largely adopted. For this purpose, the main aim of this paper is to introduce *PyGrossone*, a Python library for the Infinity Computer.

Keywords: Infinity Computer · Scientific Computing · Python Software Library

1 Introduction

The format of representation and storage of numbers, along with the set of operations that can be performed on them, are crucial aspects of traditional computer architecture that influence the accuracy of calculations. Today, almost all traditional computers use the IEEE 754-1985 binary floating point standard to represent and operate with numbers. However, the architectural limitations of traditional computers make it difficult to perform operations involving finite, infinite, and infinitesimal numbers. To overcome these restrictions, the Infinity Computer has been conceived as a new kind of supercomputer that allows one to work numerically with such numbers [20]. The existent simulators of the Infinity Computer available for the Matlab/Simulink environment are already used in several research domains to solve real-world problems, where accuracy is a crucial aspect [10,12]. To promote the adoption of the Infinity Computer, the *PyGrossone* has been developed.

The main scope of this paper is to introduce *PyGrossone*, a cross-platform and domain-independent Python library that allows one to work with finite, infinite, and infinitesimal numbers expressed in the positional numeral system with the infinite radix called grossone ① provided by the Infinity Computer. *PyGrossone* offers a set of arithmetic, elementary, and differentiation modules to perform computations with ①-based numbers with exact precision up to the machine one, since the computations on the Infinity Computer are numeric, not symbolic. The availability of a Python-based implementation of the Infinity Computer enables its exploitation in other research fields where Python is currently the reference programming language, such as Artificial Intelligence and Machine Learning.

The rest of the paper is organized as follows. Section 2 provides a brief introduction to the Infinity Computing. Section 3 presents the overall design and main modules of the *PyGrossone* library. In Sect. 4, the proposed library is used for the problem of numerical differentiation. Finally, some concluding remarks are presented in Sect. 5.

2 The Infinity Computing: A Brief Overview

The Infinity Computer is a methodology (not related to non-standard analysis, see [25]) that allows one to perform computations involving finite, infinite, and infinitesimal numbers in a unique computational framework and in accordance with Euclid's notion no. 5 *"The whole is greater than the part"* (see, e.g., [21,24]). This methodology introduces a numeral system where the infinite unit of measure is defined as the number of elements of the set $\mathbb{N}$, and it is expressed by the numeral ①, called *grossone*. In the Infinity Computer, a number C is named *grossnumber* and is expressed as follows:

$$C = d_n ①^{p_n} d_{n-1} ①^{p_{n-1}} ... d_0 ①^{p_0} ... d_{-k} ①^{p_{-k}}, \qquad (1)$$

where numerals d_i, $i = n, ..., -k,$, known as *grossdigits*, are not equal to zero and belong to a traditional numeral system that expresses finite positive or negative numbers and show how many corresponding units $①^{p_i}$ should be added or subtracted in order to form the number C. Numbers p_i, $i = n, ..., -k$, are called *grosspowers* and can be finite, infinite, and infinitesimal quantities sorted in the decreasing order:

$$p_n > p_{n-1} > ... > p_1 > p_0 > p_{-1} > ... > p_{-(k-1)} > p_k \qquad (2)$$

with $p_0 = 0^1$. The Infinity Computer numeral system based on ① offers the possibility to define, for example, a number as $6.3①^{2.3} 8.5①^{1.2}$ having grosspowers $p_2 = 2.3$, $p_1 = 1.2$ and grossdigits $c_2 = 6.3$, $c_1 = 8.5$. According to this methodology, a number C in the form (1) can be:

- *infinite*, when it has at least one term $d_i ①^{p_i}$ with positive finite or infinite grosspower p_i. For example, the simplest infinite number is $① = 1①^1$;

---

[1] In this paper, only finite grosspowers are implemented for simplicity.

- *finite*, when it has the term $d_0①^{p_0}$, $p_0 = 0$, and no infinite terms. A special case is when C contains only the term $d_0①^{p_0}$; in this case, the number C is *purely finite*. For example, any finite floating-point number can be represented in the form: $y = y①^0$ (e.g., $6.8 = 6.8①^0$);
- *infinitesimal*, when it contains only negative finite or infinite grosspowers. For example, the simplest infinitesimal is $①^{-1} = \frac{1}{①}$, which is positive, since it is the result of the division of two positive numbers 1 and $①$.

The main advantage of the Infinity Computer is that arithmetical operations of numbers in the form (1) can be performed numerically in a similar way to operations involving traditional finite numbers [24], for example:

$$①^0 = \frac{①}{①} = ① \cdot ①^{-1} = 1, \quad ① - ① = 0 = 0^①,$$

$$(3.5①^{1.5}7.2①^0 - 1.7①^{-2.1}) + (7.4①^{1.5} - 6.1①^0 - 3.25①^{-5.1}) =$$

$$= 10.9①^{1.5}1.1①^0 - 1.7①^{-2.1} - 3.25①^{-5.1}.$$

The software simulators of the Infinity Computer have been successfully exploited to solve many real-life problems in the following research areas: optimization (see, e.g., [5–7,26]), calculation of ODEs (see, e.g., [1,16,19]), numerical differentiation (see, e.g., [12,22,23]), ill-conditioning problems (see, e.g., [17]), fractals theory (see, e.g., [3,4]), handling of hybrid systems under Zeno behavior (see, e.g., [11,13,14]).

3 The PyGrossone Library: Design and Modules

With reference to the principles of Infinity Computing described in Sect. 2, in this section the *PyGrossone* library is presented. The proposed library is general-purpose and domain-independent, making it applicable to all industrial and scientific domains where computation precision represents an important aspect (e.g., Cyber-Physical Systems, formal representation and evaluation of systems' requirements, resources optimization (see [2,8,9,18])). It has been designed for Python 3 with an emphasis on flexibility, adaptability, and performance. To achieve these objectives, its design and implementation have been centered on typical software engineering methodology and, in particular, on the *Agile* one by using the built-in functionalities and data structures provided by Python 3. In this way, researchers can easily integrate the *PyGrossone* functionalities with other libraries.

Figure 1 shows the architectural design of the *PyGrossone* library and its core services. The *Core Services* layer represents the kernel of *PyGrossone* that offers a set of low-level services to manage computations according to the Infinity Computer methodology. It provides four services: (i) *Data Types*, which implements data types for representing and handling infinite, finite, and infinitesimal quantities; (ii) *Logging*, which is mainly used for debugging operations to trace the operations done during a computation; (iii) *Operations optimizer*, which manages the memory used by a *grossnumber* in an efficient and effective manner. It also optimizes the mathematical operations to speed up computations; and (iv) *Infinity Computer Engine*, which offers low-level functionalities to perform high-precision computations according to the Infinity Computer methodology.

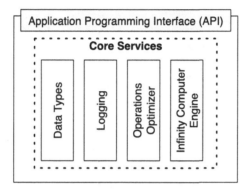

Fig. 1. The architectural design of the *PyGrossone* library and its core services.

Figure 2 presents the organization of the *PyGrossone* library through a UML class diagram that depicts the library's classes, their attributes, operations (or methods), and the relationships among classes. The library is built around two fundamental classes: (i) *GrossNumber*, which allows to represent and manage a *grossnumber* in the form (1); and (ii) *GrossToken*, which offers functionalities to handle a generic term in the form $d_i ①^{p_i}$, $i = n, ..., -k$, that make up a *grossnumber* (see Sect. 2). Since the construction and management of a *GrossNumber* is a fairly complex operation involving *GrossTokens*, the library uses a step-by-step approach based on the Builder design pattern [15]. Builder is a creational pattern whose objective is to separate the creation of a complex object from its representation so that alternative representations may be created using the same building process. It allows one to have in-depth control of the entire building process and gives the possibility to change the internal state of objects without impacting the building process. According to the Builder design pattern, the library adopts a dedicated class, named *Director*, that is responsible for building a *GrossNumber* object invoking the methods exposed by the *Builder* interface. Once the *GrossNumber* is built, the *Client Application* can directly request to the *Builder* the entirely constructed *GrossNumber* by invoking the *build()* method, which is concretely implemented by the *GrossNumberBuilder* class. This latter class implements the *Builder* interface to provide functionalities to assemble the parts that make up a *GrossNumber*, i.e., *grossTokens*. It also defines and tracks the state of the object under construction and overrides the *build()* method for retrieving it when the construction process ends. Finally, the *GrossNumber::_grossnumber_iterator* class is used to iterate over the *GrossNumber* data structure.

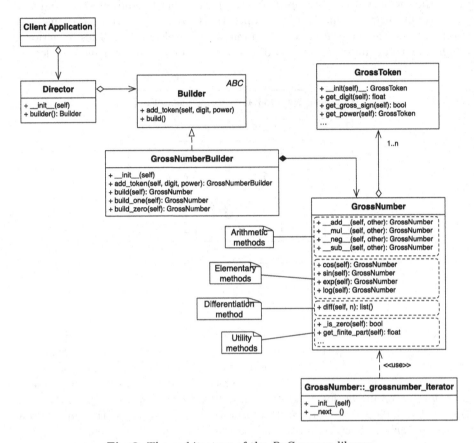

Fig. 2. The architecture of the *PyGrossone* library.

4 Using PyGrossone: An Application to Numerical Differentiation

In this section, the *PyGrossone* library is used to the problem of numerical differentiation of a function $f(x)$. Suppose to have the following function

$$f(x) = \frac{x^2 + 2}{x^2 - 2}, \qquad (3)$$

defined on the Infinity Computer as

$$g(x) = \frac{(x^2 + 2①^0)}{(x^2 - 2①^0)}, \qquad (4)$$

where x can be finite, infinite, or infinitesimal of the form (1).
Let us calculate the first four derivatives of the function $g(x)$ at the finite point $x^* = 5$. As described in [23], the Infinity Computer executes the following operations to calculate the derivatives of the function $g(x)$ at the point $y = x^* + ①^{-1} = 5①^0 1①^{-1}$

$$g(5①^0 1①^{-1}) = \frac{\left(5①^0 1①^{-1}\right)^2 + 2①^0}{\left(5①^0 1①^{-1}\right)^2 - 2①^0} = \frac{\left[\left(5①^0 1①^{-1}\right) \cdot \left(5①^0 1①^{-1}\right)\right] + 2①^0}{\left[\left(5①^0 1①^{-1}\right) \cdot \left(5①^0 1①^{-1}\right)\right] - 2①^0} = \frac{27①^0 10①^{-1} 1①^{-2}}{23①^0 10①^{-1} 1①^{-2}} = 1.17391①^0 - 0.0756①^{-1} + 0.0253①^{-2} - 0.0077①^{-3} \quad (5)$$
$$+ 0.00225①^{-4}...,$$

after obtaining the value of $g(y)$ the following formula is applied

$$f(y) = c_0, \ f'(y) = 1! \cdot c_{-1}, \ f''(y) = 2! \cdot c_{-2}, \ ..., \ f^{(k)}(y) = k! \cdot c_{-k}, \quad (6)$$

to obtain the first $four$-derivatives of $g(y)$

$$\begin{aligned} g(5) &= 1.17391, & g'(5) &= -0.0756 \cdot 1! = -0.0756, \\ g''(5) &= 0.0253 \cdot 2! = 0.0506, & g'''(5) &= -0.0077 \cdot 3! = -0.0462, \quad (7) \\ g^{(4)}(5) &= 0.00225 \cdot 4! = 0.0541 \end{aligned}$$

The scheme of the above mentioned procedures using *PyGrossone* is presented in List. 1.1.

Listing 1.1. Source code based the PyGrossone library that computes the first four derivatives of the function g(x) from (4).

```
builder = gross_number_factory.create()

gn1 = builder \
    .add_token(digit=5, power=0) \
    .add_token(digit=1, power=-1) \
    .build()


gn2 = builder \
    .add_token(digit=2, power=0) \
    .build()

numerator = (gn1 * gn1) + gn2
denominator = (gn1 * gn1) - gn2
result = numerator / denominator

derivates = result.diff(4) #Compute derivaters
```

In the *PyGrossone* library, each *grossnumber* is built by using the functionalities provided by the *Builder* class (see Fig. 2), where a *grossToken* is defined by using the *add_token(digit, power)* method. It takes two input arguments: (i) *digit*, which represents the *grossdigit*; and (ii) *power*, which represents the *grosspower*. The *GrossNumber* class provides the method *diff(n)* that allows one to differentiate a given equation $g(x)$ represented in the Infinity Computer. Using the method *diff(n)* is straightforward since it takes just one input argument, i.e., $n \geq 0$, representing the number of the n-th derivative to compute numerically. The output of the method is a list of the first n-derivatives $[f(x_0), f'(x_0), f''(x_0), f(3)(x_0), ..., f^{(n)}(x_0)]$ given as the list of IEEE 754 binary64 floating-point numbers. With reference to the source code reported in List. 1.1, the variable *derivates* stores the list: $[1.17391, -0.0756, 0.0506, -0.0462, 0.0541]$.

5 Conclusion

The new *PyGrossone* for the Infinity Computer has been developed. The proposed library offers a set of modules to perform computations with ①-based numbers with exact precision (up to the machine one). The availability of a Python-based implementation of the Infinity Computer enables its exploitation in research domains where Python is currently the reference programming language. *PyGrossone* has been successfully applied to the problem of numerical differentiation of a function.

References

1. Amodio, P., Iavernaro, F., Mazzia, F., Mukhametzhanov, M.S., Sergeyev, Y.D.: A generalized Taylor method of order three for the solution of initial value problems in standard and infinity floating-point arithmetic. Math. Comput. Simul. **141**, 24–39 (2016)
2. Bouskela, D., et al.: Formal requirements modeling for cyber-physical systems engineering: an integrated solution based on form-l and modelica. Requirements Eng. **27**(1), 1–30 (2022)
3. Caldarola, F.: The exact measures of the Sierpiński d-dimensional tetrahedron in connection with a diophantine nonlinear system. Commun. Nonlinear Sci. Numer. Simul. **63**, 228–238 (2018)
4. Caldarola, F., Maiolo, M.: On the topological convergence of multi-rule sequences of sets and fractal patterns. Soft. Comput. **24**(23), 17737–17749 (2020). https://doi.org/10.1007/s00500-020-05358-w
5. Cococcioni, M., Cudazzo, A., Pappalardo, M., Sergeyev, Y.D.: Solving the lexicographic multi-objective mixed-integer linear programming problem using branch-and-bound and grossone methodology. Commun. Nonlinear Sci. Numer. Simul. **84**, 105177 (2020)
6. De Leone, R.: Nonlinear programming and grossone: quadratic programming and the role of constraint qualifications. Appl. Math. Comput. **318**, 290–297 (2018)
7. De Leone, R., Fasano, G., Sergeyev, Y.D.: Planar methods and grossone for the conjugate gradient breakdown in nonlinear programming. Comput. Optim. Appl. **71**(1), 73–93 (2018)
8. Falcone, A., Garro, A.: Pitfalls and remedies in modeling and simulation of cyber physical systems. In: 24th IEEE/ACM International Symposium on Distributed Simulation and Real Time Applications, DS-RT 2020, Prague, Czech Republic, 14-16 September 2020, pp. 1–5. IEEE (2020)

9. Falcone, A., Garro, A., D'Ambrogio, A., Giglio, A.: Engineering systems by combining bpmn and hla-based distributed simulation. In: 2017 IEEE International Conference on Systems Engineering Symposium, ISSE 2017, Vienna, Austria, 11-13 October 2017, pp. 1–6. IEEE (2017)
10. Falcone, A., Garro, A., Mukhametzhanov, M.S., Sergeyev, Y.D.: Representation of grossone-based arithmetic in simulink for scientific computing. Soft. Comput. **24**(23), 17525–17539 (2020). https://doi.org/10.1007/s00500-020-05221-y
11. Falcone, A., Garro, A., Mukhametzhanov, M.S., Sergeyev, Y.D.: A simulink-based infinity computer simulator and some applications. In: Sergeyev, Y.D., Kvasov, D.E. (eds.) NUMTA 2019. LNCS, vol. 11974, pp. 362–369. Springer, Cham (2020). https://doi.org/10.1007/978-3-030-40616-5_31
12. Falcone, A., Garro, A., Mukhametzhanov, M.S., Sergeyev, Y.D.: A simulink-based software solution using the Infinity Computer methodology for higher order differentiation. Appl. Math. Comput. **409**, 125606 (2021)
13. Falcone, A., Garro, A., Mukhametzhanov, M.S., Sergeyev, Y.D.: Advantages of the usage of the Infinity Computer for reducing the Zeno behavior in hybrid systems models. Soft Comput., 1–20 (2022)
14. Falcone, A., Garro, A., Mukhametzhanov, M.S., Sergeyev, Y.D.: Simulation of hybrid systems under Zeno behavior using numerical infinitesimals. Commun. Nonlinear Sci. Numer. Simul. **111**, 106443 (2022)
15. Gamma, E., Helm, R., Johnson, R., Johnson, R.E., Vlissides, J.: Design patterns: elements of reusable object-oriented software. Pearson Deutschland GmbH (1995)
16. Iavernaro, F., Mazzia, F., Mukhametzhanov, M., Sergeyev, Y.D.: Conjugate-symplecticity properties of Euler–Maclaurin methods and their implementation on the Infinity Computer. Appli. Numer. Math. (2019)
17. Kvasov, D.E., Mukhametzhanov, M.S., Sergeyev, Y.D.: Ill-conditioning provoked by scaling in univariate global optimization and its handling on the Infinity Computer. In: AIP Conference Proceedings, vol. 2070, p. 020011. AIP (2019)
18. Legato, P., Mazza, R.M., Trunfio, R.: Medcenter container terminal spa uses simulation in housekeeping operations. Interfaces **43**(4), 313–324 (2013)
19. Mazzia, F., Sergeyev, Y.D., Iavernaro, F., Amodio, P., Mukhametzhanov, M.S.: Numerical methods for solving ODEs on the Infinity Computer. In: AIP Conference Proceedings, New York, vol. 1776, p. 090033 (2016)
20. Sergeyev, Y.D.: Computer system for storing infinite, infinitesimal, and finite quantities and executing arithmetical operations with them. USA patent 7,860,914 (2010), EU patent 1728149 (2009), RF patent 2395111 (2010)
21. Sergeyev, Y.D.: Arithmetic of Infinity. Edizioni Orizzonti Meridionali (2003)
22. Sergeyev, Y.D.: Numerical point of view on Calculus for functions assuming finite, infinite, and infinitesimal values over finite, infinite, and infinitesimal domains. Nonlinear Anal. Ser. A: Theory Methods & Appli. **71(12)**, e1688–e1707 (2009)
23. Sergeyev, Y.D.: Higher order numerical differentiation on the Infinity Computer. Optimization Lett. 5(4), 575–585 (2011)
24. Sergeyev, Y.D.: Numerical infinities and infinitesimals: methodology, applications, and repercussions on two hilbert problems. EMS Surv. Math. Sci. **4**, 219–320 (2017)
25. Sergeyev, Y.D.: Independence of the grossone-based infinity methodology from non-standard analysis and comments upon logical fallacies in some texts asserting the opposite. Found. Sci. **24**, 153–170 (2019)
26. Sergeyev, Y.D., De Leone, R.: Numerical infinities and infinitesimals in optimization. Springer (2022)

Towards Reproducible Research in Machine Learning via Blockchain

Ernestas Filatovas[✉], Linas Stripinis, Francisco Orts, and Remigijus Paulavičius

Vilnius University Institute of Data Science and Digital Technologies, Akademijos 4, 08663 Vilnius, Lithuania
ernestas.filatovas@mif.vu.lt

Abstract. Artificial Intelligence, particularly in Machine Learning and related research areas such as Operational Research, currently faces a reproducibility crisis. Researchers encounter difficulties reproducing key results due to lacking critical details, including the disconnection between publications and the associated codes, data, and parameter settings. Solutions that improve code accessibility, data provenance tracking, research transparency, auditing of obtained results, and trust can significantly accelerate algorithm and model development, validation, and transition into real-world applications. Blockchain technology, with its features of decentralization, data immutability, cryptographic hash functions, and consensus algorithms, provides a promising avenue for developing such solutions. By leveraging the distributed ledger working over a peer-to-peer network, a secure and auditable infrastructure can be established for sharing and controlling data in a trusted manner. Our analysis examines the current state-of-the-art blockchain-based proposals that target reproducibility issues in the Machine Learning domain. Based on the analysis of existing solutions, we propose a high-level architecture and main modules for developing a blockchain-based platform that enhances reproducible research in Machine Learning and can be adapted to other Artificial Intelligence domains.

Keywords: Machine Learning · Reproducibility · Reproducible research · Blockchain · Blockchain-based platform

1 Introduction

Today, challenging decision problems such as image analysis, voice and face recognition, planning, scheduling, and routing rely heavily on Machine Learning (ML) techniques. ML, including Deep Learning, has become the mainstream in Artificial Intelligence (AI) due to increased computational capabilities. However, the dependence on large datasets and complex models poses challenges to reproducibility. Unfortunately, with the increasing popularity of ML and other

---

This research has received funding from the Research Council of Lithuania (LMTLT), agreement No. S-MIP-21-53.

© The Author(s), under exclusive license to Springer Nature Switzerland AG 2025
Y. D. Sergeyev et al. (Eds.): NUMTA 2023, LNCS 14478, pp. 278–285, 2025.
https://doi.org/10.1007/978-3-031-81247-7_24

AI domains, researchers face a Reproducibility Crisis (RC) [9]. A significant challenge arises when attempting to reproduce various key results due to disconnection between publications and the underlying codes, data, and parameter settings, which often lack critical details.

To solve RC-related issues, the research community and industry actively explore diverse solutions. Commonly, researchers rely on depositing code and data in repositories like *GitLab* or *GitHub*. Still, this practice often falls short as it lacks crucial information, including runtime environments, contextual details, and system information [22]. Data deposited in the repositories of a journal can also be unreliable due to broken links [21]. In addition to traditional version control methods, recent advances in research technology have introduced open-access collaborative cloud-based platforms such as *Code Ocean*[1], *Whole Tale*[2], *Binder*[3], etc. These platforms enable the capture of research environments, facilitating research processes' reuse, sharing, and reproducibility. Furthermore, online platforms such as *OpenML* [25] and *ModelDB* [26] are gaining popularity in the ML community, as they provide storage and sharing capabilities for datasets and experimental results, fostering open science collaboration. However, only some required reproducibility aspects are fully covered [17], and more development is needed [6]. Moreover, these platforms are centralized and often lack operational transparency, reliable traceability, high security, and trusted data provenance features. Additionally, they carry a risk of single-point failure.

Decentralized alternatives can tackle RC issues from different perspectives, aiming to avoid the traditional issues with centralized systems and take advances such as transparency, tracebility, tokenization, incentivization, consensus mechanism, codification of trust, and decentralized infrastructure [5,20]. The research community sees great potential for blockchain [19] when dealing with RC and related challenging problems in various research fields and already develops blockchain-enhanced research workflow solutions that address provenance, reliability, and collaboration [1,3,8,16].

Several notable business-oriented projects are currently in development that combine blockchain technologies and AI (mainly focusing on ML) to enhance the research process, improve reproducibility, and foster business collaboration. *SingularityNet*[4] aims to establish a decentralized marketplace for AI algorithms and Federated Learning (FL), *PlatON*[5] is working towards building a decentralized and collaborative AI network and global brain to promote the democratization of AI for safe artificial general intelligence, and *FETCH.AI*[6] is developing a decentralized collaborative ML platform for various business applications. The research community also is increasingly interested in exploring the intersection of ML and blockchain to address reproducibility issues.

---

[1] https://codeocean.com/.
[2] https://wholetale.org/.
[3] https://mybinder.org/.
[4] https://singularitynet.io/.
[5] https://www.platon.network/.
[6] https://fetch.ai/.

This paper aims to provide a comprehensive review and evaluation of current state-of-the-art (SOTA) blockchain-based proposals that target reproducibility issues in the ML domain highlighting their main contributions, features, and progress. Additionally, leveraging the analysis of these proposals and existing solutions, we provide a high-level architecture for an efficient platform that fosters collaboration, specifically in ML, with potential adaptability to other domains of AI, such as Operations Research. This platform could harness the benefits of blockchain technology, incorporating essential features such as decentralization, incentives, traceability, consensus among the research community, and controlled data sharing while ensuring seamless integration with existing ML research solutions.

The structure of the paper is as follows: Sect. 2 presents an overview and current status of existing blockchain-based SOTA proposals to enhance reproducibility in ML research. Section 2 outlines the high-level architecture of a flexible blockchain-based platform to facilitate collaborative and reproducible research in ML while also addressing its implementation challenges. Finally, Sect. 4 provides concluding remarks.

2 Overview of Blockchain-Based Proposals for Machine Learning Research Reproducibility

This section summarizes the existing SOTA literature on blockchain-based proposals to improve reproducibility and data provenance in ML research.

We conducted a comprehensive search using backward and forward snowballing methods to identify key publications, and the searches were performed in May 2023. In Table 1, we identify the publications obtained, highlighting their main contributions and features, reproducibility aspects analyzed, application areas, blockchain platforms considered, and implementation levels.

We observe that most of the solutions focus on some specific aspects of reproducibility. The aspects of "model", "data" and "provenance" are addressed mostly, while "parameters" and "results" receive less attention in proposals. Moreover, none of the proposals emphasize the critical aspect of environmental reproducibility. Notably, the earliest related publications emerged in 2019, and most are on the "Idea" implementation level. Most of the recent proposals have progressed to higher implementation levels, including "Concept" or even "Prototype". Both public and private blockchain platforms for the implementation are considered; however, solutions with the "Idea" implementation level do not specify them at all. Another observation is the notable emphasis on solutions in Federated Learning. FL inherently relies on a decentralized approach, making blockchain an obvious choice to enhance fairness and ensure data privacy. In FL application cases, blockchain platforms specifically designed for private/consortium implementations, such as *Hyperledger Fabric*, *Corda*, or *Parity*, are frequently chosen as they facilitate FL processes involving small groups of authorized actors. Furthermore, some efforts have been made to develop systems/platforms for reproducible research in ML across various application areas (as indicated by the "General" item). Both the popular public *Ethereum* and private *Hyperledger Fabric* platforms are considered in these instances.

In summary, SOTA literature analysis showed that existing solutions are very limited in their scope, focusing primarily on only some specific aspects. This underscores the need for further consideration from the research community to effectively address research reproducibility challenges in the field of ML and, enhance the research cycle, increase trust and transparency by developing a blockchain-based decentralized platform that is efficient, scalable, interoperable, and adaptable across various AI research domains.

Table 1. Proposed state-of-the-art blockchain-based solutions to improve reproducibility in ML research.

Authors, source, year	Key contribution	Reproducibility aspect	Application area in ML	Blockchain platform	Implementation level
Harris & Waggoner [7] (2019)	Framework for training a model and collecting data on a blockchain by leveraging several incentive mechanisms	Model, data, provenance	General	N/S*	Idea
Lu et al. [14] (2019)	Data sharing architecture for privacy-preserving FL with Proof of Training Quality consensus protocol	Model, data, parameters	Federated Learning	N/S	Idea
Sarpatwar et al. [23] (2019)	Vision to build a generic blockchain library to enable trust in distributed AI applications and processing	Model, data, parameters, provenance	Federated Learning	N/S	Idea
Weng et al. [27] (2019)	Decentralized platform with an incentive mechanism and a consensus protocol for ML model training and sharing of the local gradients	Model, parameters	Federated Learning	Corda	Concept
Kannan et al. [10] (2020)	Decentralized trusted data and model platform for collaborative AI	Model, data, provenance, results	General	Hyperledger Fabric	Concept
Lüthi et al. [15] (2020)	Graph-based provenance tracking model and smart-contract for managing AI assets and their relationships	Data, provenance	General	Ethereum	Concept
Li et al. [12] (2021)	Decentralized FL framework with the committee consensus mechanism	Model, parameters, provenance, results	Federated Learning	FISCO-BCOS	Concept
Mothukuri et al. [18] (2021)	Decentralized framework for permissioned FL	Model, parameters	Federated Learning	Hyperledger Fabric	Concept
Bathen and Jadav [2] (2022)	Framework that combines AutoML techniques with blockchain to fully decentralize the design and training process	Model, data, parameters, results	General	Hyperledger Fabric	Concept
Coelho et al. [4] (2022)	Blockchain platform for collaborative research and reproducibility of experiment results	Provenance, results	Federated Learning	Hyperledger Fabric	Prototype
Khoi et al. [11] (2022)	Decentralized platform for disseminating, storing, and updating ML provenance	Model, data, parameters, provenance	General	Ethereum	Concept
Lo et al. [13] (2022)	FL architecture, including a data sampler algorithm to enhance fairness in training data and smart contract-based data-model provenance registry	Model, data, provenance, results	Federated Learning	Parity	Concept
Peregrina et al. [18] (2022)	Adaptation of data governance to FL using blockchain	Model, data, provenance, results	Federated Learning	N/S	Idea
Stodt et al. [24] (2022)	ML Birth Certificate and ML Family Tree secured by blockchain technology	Model, provenance	General	N/S	Idea

* N/S - Not Specified

3 Concept of a Blockchain-Based Platform for Reproducible Machine Learning Research

This section presents a high-level architecture for developing an interoperable, community-driven, blockchain-empowered platform to enhance reproducible research in Machine Learning. We also address the implementation challenges associated with this platform development.

3.1 Architecture

In Fig. 1, we present the layered architecture of the platform, emphasising the key modules. In the "Tools Layer," the focus is on integrating the platform with popular existing open science collaboration ML frameworks (such as *OpenML* and *MLflow*), reproducible research tools (e.g., *Code Ocean, Whole Tale*), and ML libraries (such as *TensorFlow* and *Keras*).

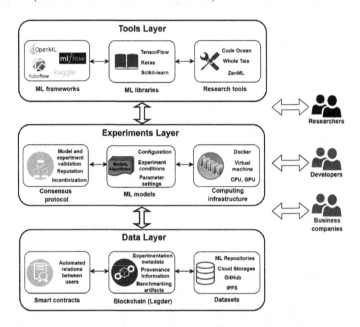

Fig. 1. Layered architecture of a blockchain-based platform for research reproducibility in ML.

The "Experiments Layer" is dedicated to conducting experimental investigations of ML models, allowing researchers to utilize various computing infrastructures. The research and developers community validate these experiments' outcomes through a specially designed community-driven consensus protocol. This protocol relies on proofs related to data quality, the nature of learning models and parameters, and the computation equipment used. The consensus protocol also considers the user's reputation and incorporates an incentivization mechanism.

The "Data Layer" focuses on data sources, automated data management, and the storage of metadata and provenance information on the blockchain. The platform considers various external data storage sources, including popular ML repositories (e.g., *Amazon, Kaggle*), code and data sharing platforms (e.g., *GitHub, GitLab*), dedicated cloud storages, and the built-in InterPlanetary File

System (IPFS) for data storage. The blockchain enhances the provenance of ML models and securely stores experimentation metadata in a decentralized manner, allowing platform participants to verify it. To consider various reproducibility aspects, the metadata includes information about the researcher, trained model, hyperparameter settings, used datasets, computing environment, obtained experimental results, benchmarking artifacts, and other relevant investigation conditions. Smart contracts automate processes and enhance participant trust by controlling user relations, managing data access, and verifying data validity before storing them on the blockchain. In particular, platform users should not be limited to researchers but also open to other developers and business companies to improve collaboration and research knowledge exchange.

3.2 Implementation Challenges

Despite the high potential of a platform built on the architecture presented, several challenges need to be addressed. Integration with other tools will require significant efforts and cooperation. Attracting participants from the research and business communities is crucial for the platform's success. A community-driven consensus protocol with reputation management must be carefully defined to ensure proper validation of results. Additionally, appropriate incentive mechanisms should be considered to encourage actors to participate in model and experiment validation and contribute their models and data. Robust protection mechanisms must be integrated into the platform to maintain control over shared data and prevent malicious usage. It is important to note that while popular blockchain platforms such as *Ethereum* have made significant progress, they still face scalability issues, low transaction throughput, and high transaction costs. Newer platforms like *Cardano* and *Polkadot* may not yet support the required functionality. Therefore, when selecting a base blockchain platform, careful consideration should be given to the required functionality and efficiency, considering the platform's specific needs and goals.

To summarize, the platform's architecture should be built on a reliable and advanced blockchain platform, such as *Ethereum* for public implementations or *Hyperledger Fabric* for consortium use. This ensures the incorporation of multiple reproducibility aspects and facilitates efficient management of data and users, storage and tracking of provenance information, and seamless operation of smart contracts. Additionally, integration with existing ML research and data storage solutions is crucial to enable effective model development, streamline the experimentation process, and foster collaboration.

4 Conclusions

This paper investigates the current landscape of blockchain-based solutions designed to address the reproducibility challenges encountered in ML. The analysis reveals that the existing solutions are rather limited in scope, primarily focusing on some specific reproducibility aspects and considering only a few blockchain features, such as consensus protocols or incentive mechanisms, while lacking a joint blockchain-based approach to enhance reproducibility in ML research. To bridge this gap, we propose a high-level layered architecture and identify crucial modules for developing a blockchain-based platform that facilitates reproducible research in Machine Learning and has the potential for adaptation to other AI domains. Moreover, the paper highlights the implementation challenges that need to be overcome to realize the envisioned platform.

Acknowledgements. This research has received funding from the Research Council of Lithuania (LMTLT), agreement No. S-MIP-21-53.

References

1. Bag, R., Spilak, B., Winkel, J., Härdle, W.K.: Quantinar: a blockchain p2p ecosystem for honest scientific research. arXiv preprint arXiv:2211.11525 (2022)
2. Bathen, L.A.D., Jadav, D.: Trustless AutoML for the age of internet of things. In: 2022 IEEE International Conference on Blockchain and Cryptocurrency (ICBC), pp. 1–3. IEEE (2022)
3. Coelho, R., Braga, R., David, J.M.N., Dantas, M., Ströele, V., Campos, F.: Blockchain for reliability in collaborative scientific workflows on cloud platforms. In: 2020 IEEE Symposium on Computers and Communications (ISCC), pp. 1–7. IEEE (2020)
4. Coelho, R., Braga, R., David, J.M.N., Stroele, V., Campos, F., Dantas, M.: A blockchain-based architecture for trust in collaborative scientific experimentation. J. Grid Comput. **20**(4), 35 (2022)
5. Filatovas, E., Marcozzi, M., Mostarda, L., Paulavičius, R.: A MCDM-based framework for blockchain consensus protocol selection. Expert Syst. Appl. **204**, 117609 (2022)
6. Gundersen, O.E., Shamsaliei, S., Isdahl, R.J.: Do machine learning platforms provide out-of-the-box reproducibility? Futur. Gener. Comput. Syst. **126**, 34–47 (2022)
7. Harris, J.D., Waggoner, B.: Decentralized and collaborative AI on blockchain. In: 2019 IEEE International Conference on Blockchain (Blockchain), pp. 368–375. IEEE (2019)
8. Hoopes, R., Hardy, H., Long, M., Dagher, G.G.: Sciledger: a blockchain-based scientific workflow provenance and data sharing platform. In: 2022 IEEE 8th International Conference on Collaboration and Internet Computing (CIC), pp. 125–134. IEEE (2022)
9. Hutson, M.: Artificial intelligence faces reproducibility crisis. Science **359**(6377) (2018)
10. Kannan, K., Singh, A., Verma, M., Jayachandran, P., Mehta, S.: Blockchain-based platform for trusted collaborations on data and AI models. In: 2020 IEEE International Conference on Blockchain (Blockchain), pp. 82–89. IEEE (2020)
11. Khoi Tran, N., Sabir, B., Babar, M.A., Cui, N., Abolhasan, M., Lipman, J.: ProML: a decentralised platform for provenance management of machine learning software systems. In: Gerostathopoulos, I., Lewis, G., Batista, T., Bureš, T. (eds.) ECSA 2022. LNCS, vol. 13444, pp. 49–65. Springer, Cham (2022). doi: https://doi.org/10.1007/978-3-031-16697-6_4

12. Li, Y., Chen, C., Liu, N., Huang, H., Zheng, Z., Yan, Q.: A blockchain-based decentralized federated learning framework with committee consensus. IEEE Network **35**(1), 234–241 (2021)
13. Lo, S.K., et al.: Towards trustworthy AI: blockchain-based architecture design for accountability and fairness of federated learning systems. IEEE Internet Things J. (2022)
14. Lu, Y., Huang, X., Dai, Y., Maharjan, S., Zhang, Y.: Blockchain and federated learning for privacy-preserved data sharing in industrial IoT. IEEE Trans. Industr. Inf. **16**(6), 4177–4186 (2019)
15. Lüthi, P., Gagnaux, T., Gygli, M.: Distributed ledger for provenance tracking of artificial intelligence assets. In: Friedewald, M., Önen, M., Lievens, E., Krenn, S., Fricker, S. (eds.) Privacy and Identity 2019. IAICT, vol. 576, pp. 411–426. Springer, Cham (2020). https://doi.org/10.1007/978-3-030-42504-3_26
16. Meng, Q., Sun, R.: Towards secure and efficient scientific research project management using consortium blockchain. J. Signal Process. Syst. **93**, 323–332 (2021)
17. Mora-Cantallops, M., Sánchez-Alonso, S., García-Barriocanal, E., Sicilia, M.A.: Traceability for trustworthy AI: a review of models and tools. Big Data Cogn. Comput. **5**(2), 20 (2021)
18. Mothukuri, V., Parizi, R.M., Pouriyeh, S., Dehghantanha, A., Choo, K.K.R.: FabricFL: blockchain-in-the-loop federated learning for trusted decentralized systems. IEEE Syst. J. **16**(3), 3711–3722 (2021)
19. Nakamoto, S.: Bitcoin: a peer-to-peer electronic cash system (2008). https://bitcoin.org/bitcoin.pdf
20. Paulavičius, R., Grigaitis, S., Igumenov, A., Filatovas, E.: A decade of blockchain: review of the current status, challenges, and future directions. Informatica **30**(4), 729–748 (2019)
21. Pimentel, J.F., Murta, L., Braganholo, V., Freire, J.: A large-scale study about quality and reproducibility of jupyter notebooks. In: 2019 IEEE/ACM 16th International Conference on Mining Software repositories (MSR), pp. 507–517. IEEE (2019)
22. Rowhani-Farid, A., Barnett, A.G.: Badges for sharing data and code at biostatistics: an observational study. F1000Research **7**, 2 (2018)
23. Sarpatwar, K., et al.: Towards enabling trusted artificial intelligence via blockchain. In: Calo, S., Bertino, E., Verma, D. (eds.) Policy-Based Autonomic Data Governance. LNCS, vol. 11550, pp. 137–153. Springer, Cham (2019). https://doi.org/10.1007/978-3-030-17277-0_8
24. Stodt, J., Stodt, F., Reich, C., Clarke, N.: Verifiable machine learning models in industrial IoT via blockchain. In: Proceedings of the 12th International Advanced Computing Conference, Hyderabad, Telangana, pp. 16–17 (2022)
25. Vanschoren, J., Van Rijn, J.N., Bischl, B., Torgo, L.: OpenML: networked science in machine learning. ACM SIGKDD Explorations Newsl **15**(2), 49–60 (2014)
26. Vartak, M., et al.: ModelDB: a system for machine learning model management. In: Proceedings of the Workshop on Human-In-the-Loop Data Analytics, pp. 1–3 (2016)
27. Weng, J., Weng, J., Zhang, J., Li, M., Zhang, Y., Luo, W.: Deepchain: auditable and privacy-preserving deep learning with blockchain-based incentive. IEEE Trans. Dependable Secure Comput. **18**(5), 2438–2455 (2019)

Dossier Classification to Support Workflow Management Optimization

Simona Fioretto[1], Elio Masciari[1,2], and Enea Vincenzo Napolitano[1]

[1] Department of Electrical and Information Technology Engineering, University of Naples Federico II, Naples, Italy
{simona.fioretto,eneavincenzo.napolitano}@unina.it
[2] Italian National Research Council (ICAR-CNR), Arcavacata, Italy
elio.masciari@icar.cnr.it

Abstract. One of the main challenges in optimizing services in various fields, including the legal sector, is the lack of an automatic system that can efficiently assign tasks based on predefined rules. In many cases, documents are still processed manually by qualified staff, which can be time-consuming and prone to errors. To address this issue, there is a need to develop a system that can automate the classification process, thereby simplifying the allocation of tasks to the most competent professionals in a specific field. This can be achieved by using Natural Language Processing (NLP) algorithms. In our Case Study the problem is to classify legal documents and suggest the most suitable judiciary branch for each dossier. By implementing such a system, it is possible to reduce processing times and enhance the overall efficiency of judicial offices. The proposed method can be supported by using a legal distributed system, which can facilitate seamless communication and collaboration among different actors.

1 Introduction

The efficient allocation of tasks based on predefined rules is a vital aspect of optimizing services in various domains, with the legal sector being no exception. However, the absence of an automatic system capable of effectively assigning tasks has posed a significant challenge in this field [6]. Consequently, the manual processing of documents by qualified personnel remains prevalent, leading to time-consuming procedures and an increased risk of errors.

To address this challenge and improve task assignment in the legal sector, it is imperative to develop a system that can automate the classification process, thereby streamlining the allocation of tasks to professionals who possess the required expertise in specific fields. Natural Language Processing (NLP) algorithms offer a promising avenue for achieving this objective [12]. By leveraging the capabilities of NLP, the classification of legal documents can be automated, enabling the system to suggest the most suitable judiciary branch for each dossier [11].

Implementing an automated system with NLP-based classification holds the potential to significantly reduce processing times and enhance the overall efficiency of judicial offices. By eliminating the need for manual sorting and categorization [8], valuable time and resources can be saved, allowing legal professionals to focus on more complex and high-value tasks. Additionally, the accuracy and consistency of task assignments can be improved, minimizing the risk of errors associated with human involvement.

Moreover, the proposed method can be complemented by employing a legal distributed system. This distributed system, tailored specifically for the legal sector, facilitates seamless communication and collaboration among different actors involved in the task assignment process. By leveraging distributed technologies, such as cloud computing and secure networks, the system can ensure efficient coordination among legal professionals, administrative staff, and judiciary branches.

In this paper, we present a comprehensive investigation into the development and implementation of an automated system for task assignment in the legal sector. We specifically focus on the classification of legal documents and the suggestion of the most appropriate judiciary branch for each dossier. The study aims to demonstrate the potential of NLP algorithms in streamlining task allocation processes, reducing processing times, and enhancing overall efficiency.

2 Judicial Systems: A Review

The functioning of the judicial system depends on multiple factors, and indeed must be carefully and deeply examined.

Firstly, as highlighted by [9] the current courtroom process is still similar to what it looked like back in 1980, and this calls for proper digital transformation. The efficiency of justice significantly influences the functioning of a country's economy. For instance, in common law countries, where the protection of shareholders and creditors is stronger, there is a greater trust on the part of economic agents to invest directly in companies and, consequently, greater development of the financial market [5].

In addition [5] shows that there is an increasing difference between the new demand of legal services and the old judicial maps which has increased processing time and backlog, therefore badly affecting judiciary efficiency. In particular, the Italian judicial system is considered one of the most inefficient in Europe, according to the Organization for Economic Co-operation and Development and the European Union, as well as the International Monetary Fund [5].

In order to solve some of the inefficiencies of judicial offices, an automatic task allocation multi agent system could be helpful, but while the use of Multi Agent systems for simulation is commonly applied in several scenarios for scheduling or prediction purposes, very little has been done in the juridical domain [2].

A real problem is represented by the classification and allocation of legal processes to judges who are responsible for studying, analyze and interpret them. In addition, law comprehends various areas of definition of the processes, from the criminal to the civil and further subdivided internally; this states two problems: one regarding the definition of the domain of the process, and one regarding the magistrate or judge interest both individuated by the legal expert.

In particular, the classification of the processes in their category could be helpful to the assignment to the judge, and therefore to support a proper task allocation system.

The problem of classifying legal documents was addressed using different methods. [13] examine the capability of a deep learning model based on CNN for binary classification (responsive and non-responsive) in real legal matters. Since the main goal of the paper is to classify legal documents into multiple categories, the research we are trying to address relates to a Multi Label Text Classification of Natural Language Processing methods (NLP). [10] use transformers models such as BERT, RoBERTa, DistilBERT, XLNet and M-BERT and combine them with a number of training strategies such as gradual unfreezing, slanted triangular learning rates and language model fine-tuning in order to solve a Multi-Label problem using legal documents from the legal information system of the European Union (Eur-Lex). [4] use two methods for sentence classification in legal documents for obtaining summaries of legal documents: LEGAL-BERT, a BERT model pretrained on legal domain data and a joint feature vector of the legal text by combining statistical features, acquired using TF-IDF, and semantic text features, extracted from LEGAL-BERT. This joint feature vector was then classified using SVM and Logistic Regression. [1] compare legal text classification using two different types of approaches: domain concept-based classification using random forests, and word embeddings-based using deep neural network showing better results for the first one.

3 Problem Statement

This section focuses on the analysis and description of the problem. The problem to be addressed in this paper is the manual assignment of tasks in the judiciary, which affects the slow speed of resolution of court cases. The current assignment process is the one used in Fig. 1.

More in details, offices currently receive folders of documents that are not assigned to a specific branch of justice. Given the large number of branches in the judiciary, the offices must manually read and analyze the folders and classify the content in order to assign the folder to a specific branch. After that, since each judge is more connected to a number of areas based on their personal experiences and interests, the offices check the compatibility between the contents of the folder and the judges and then assign the folder to the most compatible judge.

This apparently simple process, causes slowness and delays. In fact, it is a manual process that can only be performed by qualified personnel and brings some difficulties:

- the involved personnel cannot perform their main job due to this activity;
- takes a lot of time due to the difficulty and length of the content of folders;
- when a judge is not more available the process must be repeated;

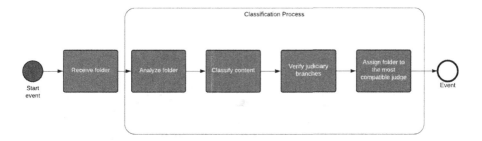

Fig. 1. Classification and Assignment process

- the personnel does not know the current interests of each judge, which can lead to numerous errors and mistakes.

This problem directly affects the workload assignment and tasks allocation, indirectly causing the problem of slow jurisdiction.

The "Classification" sub-process represents the critical part of the process. Currently it is done manually, and the main goal of this paper is to focus the attention on this activities, and do the automatic execution of this process part.

4 Automatic Dossier Classification

The proposed method aims to face with the problems mentioned in 3 in order to automatically classify a judicial case file [7] for supporting the direct assignment to the relevant judge.

The ideal flow, should be the one in which given an input dossier, it undergoes processing by an NLP model, which classifies folders with several labels, then these labels are cross-referenced with previously given associated labels of the judges. As a result of this process, the proposed method generates output associations between the dossier and the corresponding judges.

This brief explanation needs a more accurate in-deep look into the Fig. 2. From the figure we can see that some steps are needed to be completed.

First, the input dossier needs to be preprocessed to extract relevant textual information, such as case details, legal arguments, and supporting evidence. This preprocessing step helps to refine the input data and prepare it for further analysis.

Next, the preprocessed dossier is fed into an NLP model specifically trained for legal text analysis. The NLP model employs various techniques, such as tokenization and semantic analysis, to extract meaningful features from the text. These features capture the underlying semantic structure and context of the dossier [3]. The choice of the right NLP model will rely on the selected models in Sect. 2 from other authors who used this models to solve classification problem in the same domain but with different scope.

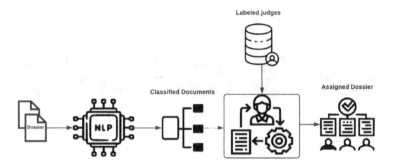

Fig. 2. Steps of proposed method

Once the NLP model has processed the dossier, it assigns the appropriate labels to classify the case. The obtained labels represent different categories of crime.

After the dossier is classified, the process proceeds to match the obtained labels with the corresponding labels associated with the judges. This matching process aims to identify the judges who possess the necessary expertise and jurisdiction to handle cases falling under the classified labels. The matching phase can be done using various techniques, such as rule-based matching, or similarity scoring.

Finally, the process gives as output the associations between the dossier and the identified judges. These associations indicate which judges are most suitable for handling the given case based on their expertise in the relevant legal domain. The output can be presented as a recommendation or an automated assignment, enabling the efficient allocation of judicial cases to the appropriate judges.

By automating the dossier classification and judge assignment process, the proposed method offers several potential benefits. It reduces the manual effort and time required for case allocation, enhances the accuracy and consistency of judge assignments, and optimizes the overall efficiency of judicial offices. Additionally, it minimizes the risk of human errors and ensures that cases are handled by judges with the relevant expertise, leading to improved legal outcomes.

5 Conclusion

In conclusion, the proposed method presented in this paper offers a promising solution to address the challenge of automatically classifying judicial case files for efficient assignment to the relevant judges. By leveraging NLP techniques and cross-referencing label associations, the method demonstrates the potential to streamline the task allocation process within the legal sector.

Through the application of NLP algorithms, the method enables the automated classification of case files based on their inherent textual features. This classification process ensures that cases are appropriately categorized into relevant legal domains, allowing for accurate matching with judges possessing the necessary expertise. By eliminating the manual effort and subjectivity involved in traditional case assignment, the method significantly reduces processing times.

Furthermore, the proposed method contributes to enhancing the overall efficiency of judicial offices. By automating the classification and assignment process, valuable time and resources can be saved, enabling legal professionals to focus on more complex and high-value tasks.

The integration of a legal distributed system further could enhance the method's effectiveness by facilitating seamless communication and collaboration among different stakeholders within the legal sector. By leveraging distributed technologies, such as cloud computing and secure networks, the system could enable efficient coordination among legal professionals, administrative staff, and judiciary branches, fostering effective task allocation and workflow management.

In conclusion, the proposed method not only addresses the existing challenges in task assignment within the legal sector but also offers significant potential for improving the overall efficiency and effectiveness of judicial processes. As future work, the method can be further refined and expanded to incorporate additional features, such as contextual analysis or add a workflow management algorithm, to enhance its accuracy and applicability in different legal domains and manage the work-assignment-process in all the aspects.

Acknowledgements. Work supported by the project "MOD-UPP" - Macroarea 4 - project PON_MDG_1.4.1_17- PON GOV grant.

We acknowledge financial support from the project PNRR MUR project PE0000013-FAIR.

References

1. Chen, H., Wu, L., Chen, J., Lu, W., Ding, J.: A comparative study of automated legal text classification using random forests and deep learning. Inform. Process. Manag. **59**(2), 102798 (2022)
2. Di Martino, B., Esposito, A., Colucci Cante, L.: Multi agents simulation of justice trials to support control management and reduction of civil trials duration. J. Ambient Intell. Humanized Comput., 1–13 (2021)
3. Flesca, S., Manco, G., Masciari, E., Pontieri, L., Pugliese, A.: Exploiting structural similarity for effective web information extraction. Data Knowl. Eng. **60**(1), 222–234 (2007). https://doi.org/10.1016/j.datak.2006.01.001
4. Furniturewala, S., Jain, R., Kumari, V., Sharma, Y.: Legal text classification and summarization using transformers and joint text features (2021)
5. Giacalone, M., Nissi, E., Cusatelli, C.: Dynamic efficiency evaluation of italian judicial system using dea based malmquist productivity indexes. Socioecon. Plann. Sci. **72**, 100952 (2020)
6. Graham, R.L., Lawler, E.L., Lenstra, J.K., Kan, A.R.: Optimization and approximation in deterministic sequencing and scheduling: a survey. Annals Dis. Math. **5**, 287–326 (1979)
7. Manco, G., Masciari, E., Tagarelli, A.: A framework for adaptive mail classification. In: 14th IEEE International Conference on Tools with Artificial Intelligence, (ICTAI 2002), Proceedings, pp. 387–392 (2002). https://doi.org/10.1109/TAI.2002.1180829

8. Masciari, E.: Trajectory clustering via effective partitioning. In: Andreasen, T., Yager, R.R., Bulskov, H., Christiansen, H., Larsen, H.L. (eds.) FQAS 2009. LNCS (LNAI), vol. 5822, pp. 358–370. Springer, Heidelberg (2009). https://doi.org/10.1007/978-3-642-04957-6_31
9. Rule, C.: Online dispute resolution and the future of justice. Annual Rev. Law Soc. Sci. **16**, 277–292 (2020)
10. Shaheen, Z., Wohlgenannt, G., Filtz, E.: Large scale legal text classification using transformer models. arXiv preprint arXiv:2010.12871 (2020)
11. Sulea, O.M., Zampieri, M., Malmasi, S., Vela, M., Dinu, L.P., Van Genabith, J.: Exploring the use of text classification in the legal domain. arXiv preprint arXiv:1710.09306 (2017)
12. Undavia, S., Meyers, A., Ortega, J.E.: A comparative study of classifying legal documents with neural networks. In: 2018 Federated Conference on Computer Science and Information Systems (FedCSIS), pp. 515–522. IEEE (2018)
13. Wei, F., Qin, H., Ye, S., Zhao, H.: Empirical study of deep learning for text classification in legal document review. In: 2018 IEEE International Conference on Big Data (Big Data), pp. 3317–3320 (2018).https://doi.org/10.1109/BigData.2018.8622157

Self-Sovereign Identification of IoT Devices by Using Physically Unclonable Functions and Blockchain

Gianluigi Folino[✉], Agostino Forestiero, and Giuseppe Papuzzo

ICAR-CNR, Via P. Bucci 8-9C, 87036 Rende, (CS), Italy
gianluigi.folino@icar.cnr.it

Abstract. The Data Sovereignty paradigm gives users full control of their digital data, allowing them to choose which parts of the data they want to share and to whom. In the case of IoT devices, a secure mechanism is necessary to verify the identity associated with the device, as typically, no trusted third-party authority can verify the identity of the device's owner. Other important issues in this context concern data provenance, efficient storing and processing while maintaining privacy and security. Physically Unclonable Functions (PUFs) could be helpful in these tasks, as they can generate secure keys or fingerprints inherently and uniquely identify and authenticate physical items or physical objects in which they are embedded. Therefore, to better overcome the above-cited issues, a framework for handling the identities associated with IoT devices, based on the PUF technology to assess their identity and a blockchain network to verify the associated transactions, is proposed in this work. More into detail, our system permits the registration and verification in real-time of the identity associated with an IoT device in a self-sovereign fashion and the traceability of the accesses and processing of the data. Indeed, the identity is checked by using the PUF associated with an IoT device, and the blockchain layer ensures the security and privacy of access to the data in a decentralised way.

1 Introduction

In the last few years, data sovereignty, particularly the Digital Sovereignty paradigm, indicated that users must have control of their digital data, allowing them to choose which parts of the data they want to share and to whom. However, in the case of IoT devices, typically, no trusted third-party authority is recognised to have the power to verify the identity of the device's owner. Therefore, a secure mechanism to verify the identity associated with an IoT device is essential for the aims of digital sovereignty. In addition, information on data, such as data provenance, processing, etc., should be efficiently stored and accessed in real-time to associate the identity with the data usage conveniently, and also privacy and security should be guaranteed. This aspect is particularly critical when a large number of IoT devices and parallel access to data must be handled. Physically Unclonable Functions (PUFs) [5] could be helpful in these tasks, as they can generate secure keys or fingerprints inherently and uniquely identify and authenticate physical items or physical objects in which they are embedded.

Blockchain distributed technology [8] permits the storage of transactions on a flexible network, guaranteeing safe authentication, management, and access to data in a distributed way with high reliability, integrity, and resilience. Its network's public databases hold transactions that authorised individuals have verified, and the transaction data stored on the blockchain are encrypted for security and, therefore, unalterable. The blockchain can be utilised in numerous financial services, such as digital assets, because it enables us to carry out payment without using a bank or intermediary; however, it is useful in many fields in which privacy and integrity of the data, together with efficient access, is essential.

To better overcome the above-cited issues, and by exploiting the advantages of the above-discussed technologies, a framework for handling the identities associated with IoT devices, based on the PUF technology to assess their identity and a blockchain network to verify the associated transactions, is proposed in this work. More into detail, our system permits the registration and verification in real-time of the identity associated with an IoT device in a self-sovereign fashion and the traceability of the accesses and processing of the data. Indeed, the identity is checked by using a biological PUF associated with an IoT device, which makes them adapt to use in the IoT scenario because of its hard reproducibility, low cost and low energy consumption; in addition, the blockchain layer ensures the security and privacy of access to the data in a decentralised way.

The rest of the paper is organised as follows. Section 2 gives background information on the PUF and on the Self-sovereign Identity and Digital Sovereignty concepts. Section 3 illustrates the software architecture of the system, the supply chain used, and the algorithm used for the authentication mechanism of the PUF.

Finally, Sect. 4 draws concluding remarks and directions for future work.

2 Background: PUF, Blockchain and Self-Sovereign Identity

2.1 Physical Unclonable Functions and IoT Devices

Physical Unclonable Functions (PUFs) [5] have emerged as a promising technology for enhancing security in various applications, including authentication and key generation. PUFs exploit inherent physical variations in physical elements to create unique, non-reproducible identities, offering a robust defence against cloning attacks and are often used for key generation. Indeed, PUF-based key generation schemes exploit their unique response patterns to generate cryptographic keys securely. Instead of storing secret keys directly in memory, PUF-based systems generate keys on the fly using their response as the seed. Since

PUF responses are unpredictable and device-specific, the resulting keys are inherently secure and resistant to attacks targeting key storage. The key advantages of this technology include their inherent tamper-resistance, low cost, and compatibility with existing manufacturing processes. PUFs require minimal additional hardware, making them suitable for resource-constrained devices. Moreover, they can be incorporated during manufacturing, eliminating the need for additional post-processing steps or modifications to existing fabrication facilities. PUFs have many interpretations and modifications, and many more have been proposed for characterizing tangible objects [5]. For example, Electric PUFs leverage the inherent manufacturing variations in electronic devices, such as transistors, to generate unique response patterns based on a challenge input. These response patterns are derived from uncontrollable physical phenomena, such as process variations, temperature, and noise, making them nearly impossible to replicate. Optical PUFs [1] exploit the behaviour of light in optical structures to generate unique responses (images). These structures' scattering or diffraction effects introduce inherent variations, resulting in distinct and unpredictable responses to a given light input. In general, by using their physical characteristics as a source of identity, they provide a hardware-based security solution that is resistant to tampering, reverse engineering, and traditional attacks on software-based security mechanisms. In addition, they can be integrated/embedded into devices, circuits, smart cards, RFID tags, or any other physical component to establish a secure identity that cannot be duplicated. The challenge-response mechanism ensures that each object responds uniquely to a given challenge, enabling robust device identification and preventing unauthorized access or counterfeit products.

In our context, Biological PUFs (Bio-Puf) [7] present many advantages in comparison with the other technologies, such as hard reproducibility and especially low cost and low energy consumption, which make them adapt to use in the IoT scenario.

2.2 Self-Sovereign Identity, Digital Sovereignty and DSAs

We refer to Self-sovereign Identity, indicating that the data owner must fully control their information and related digital *identity* once they are exchanged and shared. This requires the definition of constraints in the use of data, clarifying who is allowed to do what and in what context with the data shared by those who own it. Data sovereignty is more specific on the digital data and the owner's control of these data, even specifying which parts of the data the owners want to share and to whom. At the European level, and with explicit reference to the digital world, Data Sovereignty follows two main areas: Cloud Sovereignty, i.e., the acquisition of federated cloud services and infrastructures compliant with existing regulations, and Secure online exchange of data between multiple participants in a consortium or group of companies.

Data Sharing Agreements (DSAs) [6] constitute a formal contract regulating the usage and access to data and also define how these data can be shared with other entities/organisations. Therefore, they can be profitably used in the context of Digital Sovereignty to define the rules that producers and consumers of data must respect (i.e., business, legal and Cloud-based rules.). To better clarify, imagine a scenario in which a producer/consumer of data, i.e., a health operator, supplies some data and can also use the results of some analytics/machine learning operations. In this scenario, several operations must be performed and, depending on the operator (i.e., patient, paramedical or medical doctor), not all the data features can be accessed, and part of them also needs to be anonymised. In addition, not all the data can be transferred outside the country of origin or the European Community. Therefore, without using different DSAs, this scenario could hardly be handled adequately.

3 Software Architecture and Supply Chain

In this section, first, we describe the software architecture defined in this work for the secure registration and identification of IoT Devices, and then the computer vision-based algorithm used for the authentication mechanism of the biological PUF is illustrated.

3.1 Software Architecture for the Secure Registration and Identification of the IoT Devices

Biological PUFs have been introduced in our system for efficiently handling the authentification of the users, which operate on IoT devices, mainly for their low cost and low energy consumption. Figure 1 illustrates the overall software architecture for the secure registration and identification of the IoT Devices in a scenario of Self-sovereign Identity.

The first step of the supply chain (on the bottom-left) considers the association of an IoT device with the identity of its owner or user. A biological PUF is associated with this device; then, the features (key points), computed with

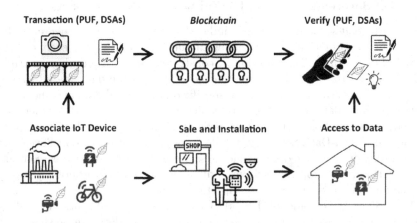

Fig. 1. Software Architecture for the secure registration and identification of the IoT Devices.

the procedure detailed in Subsect. 3.2, of the biological PUF and some DSAs associated with data usage are stored in the blockchain. In a second moment, when the devices are installed in the proper place and need to access some data, their associated PUF is verified by following the procedure explained in the next subsection and in accordance with the particular DSAs associated with the data they can access, the relative permissions are accorded.

The adoption of blockchain guarantees immutability and distributed trust and makes the device authentication independent of the manufacturer. The information saved in the blockchain is used to validate both device signatures and associated DSAs and, in addition, permits ensuring the authenticity and integrity of all the transactions. Replacing the central database operated by a manufacturer with a blockchain makes the system independent of the manufacturer. Even an organization or user can introduce entries on the distributed ledger; all the information is distributed to all the participants; therefore, its availability is guaranteed, differently from a central database, with their relative problems.

Currently, our framework does not provide systems for handling different types of accounts and different organizations. However, the system could be extended to provide mechanisms of accountability, including, for instance, the methodology proposed in [2], in which an infrastructure to offer decentralized accountability services for IoT devices participating in cooperation processes that involve different organizations is proposed.

3.2 PUF-Based Authentication Mechanism

In this subsection, the computer vision-based algorithm used for the authentication mechanism of the biological PUF is illustrated in more detail.

A set of unique and trusted images, generated through a collection of Optical-PUFs, enables the generation of an authentication mechanism so that only PUF-generated images (photos) contained in the dataset will be authenticated and/or authorized. The photos are examined using tailored software that recognises distinctive elements in the images, named key points. The Scale Invariant Feature Transform (SIFT) technique is the base for picture identification and feature-matching software programs, according to [3]. With the SIFT method, pattern identification may be made on 2D images regardless of the angle of acquisition, scale zooming, or brightness changes.

More in detail, the algorithm's main steps are described in the following: (i) creating a scale space to guarantee that the features are independent of scale; (ii) locating appropriate features (key points); (iii) assigning an orientation to guarantee rotation-invariant key points; and (iv) creating a key point descriptor to give each key point a distinct fingerprint [4]. Figure 2 reports an example of the key points detected through the implemented software, Fig. 2a, when the image reported in Fig. 2b is given as input.

Identifying positions and sizes that may be associated with the same object from various viewpoints constitutes detecting the proper features (key points). The position can be determined by examining features with constant values across all relevant dimensions and independent of changes in the image's scale.

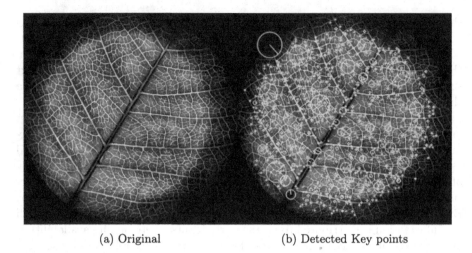

(a) Original (b) Detected Key points

Fig. 2. Images of (a) the original leaf and (b) the map of the key points detected by the computer vision tool.

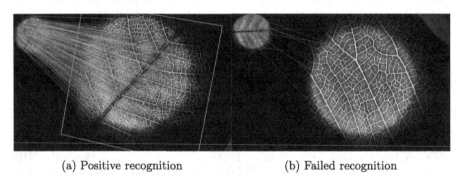

(a) Positive recognition (b) Failed recognition

Fig. 3. Image recognition

In particular, to analyze the considered images, a tool that uses open-source software libraries[1] exploiting the SIFT algorithm written in C++ was designed and implemented.

Figure 3 illustrates a feature comparison between two photos, where a colour line connects each similar attribute. The high number of interconnected lines in Fig. 3a visually represents how many features may be identified when comparing two photos taken over the same area. Conversely, comparing two pictures from different locations yields extremely few "recognitions", as evidenced by the small number of connecting lines between the two images in Fig. 3b. In order to demonstrate the correctness of the approach, the validation method was applied to a set of pictures acquired from a collection of PUFs, and the outcomes are reported in Table 1. The image of the PUF, $P_x, x = \{1..3\}$, was acquired with

---
[1] https://opencv.org.

Table 1. Verification matrix

	$P_1\alpha$	$P_1\beta$	$P_1\gamma$	$P_1\delta$	$P_2\alpha$	$P_2\beta$	$P_2\gamma$	$P_2\delta$	$P_3\alpha$	$P_3\beta$	$P_3\gamma$	$P_3\delta$
$P_1\alpha$	481	360	357	348	12	12	13	14	6	10	7	14
$P_1\beta$	419	482	385	427	9	13	11	14	12	5	6	8
$P_1\gamma$	288	279	474	370	13	8	7	10	14	11	6	14
$P_1\delta$	315	316	458	472	11	11	8	6	11	10	7	13
$P_2\alpha$	6	5	5	8	532	402	354	405	7	12	5	13
$P_2\beta$	8	12	12	8	483	528	480	515	15	13	8	12
$P_2\gamma$	12	11	6	5	521	532	536	529	5	6	10	5
$P_2\delta$	14	7	10	6	521	511	511	509	6	15	5	14
$P_3\alpha$	12	9	8	6	9	5	13	10	429	384	443	501
$P_3\beta$	14	6	15	8	7	8	15	13	275	272	319	327
$P_3\gamma$	14	5	8	6	5	13	8	13	318	291	363	392
$P_3\delta$	6	5	6	5	11	6	5	8	348	333	373	490

four different angles, i.e. $\{\alpha, \beta, \gamma, \delta\}$ and the number of matching recognized features for each couple of images were computed. As expected, a higher number of detected points occurs when the same PUF is compared, regardless of the acquisition angle. That can be observed by examining the values corresponding to the number of key points matching the stored image on the blocks belonging to the diagonal of the matrix and the block outside the diagonal. As expected, the values on the diagonal blocks overcome the other blocks by more than an order of magnitude, confirming the goodness of the authentication method and its robustness to changes in scale, angles of observation, etc.

4 Conclusions and Future Works

Our system permits the registration and verification in real-time of the identity associated with an IoT device in a self-sovereign fashion and the traceability of the accesses and processing of the data. Indeed, the identity is checked by using the PUF associated with an IoT device, and the blockchain layer ensures the security and privacy of access to the data in a decentralised way. The efficient authentication system, based on computer vision and low cost and low energy consumption associated with a biological PUF, makes our approach ideal for an IoT scenario.

Future works aim to design and implement a security authentication schema using PUF-based hardware security elements boasting unclonability, uniqueness, and anti-tampering qualities provided by the random irregularities created during the manufacturing process. Furthermore, we want to extend our framework with accountability mechanisms for handling different types of accounts and different organizations.

Acknowledgment. This work was partially supported by project SERIES (PE00000014) under the NRRP MUR program funded by the EU - NGEU.

References

1. Caligiuri, V.: Hybrid plasmonic/photonic nanoscale strategy for multilevel anticounterfeit labels. ACS Appli. Mater. Interfaces **13**(41), 49172–49183 (2021)
2. Felicetti, C., Furfaro, A., Saccà, D., Vatalaro, M., Lanuzza, M., Crupi, F.: Making IoT services accountable: a solution based on blockchain and physically unclonable functions. In: Montella, R., Ciaramella, A., Fortino, G., Guerrieri, A., Liotta, A. (eds.) IDCS 2019. LNCS, vol. 11874, pp. 294–305. Springer, Cham (2019). https://doi.org/10.1007/978-3-030-34914-1_28
3. Lowe, D.G.: Object recognition from local scale-invariant features. In: Proceedings of the Seventh IEEE International Conference on Computer Vision,. vol. 2, pp. 1150–1157 (1999). https://doi.org/10.1109/ICCV.1999.790410
4. Lowe, D.G.: Distinctive image features from scale-invariant keypoints. J. Comput. Vis., 91–110 (2004)
5. McGrath, T., Bagci, I.E., Wang, Z.M., Roedig, U., Young, R.J.: A PUF taxonomy. Appli. Phys. Rev. **6**(1) (02 2019)
6. Sheikhalishahi, M., Saracino, A., Martinelli, F., Marra, A.L.: Privacy preserving data sharing and analysis for edge-based architectures. Int. J. Inf. Sec. **21**(1), 79–101 (2022)
7. Wali, A., et al.: Biological physically unclonable function. Commun. Phys. **2** (04 2019)
8. Zheng, Z., Xie, S., Dai, H.N., Chen, X., Wang, H.: Blockchain challenges and opportunities: a survey. Inter. J. Web Grid Serv. **14**, 352 (2018). https://doi.org/10.1504/IJWGS.2018.095647

Introducing *Nondum*, A Mathematical Notation for Computation with Approximations

Francesco La Regina[✉] and Gianfranco d'Atri

University of Calabria, 87036 Arcavacata di Rende, CS, Italy
laregina@ieee.org, datri@mat.unical.it

Abstract. In this paper, we are introducing a symbol representing a not-yet-known value. This symbol can be used to represent one or more unknown digits within a number. A mathematical notation "extended" in this way makes it easier to store numerical approximations. It also allows you to perform computations between approximate values or to perform partial or approximate computations. We called this symbol *Nondum*, the Latin word meaning not yet (known, measured, or calculated). Furthermore, on the level of symbolic computing, we will show some analogies with infinite computing, *grossone* and the so-called unimaginable numbers.

Keywords: Approximation · Grossone · Unimaginable numbers

1 Introduction

The word *approximation* is derived from Latin *approximatus*, from *proximus* meaning *very near* and *approximare* meaning *come near*. In mathematics, an approximation is often used when an exact numerical form is unknown or difficult to obtain.

Two types of notations are mainly used to represent numerical approximations. Both exploit the concept of margin of error, adding it to the value of the approximated number. In this way a range of acceptable values between a lower and an upper bound is obtained. This choice, although formally impeccable, becomes a difficult model to apply in the computation.

Another approach used to perform calculations with incompletely defined values is the Interval Analysis (See [2]). Also this model, being mainly symbolic, is difficult to apply in computation.

On the other hand, the paradigm of *Approximate computing* is rapidly emerging. In this case, a greater obtainable precision is sacrificed in order to obtain better computational performances. This approach is useful for many application fields in which approximate results are still tolerated and valid, such as scientific computing, machine learning, or multimedia processing.

Table 1. Principle of excluded middle

A	¬ A	A ∨ (¬ A)
0	1	1
1	0	1

2 Intuitionist Mathematics and the Principle of Excluded Middle

In the first decades of the twentieth century there was a great debate on the existence of a continuous set of real numbers, most of which have infinite digits after the decimal point.

The great German mathematician David Hilbert accepted the now established view that real numbers exist and can be manipulated as complete entities. This is what we usually do today.

But, this idea was opposed by the "intuitionist" mathematicians led by the great Dutch topologist L.E.J. Brouwer, who saw mathematics as a construct. According to Brouwer, numbers had to be constructible, with the digits calculated or chosen or randomly determined one at a time because numbers are limited and are processes: they can become all the more exact the more digits are revealed in what he called a sequence of choice, a function to produce values with ever greater precision (see [10,29]).

By basing mathematics on what can be constructed, intuitionism has far-reaching consequences for the practice of mathematics, the most radical of which is that the principle of the excluded middle no longer holds. The principle of excluded middle states that either a proposition is true or its negation is (see Table 1). But, in Brouwer's approach, statements about numbers may be neither true nor false at any given moment, since the exact value of the number has not yet been revealed (see [11]).

2.1 "Constructivist" Principle of Numbers

The consequences are well explained by Carl Posy, a philosopher of mathematics at the Hebrew University of Jerusalem, a leading expert in intuitionist mathematics. Consider a number x, very close to 1. Let's say the value of x is 0.9999 with more decimal places emerging from a sequence of choices. Perhaps the sequence of 9s continues forever, in which case x converges exactly to 1. So far there is no difference with standard mathematics. But if at some future point in the sequence a digit other than 9 appears - for example, if the value of x becomes 0.9999999999997... - then, whatever happens after that, x is less than 1. Before that happens though, i.e. when we know only 0.9999, we do not know if a digit other than 9 will ever appear, therefore we cannot say either that x is less than 1, or that it is equal to 1. The proposition "x equals 1" is not true, but neither is its denial. The principle of excluded middle does not apply.

This approach, however limiting in the fields of pure mathematics, can be very interesting for application fields in which mathematics is used as a tool to manipulate information about the physical world. In these fields, knowledge of a number is effectively limited by physical knowledge of what that number represents. This concept was expressed in a very interesting way by the experimental physicist Nicolas Gisin to express the time we define as "present" (see [20,28]). In fact, we are all aware of living immersed in a flowing time dimension, in which the "future" becomes "present" before our eyes and therefore it seems almost obvious to conclude that the measurement of time must be built moment by moment.

But this approach is also valid for other aspects of physics. For example, if my number expresses a measure of length in metres, the number of decimal digits that I will be able to define cannot be infinite, but will be a direct consequence of the precision of the measuring instrument I am using. On the other hand, if number is a tool for me, very often I don't need to use the absolute precision formally required of mathematicians. For example, NASA for space mission trajectory calculations does not use π, but uses 3.141592653589793 - with exactly 15 decimal digits - as if it were not an irrational number, because adding further precision would become a useless waste of computing resources. It is not necessary to have infinite precision if I am doing calculations that have a practical purpose.

3 Embrace the Uncertainty

In 1976, Donald E. Knuth noted that our computing capabilities were growing so dramatically that it was possible to calculate really huge numbers, albeit always finite. And, for really large numbers, the practical difference between finite and infinite numbers turned out to be already irrelevant (see [22]). Like Knuth suggested to "copying with finiteness" in the middle ground between mathematics and computer science, we suggest to "embrace the uncertainty" in the land of automatic computation.

We take into account two aspects. Numbers with absolute precision are used only in symbolic calculation, which covers a tiny part of the use we make of the numbers themselves. Very often we have limited knowledge of the values to be treated, due to measurement limits or similar reasons.

At this point, why shouldn't a calculation tool be able to handle even partially and temporarily unknown numbers? It can certainly treat the known and the unknown parts separately (it is already done with the margin of error) but it means doubling the complexity of the calculation.

If we were able to define the basic operations by including the unknown part directly in the number we could obtain a more efficient calculation tool, even at the cost of sacrificing part of the precision, of the knowledge. Embrace the uncertainty precisely, and giving it a name and a dignity equal to the known figures.

4 Introducing *Nondum* and Its Symbol ŋ

In this paper, we are introducing a symbol representing a not-yet-known value. This symbol can be used to represent one or more unknown digits within a number.

We called this symbol *Nondum*, the Latin word meaning *not yet* (known, measured, or calculated). At the same time we had the need to choose the symbol to write *Nondum* instead of a digit within the numbers. The choice fell on the letter ŋ. This symbol is used in the international phonetic alphabet to indicate the sound *ng* (as in English *singing*) and was supposed to be printed as the union of the lowercase *n* with the hooked tail of a lowercase *g* in italics, but it was not printed correctly; in the corrigendum of [1] there is a typographical error that led to its reproduction as a union of an *n* with a *y*.

The initials of *not yet*, joined by an error in the monogram ŋ, could only become the symbol of *Nondum*.

4.1 Using ŋ

As we mentioned at the beginning, the margin of error is usually used to manipulate uncertain values. When I need to sum two uncertain values, I can do it by adding the finite part and also adding the two margins of error. This procedure obviously doubles the cost of the calculation. By introducing *Nondum* instead of the margin of error, the calculation could be performed in a single solution, but at the cost of sacrificing the information contributed by the margin of error.

Let's go back to the example of the number x very close to 1. While the new digits gradually emerge from the selection sequence, we can "copying with finiteness" and express it and treat it as a finite number - for example as 0.99 - but, if we do that is, when we use it in a calculation we lose the information on the fact that the number is "incomplete" and we have not yet defined whether it is less than or equal to 1.

Let us consider the case in which we need to add this number to a known finite value, for example 4, we will have $0.99 + 4 = 4.99$ and we will have obtained a number that is certainly less than 5. We have thus lost the information on the uncertainty.

We rewrite the same number using *Nondum* and then adding it to 4. We will have $0.99ŋ + 4 = 4.99ŋ$ which remains a number for which we have not yet defined whether it is less than or equal to 5.

But let's also consider the case in which we need to double this number, we will have $0.99 + 0.99 = 1.98$ and we will have obtained a number that is certainly less than 2. We will thus have lost the information on the uncertainty.

Let's rewrite the same number using *Nondum* and then double it. We will have $0.99ŋ + 0.99ŋ = 1.9ŋ$ which remains a number for which we have not yet defined if it is less than or equal to 2.

Note how, by keeping the uncertainty information, in this second case we have sacrificed a digit of precision.

Table 2. Logical operators involving η

A	B	A ∧ B	A ∨ B	¬ A
0	0	0	0	1
0	1	0	1	1
0	η	0	η	1
1	0	0	1	0
1	1	1	1	0
1	η	η	1	0
η	0	0	η	η
η	1	η	1	η
η	η	η	η	η

4.2 Mathematical Properties of *Nondum*

Only a few finite mathematical properties of *Nondum* can be defined: $0 \cdot \eta = 0$; $1 \cdot \eta = \eta$; $\eta^0 = 1$; $\eta^1 = \eta$; $1^\eta = 1$.

These properties show immediately how *Nondum* is different from *NaN (Not a Number)*. In fact NaN is a particular value of a numeric data type which is undefined or cannot be represented, such as the result of 0/0 (about limits of computational representation see [25]), while *Nondum* is a number *not yet* defined.

4.3 Logical Operations with *Nondum*

We can define the results of logical operations by adding Nondum to the binary values 1 and 0 (see Table 2). It is interesting to note that the AND and OR operators continue to give results that allow practical use, while the NOT operator becomes useless for practical purposes.

The statement of the principle of the excluded middle becomes irrelevant: if $A = \eta$ then $A \vee (\neg A) = \eta$.

5 *Nondum*, Unimaginable Numbers and Grossone

In this section we begin to investigate relations and connections between the newly introduced *nondum*, the unimaginable numbers (see [23]) and the grossone (see [30,31]). As we have already said, the purpose of this work is to propose the idea of *nondum* and establish some of its fundamental and basic characteristics. With other future works we will better investigate its properties. At the moment we limit ourselves, in the few space available here, to starting some reflections that seem very interesting and scientifically promising.

5.1 *Nondum* and Unimaginable Numbers

A positive integer is usually called an *unimaginable number* if it is greater than 1 googol, i.e. 10^{100} (see [8,15,22]). Unimaginable numbers are so called because they cannot be represented through an ordinary decimal expression, and very often not even with exponential or engineering notations. Special uncommon notations are necessary to express very large numbers: some of them are *Knuth's up-arrow notation*, *Steinhaus-Moser notation*, *hyperoperation symbols*, *Conway chained-arrow notation*, and many others in relation to many growing size levels. Anyway, probably, the most known way to write unimaginable numbers is to use Knuth's up-arrow notation, which is of the kind

$$B \uparrow^d T,$$

where B, d, T are non-negative integers called respectively *base*, *depth* and *tag* (see [8,15,22,23] for details and new results on Knuth's powers and the so-called *kratic representations*).

As said above, unimaginable numbers cannot be written in decimal expansions, so, an undetermined quantity like our *nondum* could be very useful. It (or η^{-1}) can also be used to replace indeterminate digits on the left of a decimal expression and not only some digits after the point, on the right of a decimal expression as seen in the previous sections.

The following simple example of use can be clarifying for the reader. Assume that we want to investigate modular congruences of some family of unimaginable numbers modulo some positive integer m, as the authors do in some cases in [9]. Then the opportunity to use a quantity as our *nondum* will naturally arise.

The same convenience can be found in many other cases involving operations with very large numbers, including their inverses. Another promising application of η seems to be to *n*-sets (see [16]) and the difficult exact enumeration problems related to them (both with ordinary and unimaginable sizes).

5.2 *Nondum* and Grossone

The mathematical properties of *nondum* also hold for a recent proposed symbol ① called *grossone*. It was introduced in the early 2000s by Y. Sergeyev and it represents the fundamental infinite unit in the grossone-based numeral system. For more details, some applications and related topics, the reader can see [3,6,7,12,13,17–19] and [14,24,26,27,32,33] for some discussions on foundations and motivations of the new system, etc. A very relevant thing is the fact that, recently, the grossone has also been used on an educational level, experimentally, in some high schools (see [5,21] and also [4] in this same volume). This is very interesting because the authors of this paper are also thinking about similar researches involving the *nondum*, also in connection with the grossone and unimaginable numbers.

At the moment we point out to the reader that Sergeyev himself talks about new positional systems constructed on infinite base or *grossbase*. In this view *grossdigits* and *grosspowers* are also introduced and the usefulness or application of *nondum* in this context seems very interesting from many points of view. But not only: the new capability of writing in a simple way different sizes of infinities and infinitesimals through the grossone-system, spontaneously calls into question the applicability of our *nondum* as we plan to do in future works.

Acknowledgments. This research was partially supported by Seeweb s.r.l., Cloud Computing provider based in Frosinone, Italy, and part of DHH Group.

References

1. Albright, R.: The international phonetic alphabet: its backgrounds and development. Int. J. Am. Linguist. Indiana Univ (1958)
2. Alefeld, G., Mayer, G.: Interval analysis: theory and applications. J. Comput. Appl. Math. **121**(1–2), 421–464 (2000)
3. Amodio, P., Iavernaro, F., Mazzia, F., Mukhametzhanov, M.S., Sergeyev, Y.D.: A generalized Taylor method of order three for the solution of initial value problems in standard and infinity floating-point arithmetic. Math. Comput. Simul. **141**, 24–39 (2017)
4. Antoniotti, L., Astorino, A., Caldarola, F.: Unimaginable numbers and infinity computing at school: an experimentation in northern Italy. In: Sergeyev, Y.D., Kvasov, D.E., Astorino, A. (eds.) NUMTA 2023, LNCS, vol. 14478, pp. xx–yy. Springer, Cham (2025)
5. Antoniotti, L., Caldarola, F., d'Atri, G., Pellegrini, M.: New approaches to basic calculus: an experimentation via numerical computation. In: Sergeyev, Y.D., Kvasov, D.E. (eds.) NUMTA 2019. LNCS, vol. 11973, pp. 329–342. Springer, Cham (2020). https://doi.org/10.1007/978-3-030-39081-5_29
6. Antoniotti, L., Caldarola, F., Maiolo, M.: Infinite numerical computing applied to Peano's, Hilbert's, and Moore's curves. Mediterr. J. Math. **17**, 99 (2020)
7. Astorino, A., Fuduli, A.: Spherical separation with infinitely far center. Soft. Comput. **24**(23), 17751–17759 (2020). https://doi.org/10.1007/s00500-020-05352-2
8. Blakley, G.R., Borosh, I.: Knuth's iterated powers. Adv. Math. **34**, 109–136 (1979)
9. Blakley, G.R., Borosh, I.: Modular arithmetic of iterated powers. Comput. Math. Appl. **9**, 567–581 (1983)
10. Brouwer, L.E.J.: Historical background, principles and methods of intuitionism. J. Symb. Log. **19**(2), 125 (1954)
11. Brouwer, L.E.J.: 1908 C - the unreliability of the logical principles. In: Heyting, A. (ed.) Philosophy and Foundations of Mathematics, pp. 107–111. North-Holland (1975)
12. Caldarola, F.: The exact measures of the Sierpiński *d*-dimensional tetrahedron in connection with a Diophantine nonlinear system. Commun. Nonlinear Sci. Numer. Simul. **63**, 228–238 (2018)
13. Caldarola, F.: The Sierpiński curve viewed by numerical computations with infinities and infinitesimals. Appl. Math. Comput. **318**, 321–328 (2018)
14. Caldarola, F., Cortese, D., d'Atri, G., Maiolo, M.: Paradoxes of the infinite and ontological dilemmas between ancient philosophy and modern mathematical solutions. In: Sergeyev, Y.D., Kvasov, D.E. (eds.) NUMTA 2019. LNCS, vol. 11973, pp. 358–372. Springer, Cham (2020). https://doi.org/10.1007/978-3-030-39081-5_31

15. Caldarola, F., d'Atri, G., Mercuri, P., Talamanca, V.: On the arithmetic of Knuth's powers and some computational results about their density. In: Sergeyev, Y.D., Kvasov, D.E. (eds.) NUMTA 2019. LNCS, vol. 11973, pp. 381–388. Springer, Cham (2020). https://doi.org/10.1007/978-3-030-39081-5_33
16. Caldarola, F., d'Atri, G., Pellegrini, M.: Combinatorics on n-sets: arithmetic properties and numerical results. In: Sergeyev, Y.D., Kvasov, D.E. (eds.) NUMTA 2019. LNCS, vol. 11973, pp. 389–401. Springer, Cham (2020). https://doi.org/10.1007/978-3-030-39081-5_34
17. Caldarola, F., Maiolo, M.: On the topological convergence of multi-rule sequences of sets and fractal patterns. Soft. Comput. **24**(23), 17737–17749 (2020). https://doi.org/10.1007/s00500-020-05358-w
18. Caldarola, F., Maiolo, M., Solferino, V.: A new approach to the Z-transform through infinite computation. Commun. Nonlinear Sci. Numer. Simul. **82**, 105019 (2020)
19. D'Alotto, L.: Infinite games on finite graphs using Grossone. Soft. Comput. **24**(23), 17509–17515 (2020). https://doi.org/10.1007/s00500-020-05167-1
20. Gisin, N.: Mathematical languages shape our understanding of time in physics. Nat. Phys. **16**(2), 114–116 (2020)
21. Ingarozza, F., Adamo, M.T., Martino, M., Piscitelli, A.: A Grossone-based numerical model for computations with infinity: a case study in an Italian high school. In: Sergeyev, Y.D., Kvasov, D.E. (eds.) NUMTA 2019. LNCS, vol. 11973, pp. 451–462. Springer, Cham (2020). https://doi.org/10.1007/978-3-030-39081-5_39
22. Knuth, D.E.: Mathematics and computer science: coping with finiteness. Science **194**, 1235–1242 (1976)
23. Leonardis, A., d'Atri, G., Caldarola, F.: Beyond Knuth's notation for unimaginable numbers within computational number theory. Int. Electron. J. Algebra **31**, 55–73 (2022)
24. Lolli, G.: Metamathematical investigations on the theory of Grossone. Appl. Math. Comput. **255**, 3–14 (2015)
25. Mazzia, F.: A computational point of view on teaching derivatives. Inform. Educ. **37**(1), 79–86 (2022)
26. Nasr, L.: Students' resolutions of some paradoxes of infinity in the lens of the grossone methodology. Inform. Educ. **38**(1), 83–91 (2023)
27. Rizza, D.: Primi passi nell'aritmetica dell'infinito. Preprint (2019)
28. Santo, F.D., Gisin, N.: Physics without determinism: alternative interpretations of classical physics. Phys. Rev. A **100**(6) (2019)
29. Scedrov, A.: Markov A. A. and Nagorny N. M.. The theory of algorithms. English translation by Greendlinger. Kluwer Academic Publishers, Dordrecht, 1988, xxiv + 369 pp. J. Symbol. Logic **56**, 336–337 (2014)
30. Sergeyev, Y.D.: Arithmetic of Infinity. Edizioni Orizzonti Meridionali, Cosenza (2003). 2nd ed 2013)
31. Sergeyev, Y.D.: Lagrange Lecture: methodology of numerical computations with infinities and infinitesimals. Rendiconti del Seminario Matematico dell'Università e del Politecnico di Torino **68**, 95–113 (2010)
32. Sergeyev, Y.D.: Some paradoxes of infinity revisited. Mediterr. J. Math. **19**, 143 (2022)
33. Sergeyev, Y.D., De Leone, R.: Numerical Infinities and Infinitesimals in Optimization. Springer, Cham (2022)

Visualization of Multilayer Networks

Ilaria Lazzaro[1](✉) and Marianna Milano[2]

[1] Department of Medical and Surgical Sciences, University "Magna Græcia", 88100 Catanzaro, Italy
[2] Department of Experimental and Clinical Medicine, University Magna Græcia, 88100 Catanzaro, Italy
{ilaria.lazzaro,m.milano}@unicz.it

Abstract. Network theory, in particular complex networks, has undergone considerable development finding its way into many real-world applications. However, they have several different types of relationships that cannot be represented by a mono layer network [11]. In fact, multilayer networks explicitly incorporate multiple channels, creating the right context to describe interconnected systems through different related layers, where nodes are the entities of the system and the edges represent the interactions between them. The application fields that may be modelled by multilayer networks range from human and social systems [8], to technological and transport systems, including biology and medicine [4]. Although various approaches for the visualization of multilayer networks have been suggested in recent years, this is an evolving field. In this work, we present an overview of the tools exploited for the visualization of multilayer networks. Then, we present a comparison among different tools through a case study represented by a real multilayer network. The network under investigation is a multilayer network composed of 8, 392 nodes and 128, 199 edges distributed over 2 layer that relate disease and drug. We propose methods for visualisation of the network that provide a topological analysis, attempting to derive the main measurement metrics. Through the visualisation carried out on the dataset, we will attempt to demonstrate the different layout configurations that a network can assume and the parameters that can be associated to each tool used in order to derive real scientific information.

Keywords: Network Visualization · Multilayer Network

1 Introduction

The study of multilayer networks makes possible to analyse the relationships between the elements of a system. Although traditional network connections only allow interactions between pairs, a multilayer network is characterised by entities that contain multiple types of connections and can exhibit several simultaneous relationships between interacting layers. The study of this model also makes it possible to encode richer information than is possible using single-layer models.

The mathematical concept of a simple network has been extensively studied and represented, for multilayer networks however, there is no conformity in structure and it assumes different terminology depending on the type of implementation [5]. The main existing multilayer network model can be distinguished between Multiplex Networks, Multilevel Networks, Hypergrids or Hypergraphs, interconnected or interdependent networks, a classification that takes into account the type of connection of a node to a layer, its structure or the type of graph. However, in the most general form, a multilayer network is defined as a tuple (A, L, V, E) where A denotes the set of actors, L the set of layers, and (V, E) a graph on $V \subseteq A \times L$ [6]. In this paper, we consider different scenarios of multilayer network visualisation, these methods offer new opportunities to study intractable systems [10]such as transport networks, social networks and in particular biological networks. In fact, the development of available software resources has become relevant and is constantly evolving, the comparison of layouts allows the study of the relationships between entities in the system. A case study based on the visualization of a specific multilayer network is then examined. The rest of the paper is organised as follows. Section 1 defines the network visualisation tools and refers to their structure and applicability. Section 2 describes the multilayer network used as a case study and the visualisation techniques applied on it. Section 3 presents and demonstrates the results obtained on the network under investigation.

2 Background

This section describes the contributions used for visualisation of multi-layer networks, explains the software architecture and then describes the individual components. The most common methods used for networks can be divided into two main groups, the first being GUI-based and the second API-based, which involves the use of high-level functions when the networks under consideration are large or the number of approaches is high, deriving the desired output.

2.1 GUI-Based Tool

Cytoscape and Gephi are two of the GUI-based solutions that support custom network manipulation. They are both two open source software projects that are flexible in terms of both visualisation and analysis.

2.1.1 Cytoscape. Cytoscape is defined as one of the most powerful high-performance network analysis projects [9]. Cytoscape provides the basic functionality for visualising and querying the network and for linking the network to functional annotation databases. Various extensions can be installed within the software. These are specific applications for multi-layer networks, which together allow the construction, management and visualisation of multi-layer networks, as well as the extraction of active multi-layer sub-networks.

Visualization of Multilayer Networks 311

2.1.2 Gephi. Gephi offers a similar set of functionalities but based on a 3D rendering engine for visualising large networks [1]. It allows the personalization of the node designer through a texture, panel or photo and is capable of handling large networks (over 20,000 nodes). Within the software, it is possible to add an algorithm, filter or tool even with little programming experience, Gephi allows multiple algorithms to run simultaneously in separate workspaces. Integrated layout visualisation plug-ins modify parameters such as colour or size, for the understanding of structure or network content. Techniques have been developed to increase clarity and readability; the architecture supports graphs of different structures. Although the networks can be explored interactively, they can also be exported and saved in PDF or SVG format.

2.2 API-Based Tool

On the other hand, API-based methods provide access from popular programming languages such as Pyhtom, JavaScript, C++ and R, which are particularly used when the entire network analysis is performed in the same environment. In particular, in this paper we focus on R, an environment consisting of a set of libraries, objects that allow, in this case, the modelling and creation of both real and virtual networks. In the specific case of networks, there are two particular packages, IGraph and Multinet, whose interfaces permit the manipulation and visualisation of structural properties, ensuring export in various file formats. A particular platform for visualization and analysis of multilayer networks available within R is Muxviz. The implementation exploits a graphical user interface for rendering networks and abstracting relationships by representing them in a visual and analytical format.

2.2.1 Igraph Package. Igraph is a package developed to efficiently manage large graphs, is open source and uses efficient data structures. It allows sophisticated information to be extracted from a network, from simple analysis metrics to active community extraction. It does not have a graphical interface; in fact, it is embedded within high-level languages (R and Python) and supports rapid development and fast prototyping [2]. The Igraph library has a three-layer architecture, connected by well-defined interfaces. A disadvantage is the lack of functionality in some areas compared to other existing network analysis packages; one of these is the visualisation of graphs.

2.2.2 Multinet Package. Multinet is an R package that was created for the study of multilayer networks [7]. It defines classes for storing multilayer networks and functions for analysing and extracting them. This package consists of different methods that include reading and writing multilayer networks, functions for extracting information on network components (actors, layers, vertices and edges), attribute management and analysis measures for extracting measures based on single actor or single degree, and functions for comparing different layers.

2.2.3 MuxViz.

MuxViz is used for the visualization and analysis of multilayer networks [3]. The tool provides a range of features to explore multilayer networks, including multilayer correlation analysis, community detection, multilayer centrality analysis as well as various parameters characterizing nodes and edges to evaluate network structure and dynamics. MuxViz focuses mainly on multiplex networks through the production of animated visualizations of dynamic network processes. It allows the choice of different layouts, to determine the position of nodes and multilayer edge structure. It is open source, specifically R allows the creation of a graphical user interface that can be used both locally and through browsers.

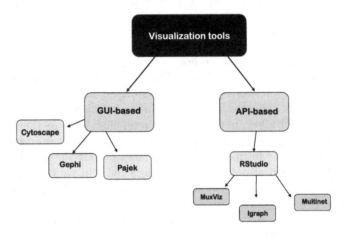

Fig. 1. Outline of the main methods used for visualisation and analysis of multilayer networks.

3 Visualization of Multilayer Network

3.1 The Multilayer Network

The case study under consideration is a multilayer network consisting of 8392 nodes and 128199 edges distributed over 2 layers. The initial database is available on https://snap.stanford.edu/ and consists of a dataset of diseases (sent by the acronym DOID), a dataset of drugs (sent by the acronym DB) and a dataset linking diseases to drugs.

3.2 Visualization

The visualization tool used is Gephi, the network under consideration is imported through a list of the actors (disease and drug) that make it up and a list of the relationships (the interactions between the actors). Within the software, it is possible to change the designer of the nodes and the colour of the two categories. However, in the 'Ranking' panel, the nodes are sized proportionally to their rank (number of connections). The output shows a visual distinction between Drug (green nodes) and Disease (pink nodes) and the degree to which the nodes are connected.

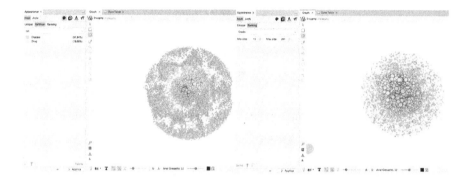

Fig. 2. Layout of visualisation of disease-drug categories, and degree of connection between nodes in the network.

Specific plug-ins allow you to go beyond the basic functionality. The one used for the Multilayer network is Network Splitter 3D whose function is to display

Fig. 3. Network output using the Network Splitter 3D plug-in, is used to divide a network layout into distinct z-levels. The network is displayed according to category (drug and disease).

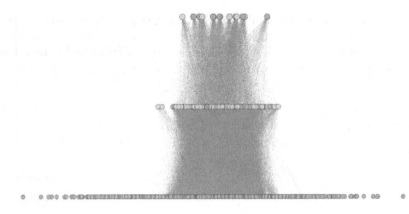

Fig. 4. Proposed network layout with variation of z-maximum level parameters and rotation angle Alpha

nodes on different network levels depending on the selected column. For the different configuration, it is necessary to change the parameters of Z- MAximun level (displays nodes in different levels according to the selected column), Z-scale (adjusts the vertical z-scale), Z-Distance Factor (adjusts the distance between each z-level layout), Alpha (rotates the network angle) which act on the network and reproduce the desired output.

4 Conclusion

Multilayer networks are frameworks widely used and studied for the analysis of complex systems and entities interconnected in different layers. Although they are extremely useful tools for real-world information extraction, visualization of multilayer networks remains one of the open problems in research due to computational limitations and costs. Modeling networks for analysis purposes may be impractical on large networks with a large number of interconnected layers. Although several network processing techniques have been proposed, however, in addition to visualization limitations, topological network analysis also involves the use of techniques to simplify and manage these particular structures. To solve these problems, as future work we plan to analyze the topology structure of the network under consideration and implement a parallel architecture for visualization of multilayer networks.

Acknowledgements. This work was funded by the Next Generation EU - Italian NRC, Mission 4, Component 2, Investment 1.5, call for the creation and strengthening of 'Innovation Ecosystems', building 'Territorial R&D Leaders' (Directorial Decree n. 2021/3277) - project Tech4You - Technologies for climate change adaptation and quality of life improvement, n. ECS0000009. This work reflects only the authors' views and opinions, neither the Ministry for University and Research nor the European Commission can be considered responsible for them.

References

1. Bastian, M., Heymann, S., Jacomy, M.: Gephi: an open source software for exploring and manipulating networks. In: Proceedings of the international AAAI Conference on Web and Social Media, vol. 3, pp. 361–362 (2009)
2. Csardi, G., Nepusz, T., et al.: The igraph software package for complex network research. Inter. J. Complex Syst. **1695**(5), 1–9 (2006)
3. De Domenico, M., Porter, M.A., Arenas, A.: MuxViz: a tool for multilayer analysis and visualization of networks. J. Complex Netw. **3**(2), 159–176 (10 2014). https://doi.org/10.1093/comnet/cnu038
4. De Toni, A.F., Marcovig, M., Nonino, F.: Lâ'organizzazione informale nella prospettiva dell'âanalisi dei network. Economia & Manag. **4**, 93–114 (2007)
5. Hammoud, Z., Kramer, F.: Multilayer networks: aspects, implementations, and application in biomedicine. Big Data Analy. **5**(1), 2 (2020)
6. Magnani, M., Hanteer, O., Interdonato, R., Rossi, L., Tagarelli, A.: Community detection in multiplex networks. ACM Comput. Surv. **54**(3) (2021). https://doi.org/10.1145/3444688
7. Magnani, M., Rossi, L., Vega, D.: Analysis of multiplex social networks with r. J. Stat. Softw. **98**(8), 1–30 (2021)
8. Murase, Y., Török, J., Jo, H.H., Kaski, K., Kertész, J.: Multilayer weighted social network model. Phys. Rev. E **90**(5), 052810 (2014)
9. Shannon, P., et al.: Cytoscape: a software environment for integrated models of biomolecular interaction networks. Genome Res. **13**(11), 2498–2504 (2003)
10. Škrlj, B., Kralj, J., Lavrač, N.: Py3plex toolkit for visualization and analysis of multilayer networks. Appli. Netw. Sci. **4**(1), 1–24 (2019). https://doi.org/10.1007/s41109-019-0203-7
11. Zhong, L., Zhang, Q., Yang, D., Chen, G., Yu, S.: Analysing motifs in multilayer networks. arXiv preprint arXiv:1903.01722 (2019)

Legal Systems and Fractals, Towards Infinity Computing

Maria Rita Maiolo[1](✉) and Mària Ivano[2]

[1] Department of Legal and Business Sciences, University of Calabria, 87036 Arcavacata di Rende, CS, Italy
mariarita.maiolo@unical.it
[2] Liceo Scientifico "G.B. Scorza", 87100 Cosenza, Italy

Abstract. The focus of this work is to take an approaching step, or rather, try to create a stronger connection than those existing between jurisprudence and mathematics. In particular, between legal systems in the broadest generality and fractal structures, using in particular the von Koch curve and infinity computing which allows a precise measurement of infinite quantities, and therefore of stretches of fractal curves with infinite length.

Keywords: Legal systems · Fractals · Hausdorff dimension · Fractal curves · Von Koch curve · Grossone

1 Introduction

In the Modern Era we have witnessed the consolidation of an orientation towards distinct disciplines of study and different fields of research with well-defined boundaries, in contrast to the ancient view that sought the unity of knowledge. Today, we are beginning to see a new reversal of this trend, aimed at recovering a community of views and an exchange of tools between the technical-scientific and humanistic disciplines. This article aims to make a small contribution to this process by proposing a rapprochement between two fields of knowledge that are generally considered to be distant: law and mathematics. Specifically, it aims to propose an unusual conference between one of the most recent topics in mathematics, namely fractals and the fractal geometry developed by them, and the intimate essence of the legal reasoning and argumentation. The first to apply a geometric view to jurisprudence was J. Balkin in 1986 in an article entitled *"The Crystal Structure of Legal Thought"* [6]. The author speaks of "the crystalline structure" because in 1986 he was not yet aware of the fractals and the results that Mandelbrot and others had achieved in the previous two decades, as he himself specifies in a subsequent article of over 120 pages in 1991 *"The promise of legal semiotics"* [7]. In fact, in 1967, B. Mandelbrot, with the publication of the article *"How long is the coast of Britain?"* [20] had opened up a new way of interpreting many physical objects and structures found in nature.

Many other papers and books followed [20] in the next two or three decades, for example the very famous book [21] by Mandelbrot himself or [15,18], but a real explosion in the research and applications of fractals has occurred in the last quarter of a century.

As regards fractals and legal systems, in 2000, two eminent researchers, D. Post and M. Eisen, the first a legal scholar, the second a computational biologist, published an article clearly inspired by [20] entitled *"How long is the coastline of the law? Thoughts on the fractal nature of legal systems"* [25]. More recently another single article [34] was published in 2013 by A. Stumpff on the relations between fractals and legal systems. To our knowledge, no other works different from the quoted ones exist until [19] in 2022 by one of the author of this paper.

The purpose of this work is to take it one step further into investigating the possibility of study legal systems with metric and geometric tools. We continue the ideas proposed in [6,7,25,34] and [19] into the direction of using fractals to interpret the internal structure of jurisprudence and legal systems. In particular we use the von Koch curve as fractal model to represent and, in some sense, encode internal paths joining different points inside legal systems, bodies of law, legal argumentation, and jurisprudence in general. A further powerful tool that we can use to push on the ideas of introducing concrete and numerical metrics into legal systems is the infinity computing introduced by Sergeyev. His new methodology allows to assign precise infinite numerical values, written with the help of a new infinite numerical unit called *grossone*, to generic infinite quantities and sizes. This is just what we need to numerically evaluate the distance of two points on the von Koch curve, where by distance we mean the length of the path obtained by walking along the curve and not the Euclidean distance in the plane. We conclude that, by measuring sections of fractal curves we can find somewhat a metric inside legal systems: using the expression of Post and Eisen, we can precisely measure all pieces of "the coastline of the law" in the same way we can give a precise infinite numerical value to a piece of real coastline or to the perimeter of an island (if we are able to mathematically model its shape).

Furthermore, the capacity to introduce a metric into legal systems allows to study them through the use of fractal or Hausdorff dimension as proposed in [19].

As regards the structure of this paper, Sect. 2 gives to the reader some basic notations and examples how to write infinite numbers through the grossone-based system and some basic references, with particular attention to fractal applications. In Sect. 3 we recall the construction of the von Koch curve and use it to explain and represents paths inside legal systems.

In this paper we use the symbols $\mathbb{N}, \mathbb{N}_0$ and $\mathbb{R}$ to denote the set of positive, non-negative integers and real numbers, respectively.

2 Infinity Computing and Fractals

In the early 2000 Y. Sergeyev introduced a new numerical system which allows computations not only with ordinary finite numbers (e.g., natural, rationals, reals, etc.) but also with infinite and infinitesimal ones. Sergeyev's system is built on two fundamental units: the ordinary 1 to generate finite numbers, and a new infinite unit① called *grossone* which generates infinite numbers and infinitesimal ones (through its reciprocal $①^{-1} = 1/①$). Examples of infinite and infinitesimal numbers in the new system are the following

$$2①, \quad 5①/9, \quad 4①^2 - 5①, \quad -6①/11 + 8①^{1/2}, \quad 6①^{-2}, \quad -4①^{-1} + 5①^{-3}.$$

Consider now the following number

$$a = -\frac{2}{3}①^3 + 5①^{2/3} - 7 + 3\sqrt{2} - 6①^{-1/2} + \frac{5}{4}①^{-2};$$

it has an infinite part $(-\frac{2}{3}①^3 + 5①^{2/3})$, a finite one $(-7 + 3\sqrt{2})$ and an infinitesimal part $(-6①^{-1/2} + \frac{5}{4}①^{-2})$. To perform computations using the common operations is very easy and intuitive in the grossone system. This is also proved by several researches which propose Sergeyev's system in some Italian high schools, using even zero knowledge tests with very satisfactory results: see for instance [3,16] and [2] in this volume. See also the very recent papers [22,23] and the book [26]. The reader can find extensive details on the new system into introductory surveys by Sergeyev himself as [27,32] or [29] also in Italian.

The new system, or methodology, has been successfully applied to a number of areas both of mathematics, physics, biology and applied sciences. For example, [1,5,10,13,14,17,32,33] apply the grossone-based systems to ordinary differential equations and optimization, cellular automata, game theory, and logic paradoxes. But for the arguments discussed in this paper, the mist important applications are those relative to fractals and fractal curves (see [4,8,9,11,12,24,28,30]), and in particular to von Koch curve (see [31]) which occupies a central role for our discussion, see Sect. 3.

3 The Von Koch Curve and Legal Constructions

3.1 Construction of the Von Koch Curve

Let K_0 be the unitary interval $[0,1] \times \{0\}$ contained in $\mathbb{R}^2$. We obtain K_1 by dividing K_0 into three parts and by replacing the central one (i.e. $[1/3, 2/3] \times \{0\}$) by the two sides of the equilateral triangle with base $[1/3, 2/3] \times \{0\}$ and a vertex in $(1/2, \sqrt{3}/6)$. In other words K_1 is the polygonal shown in Fig. 1(a) whose vertices, in order from left, are

$$(0,0), \quad \left(\frac{1}{3}, 0\right), \quad \left(\frac{1}{2}, \frac{\sqrt{3}}{6}\right), \quad \left(\frac{2}{3}, 0\right), \quad (1,0).$$

To obtain K_2 we apply the same procedure used to get K_1 from K_0 to each of the four line segments constituting the polygonal K_1: the result is shown in Fig. 1(b).

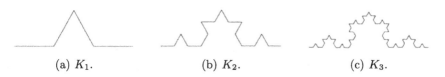

(a) K_1. (b) K_2. (c) K_3.

Fig. 1. The first three steps starting from K_0 in the construction of von Koch curve K.

Iterating this process n times we obtain a polygonal K_n composed of 4^n line segments all with the same length equal to $(1/3)^n$. Therefore we can deduce the following formula for the length $l(K_n)$ of the polygonal K_n,

$$l(K_n) = \left(\frac{4}{3}\right)^n \qquad (1)$$

for all non-negative integers n. For instance, K_3 is a polygonal line composed by $4^3 = 64$ line segments of length $(1/3)^3 = 1/27$, as shown in Fig. 1(c). The only thing you have to pay attention to is that of constructing the equilateral triangles on the correct part every time, that is, an observer who travels the segment K_0 from $(0,0)$ to $(1,0)$, he sees the curve K_1 to his left in the middle third $[1/3, 2/3] \times \{0\} \subset \mathbb{R}^2$. And the same when an observer travels along the curve K_n from $(0,0)$ to $(1,0)$ for all $n \in \mathbb{N}_0$: he sees the curve K_{n+1} to his left where K_n does not coincide with K_{n+1}.

Using the Hausdorff distance as usual in fractal geometry, the sequence $\{K_n\}_{n \in \mathbb{N}_0}$ of closed compact sets contained in $\mathbb{R}^2$ is proved to be a Cauchy sequence, hence it will converge to a set K called the *von Koch curve*. Hence K is the limit

$$K := \lim_{n \in \mathbb{N}} K_n.$$

3.2 Paths Inside Legal Systems

Legal arguments, the typical constructs of legal disciplines, legal systems and legal corpus, certainly do not follow a straight line to go from one point to another, however one understands them and whatever the two points in question represent. Similar observations have been made in the past and from different points of view in [6,7,25,34] and [19], as we mentioned in the Introduction.

We shall now try to examine, with an example, the tortuous ramifications typical of legal arguments. By way of example, we shall consider the case of a dispute concerning the exploitation of a patent for an invention and try to show how legal arguments can branch off, in this situation, *ad infinitum*, without ever potentially reaching an end point or an end (see [19]). This will result in the approximation of this logical-spatial construct to the fractal geometry and in particular to the von Koch curve, or rather to its approximations K_n.

Let us suppose that plaintiff A accuses defendant B of improperly using a certain patent or reproducing its contents without having the rights to do so.

At a first level, the level of maximum generality, one may ask whether or not B is liable for infringing ah intellectual property or copyright covered by a patent. But immediately thereafter, at the second level the question branches off or deviates from an ideal straight line like successive approximations K_n of the von Koch fractal curve.

Consider, for example, four branches or four changes of direction such as the following:

1. Is applicant A really the owner of the copyright or the invention?
2. Is the object protected by the patent or the patented process really subject to a regular patent?
3. Are the proceeds derived from that invention or innovation really subject to a patent or patentability?
4. Has B infringed one or more rules against A?

Defendant B could at this point argue along one, two, three or all four of the branches exemplified above, producing other strands and other jagged line. For example, B could divide issue 1 into three sub-issue as follows:

1.1. Is the invention or innovative process in question truly original? Or is it a development of something already existing or known?
1.2. In the patented product fully and incontrovertibly result of A's ingenuity or work?
1.3. Do the relevant laws permit the product in question to be protected or covered by a patent?
1.4. Is the patent temporally valid and in existence at the time or during the alleged infringements?
1.5. Have the rights in question been transferred?

The list could continue with further branches or deviation of "level two" and, in addition, each of the points in the list could give rise to further branches or deviations of "level three".

For example, item 1.3 could give rise the following branches:

1.3.1. Does the patent in question fully comply with the statutory indications and requirements?
1.3.2. Does what A would like to be protected coincide with what is covered by the patent?
1.3.3. Does the benefit obtained by B derive directly, according to the applicable legal interpretation, from what is protected?

In turn, point 1.3.1 opens up a series of considerations and further sub-branches that go into the merits of the specific legislation in force in Italy (or in another country), which in turn refer to international conventions.

Now it should be quite clear how it is possible to continue reasoning *ad infinitum* by successive, more or less complex articulations, which we can think of as deviations from a simple ideal straight line. And it is precisely the infinite branching or jaggedness that allows the leap towards fractal figures and fractal geometry. The above iterations, therefore, are to be understood as unlimited and with endless descending ramifications. Thus, to measure or evaluate the distance between any two points, however understood (two laws, two normative entities, or two competing positions A and B as in our case), the length of a piece of the von Koch curve seems very appropriate.

To show a numerical example and, at the same time, the great usefulness, or necessity in this context, of grossone methodology, let us consider the following two points:

$$A = (0,0) \quad \text{and} \quad B = \left(\frac{1}{3}, \frac{\sqrt{3}}{9}\right)$$

belonging to all K_n, $n \geq 2$, and so also to K. The distance between A and B along K_2 is $2/3$, along K_3 is $8/9$ and in general, along K_n is

$$\frac{1}{2} \cdot \left(\frac{4}{3}\right)^{n-1}$$

for all $n \geq 2$. This means that the distance between A and B along the von Koch fractal curve after is an infinite distance. For instance, considering ① steps in the construction (i.e. $K_①$), such infinite distance is expressible through the grossone-based system as

$$\frac{1}{2} \cdot \left(\frac{4}{3}\right)^{①-1}.$$

In conclusion, it is evident that the same procedure described above can be applied, at least in theory, in many different situations that arise in the legal field.

4 Future Work

It is clear that there is still much to be studied and developed along the lines sketched out in this article. There are many legal contexts to be examined in the future. Furthermore, from a mathematical point of view, the von Koch fractal curve has many variants of different types, which can be considered. For instance, to give a very simple example, one can consider unequal tripartitions of the original interval $[0,1] \times \{0\}$ and constructions with triangles other than the equilateral one. This would mean introducing construction parameters, even variable, that could be calibrated *ad hoc* to best interpret different contexts and situations.

Acknowledgments. This research was supported by the following grants: PON R & I 2014-2020 Action IV.4, Green Home s.c.ar.l., and NRRP mission 4 for Doctoral research.

References

1. Amodio, P., Iavernaro, F., Mazzia, F., Mukhametzhanov, M.S., Sergeyev, Y.D.: A generalized Taylor method of order three for the solution of initial value problems in standard and infinity floating-point arithmetic. Math. Comput. Simul. **141**, 24–39 (2017). https://doi.org/10.1016/j.matcom.2016.03.007
2. Antoniotti, L., Astorino, A., Caldarola, F.: Unimaginable numbers and infinity computing at school: an experimentation in Northern Italy. In: Sergeyev, Y.D., Kvasov, D.E., Astorino, A. (eds.) NUMTA 2023, LNCS 14478, pp. xx–yy. Springer, Cham (2025). https://doi.org/10.1007/978-3-031-81247-7_17
3. Antoniotti, L., Caldarola, F., d'Atri, G., Pellegrini, M.: New approaches to basic calculus: an experimentation via numerical computation. In: Sergeyev, Y.D., Kvasov, D.E. (eds.) NUMTA 2019. LNCS, vol. 11973, pp. 329–342. Springer, Cham (2020). https://doi.org/10.1007/978-3-030-39081-5_29
4. Antoniotti, L., Caldarola, F., Maiolo, M.: Infinite numerical computing applied to Hilbert's, Peano's, and Moore's Curves. Mediterr. J. Math. **17**(3), 1–19 (2020). https://doi.org/10.1007/s00009-020-01531-5
5. Astorino, A., Fuduli, A.: Spherical separation with infinitely far center. Soft. Comput. **24**(23), 17751–17759 (2020). https://doi.org/10.1007/s00500-020-05352-2
6. Balkin, J.: The crystalline structure of legal thought. Rutgers Law Rev. **39**, 1–103 (1986)
7. Balkin, J.: The promise of legal semiotics. Univ. Texas Law Rev. **69**, 1831–1852 (1991)
8. Caldarola, F.: The exact measures of the Sierpiński d-dimensional tetrahedron in connection with a Diophantine nonlinear system. Commun. Nonlinear Sci. Numer. Simul. **63**, 228–238 (2018). https://doi.org/10.1016/j.cnsns.2018.02.026
9. Caldarola, F.: The Sierpiński curve viewed by numerical computations with infinities and infinitesimals. Appl. Math. Comput. **318**, 321–328 (2018). https://doi.org/10.1016/j.amc.2017.06.024
10. Caldarola, F., Cortese, D., d'Atri, G., Maiolo, M.: Paradoxes of the infinite and ontological dilemmas between ancient philosophy and modern mathematical solutions. In: Sergeyev, Y.D., Kvasov, D.E. (eds.) NUMTA 2019. LNCS, vol. 11973, pp. 358–372. Springer, Cham (2020). https://doi.org/10.1007/978-3-030-39081-5_31
11. Caldarola, F., Maiolo, M.: On the topological convergence of multi-rule sequences of sets and fractal patterns. Soft. Comput. **24**(23), 17737–17749 (2020). https://doi.org/10.1007/s00500-020-05358-w
12. Caldarola, F., Maiolo, M., Solferino, V.: A new approach to the Z-transform through infinite computation. Commun. Nonlinear Sci. Numer. Simul. **82**, 105019 (2020). https://doi.org/10.1016/j.cnsns.2019.105019
13. Cococcioni, M., Cudazzo, A., Pappalardo, M., Sergeyev, Y.D.: Solving the lexicographic multi-objective mixed-integer linear programming problem using branch-and-bound and grossone methodology. Commun. Nonlinear Sci. Numer. Simul. **84**, 105177 (2020). https://doi.org/10.1016/j.cnsns.2020.105177
14. D'Alotto, L.: Infinite games on finite graphs using grossone. Soft. Comput. **24**(23), 17509–17515 (2020). https://doi.org/10.1007/s00500-020-05167-1
15. Hastings, H.M., Sugihara, G.: Fractals: A User's Guide for the Natural Sciences. Oxford University Press, Oxford (1994)
16. Ingarozza, F., Adamo, M.T., Martino, M., Piscitelli, A.: A Grossone-based numerical model for computations with infinity: a case study in an Italian high school. In: Sergeyev, Y.D., Kvasov, D.E. (eds.) NUMTA 2019. LNCS, vol. 11973, pp. 451–462. Springer, Cham (2020). https://doi.org/10.1007/978-3-030-39081-5_39

17. Iudin, D., Sergeyev, Y.D., Hayakawa, M.: Infinity computations in cellular automaton forest-fire model. Commun. Nonlinear Sci. Numer. Simul. **20**, 861–870 (2015). https://doi.org/10.1016/j.cnsns.2014.06.031
18. Kaandorp, J.A.: Fractal Modelling: Growth and Form in Biology. Springer, Heidelberg (1994)
19. Maiolo, M.R.: Frattali e sistemi giuridici. Queste Istituzioni **2**(2022), 135–147 (2022)
20. Mandelbrot, B.B.: How long is the coast of Britain? Science **156**, 636–638 (1967)
21. Mandelbrot, B.B.: The Fractal Geometry of Nature. W. H. Freeman and Co., New York (1982)
22. Mazzia, F.: A computational point of view on teaching derivatives. Inform. Educ. **37**, 79–86 (2022)
23. Nasr, L.: Students' resolutions of some paradoxes of infinity in the lens of the grossone methodology. Inform. Educ. **38**, 83–91 (2023)
24. Pepelyshev, A., Zhigljavsky, A.: Discrete uniform and binomial distributions with infinite support. Soft. Comput. **24**(23), 17517–17524 (2020). https://doi.org/10.1007/s00500-020-05190-2
25. Post, D.G., Eisen, M.B.: How long is the coastline of the law? Thoughts on the fractal nature of legal systems. J. Leg. Stud. **29**, 545–584 (2000)
26. Rizza, D.: Primi passi nell'Aritmetica dell'Infinito. Bonomo Editore (2023)
27. Sergeyev, Y.D.: Arithmetic of Infinity. Edizioni Orizzonti Meridionali, Cosenza (2003). 2nd ed 2013)
28. Sergeyev, Y.D.: Evaluating the exact infinitesimal values of area of Sierpinski's carpet and volume of Menger's sponge. Chaos Solit. Fractals **42**, 3042–3046 (2009). https://doi.org/10.1016/j.chaos.2009.04.013
29. Sergeyev, Y.D.: Lagrange Lecture: methodology of numerical computations with infinities and infinitesimals. Rendiconti del Seminario Matematico dell'Università e del Politecnico di Torino **68**, 95–113 (2010)
30. Sergeyev, Y.D.: Using blinking fractals for mathematical modelling of processes of growth in biological systems. Informatica **22**, 559–576 (2011)
31. Sergeyev, Y.D.: The exact (up to infinitesimals) infinite perimeter of the Koch snowflake and its finite area. Commun. Nonlinear Sci. Numer. Simul. **31**, 21–29 (2016). https://doi.org/10.1016/j.cnsns.2015.07.004
32. Sergeyev, Y.D.: Numerical infinities and infinitesimals: methodology, applications, and repercussions on two Hilbert problems. EMS Surv. Math. Sci. **4**, 219–320 (2017). https://doi.org/10.4171/EMSS/4-2-3
33. Sergeyev, Y., De Leone, R. (eds.): Numerical Infinities and Infinitesimals in Optimization. Springer, Cham (2022)
34. Stumpff, A.M.: The law is a fractal: the attempt to anticipate everything. Loyol Univ. Chicago Law J. **44**, 649–681 (2013)

Exploit Innovative Computer Architectures with Molecular Dynamics

Filippo Marchetti[1,2](✉) and Daniele Gregori[1]

[1] E4 Computer Engineering, via Martiri della Libertà 66, Scandiano, Italy
{filippo.marchetti-ext,daniele.gregori}@e4company.com
[2] INSTM, Via G. Giusti 9, Firenze, Italy

Abstract. Atomistic simulations are a powerful tool to analyse the structure and the behavior of various biological molecules and are widely used in biochemistry studies. Currently, there are several software packages available which provide the algorithms for the numerical resolution of the physical equation necessary for the simulations. Those softwares are demanding and need to handle a large amounts of resources, therefore the hardware in use has to be efficient in terms of performance and costs. Here we tested the Molecular Dynamics toolkit Gromacs on different platforms: Intel, AMD and ARM clusters are considered, also with the support of NVIDIA GPUs. During the tests energy consumption is measured in both idle and working case obtaining the energy to solution for every machine. While the highest performances are obtained with Intel we found that AMD and ARM servers are valid option if energy savings is considered alongside performance, the use of GPU gives a remarkable contribution to reduce consumption.

Keywords: Gromacs · High Performance Computing · Molecular Dynamics

1 Introduction

Proteins, given their relevant position in many biological pathways, are often chosen as target of biomedical and pharmaceutical studies. In order to better comprehend the behavior and function of proteins various studies have focused on the computational analysis of the structure and the motion of the molecules through Molecular Dynamic (MD) simulations, a tool that enables scientists to follow conformational variation of the molecule using a simplified model where all or most of the atoms are considered. Since the first studies on proteins in 1970s [1] the continuous development of computers through the years has contributed to the rise in popularity of MD in structural biology and biophysics community [2].

The simulation of a protein in explicit solvent requires, even for small sized systems, a large amount of atoms to be considered and the computational power needed to complete a simulation in a reasonable timespan scales with the size of the system. The majority of softwares used for MD keeps developing in terms of parallelization efficiency and GPU acceleration, improving the accessibility of these simulations, considering that usually the systems studied range in size from 10^4 to 10^5 atoms. When very large systems are studied the resources needed increase dramatically and normally are performed in High-Performance Computing (HPC) clusters, where great amount of CPUs or GPUs are grouped together. We have admirable examples of supercomputing centers listed in the top 500 [3]. Since the required resources are demanding in terms of computational and energetic costs the choice of the platforms for a dedicated HPC system should be supported by tests of the behaviour of the hardware with a real-life use case. An example of this type of approach could be observed in a recent work were different cloud resources are tested with an MD suite [4]. In our work we test Intel, AMD and ARM based platforms with an MD software, in particular we ask if exotic ARM based machines are a reasonable alternative to common x86 servers.

2 Materials and Methods

2.1 Description of the Machines Used

For this work we probe five different systems: an Intel base, one AMD base and three different ARM based servers. All tests were conducted in the E4 R&D unit allowing the maintenance of costant environment temperature and power supply. A list of all the equipment is provided.

1. **Intel**, The overall system is a supermicro twinsquare. Each node is a dual socket Intel Xeon Gold 6226R CPU (2.90 GHz) 16-core processor, 192 GB RAM. GPU NVIDIA A100.
2. **AMD**, The overall system is a supermicro twinsquare. Each node is a dual socket AMD EPYC 7313 (3.0 GHz) 16-core processor, 512 GB RAM.
3. **ARM-1**, Seven nodes cluster. Each server is a dual socket Cavium ThunderX2 (2.5 GHz) 32-core processor, 256 GB RAM. GPU NVIDIA V100.
4. **ARM-2**, Two nodes, dual socket Neoverse-N1 (3.0 GHz) 128-core processor, 512 GB RAM.
5. **ARM-3**, Two nodes, single socket Neoverse-N1 (1.0 GHz) 80-core processor, 256 GB RAM. GPU NVIDIA A100.

2.2 Software

The core idea of Molecular Dynamics is to simulate the evolution of a many-body system using classical mechanics, i.e. integrating Newton's equation of motion (1):

$$\frac{d^2 x_i}{dt^2} = \frac{F_x}{m_i} = -\frac{\partial V}{m_i \partial x_i} \tag{1}$$

The potential V is the sum of the interactions between all the particles and, for a all-atom protein simulation, can be expressed as:

$$V(\bar{r}) = \sum_{bond} \frac{1}{2}K_b(b-b_0)^2 + \sum_{angles} \frac{1}{2}K_\theta(\theta-\theta_0)^2 + \sum_{impdihed} \frac{1}{2}K_\phi(\phi-\phi_0)^2 \\ + \sum_{dihedral} K_\psi(1+cos(n\psi-\delta)) + \sum_{(i,j)}(\frac{C_{12}(i,j)}{r_{ij}^{12}} - \frac{C_6(i,j)}{r_{ij}^6} + \frac{q_iq_j}{4\pi\epsilon_0\epsilon_r r_{ij}}) \quad (2)$$

In the potential there is a contribution of long range non-bonded functions, Lennard-Jones and Coulombian potentials, and bonded, angular and dihedral elements. The potential is computed using a set potential functions and parameters that are collected in force fields, modern molecular dinamics suites provide different types of force fields.

The molecular simulation was conducted with Gromacs package [10,11] using the amberff99SB-ildn forcefield [6]. After an initial minimization, the system equilibration was conducted at a constant temperature of 300 K for 100 ps and, successively, at constant temperature and pressure (300 K, 1 bar) for 100 ps. For temperature and pressure coupling, the modified Berendsen thermostat (v-rescale) and Parrinello-Rahman barostat were used [7,8] with τT of 0.1 ps and τP 2 ps. Electrostatic forces were evaluated by Particle Mesh Ewald method [9] and Lennard-Jones forces by a cutoff of 0.9 nm.

2.3 Models

In this work two different molecular systems are used as a benchmark, the esterase enzyme and the Spike protein complex. Two different models permit to compare the behavior of the platforms with a small system (Esterase enzyme) and with a huge systems (Spike protein).

Spike. This model consists of the SARS-CoV-2 spike (S) protein in complex with C102 antibody. The spike protein is present on the viral surface and is the target of most of the antibody known to neutralize SARS-CoV-2. The study of the spike-antibody interaction could improve the knowledge about antibody response that can be important for therapeutic and vaccine development.

The structure model of the spike protein was obtained from the wild type protein used in the simulation describe by Grant et al. [12], without glycan envelope, based on the electron microscopy structure with PDB entry 6VSB. The interface with the antibody was modelled from the PDB entry 7K8M. The complex obtained has a total residue count of 3836 and it was solvated with tip3p water model in a cubic box (1.2 nm of margin), Cl- ions were added to ensure electroneutrality. The final system consists of 1604163 atoms.

Esterase. This model is a cold-adapted esterase EstA, a lypolitic enzyme from bacteria that lives in permanently cold environments, these kind of enzymes can be employed as additives to detergents that require low temperatures or as biocatalyst for low temperature reactions [13,14].

The enzyme structure was modelled from the PDB entry 3HP4, for a total residue count of 162 and it was solvated with tip3p water model in a cubic box (1 nm of margin), electroneutrality was reached adding Na+ atoms. The final system consists of 35182 atoms.

3 Results and Discussion

The two molecular models previously described are used as a benchmark to test molecular dynamics performances on different architectures. Gromacs package comes with various algorithms for integration of the equation of motion, in our case leap-frog [5] is used. The length of the integration step has to be chosen wisely considering a trade-off between performance and computation errors, the time interval is fixed to 0.002 ps. While this parameter refers to the time of a single step integration of the model the real time spent in the computation of the single step goes to an order of ms for Esterase to tens of ms for Spike. For comparing the behaviour of the simulation on different machines the results are reported in the common measure for Gromacs performances, i.e. the length of the simulation time covered in a day (ns/day).

3.1 CPU Parallelization

The first test is conducted without the use of GPU acceleration to check the efficiency of the CPU parallelization. The value of the performance is obtained taking the average of the measure of four independent runs.

For the Esterase model the tests reveal that the best absolute performance is reached with Intel cases. When per-core performance is considered Intel and AMD machines give the highest results for single node cases, when the number of nodes is increased both Intel and AMD systems obtain a linear speed up with Intel case have a better scaling. For ARM cases we have ARM-1 cluster that gives the lowest performances, with a worse scaling respect to Intel and AMD machines, and ARM-2 that shows the best performances when a single node is considered. With the spike protein case we observe a behaviour similar to the Esterase case, where the best absolute is gained with Intel machines (Fig. 1b). For a single node comparison ARM machines have the highest performance. When more nodes are considered Intel and ARM have a better scaling than AMD, especially the four nodes case shows a significant drop in performance (respect to a linear scaling). In both Esterase and Spike cases it is worth notice that ARM platform used in ARM-2 node could be an interesting alternative for Intel clusters, also AMD machines gives good performances but our results suggest that may rise a worse efficiency in scaling, especially when system size is increased.

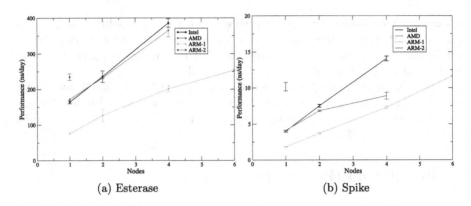

(a) Esterase (b) Spike

Fig. 1. In the Figure are reported the values for the performance in terms of ns/day scaling from 1 to 6 nodes obtained without the use of GPUs. In figure a there are data for Esterase model and in figure b for the Spike model.

3.2 GPU Acceleration

The measurements show the relevant impact of the GPU acceleration on simulation performances. In Fig. 2 we observe that with the use of both A100 and V100 the performance scale more than four times respect to the CPU only single node

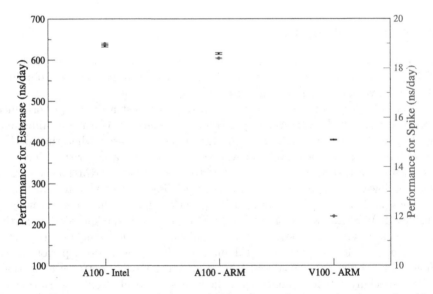

Fig. 2. In Figure are presented the results for the performance in terms of ns/day. The black/left axis is referred to Esterase case while the red/right axis to Spike case. For both cases there are values obtained in three different configurations: Intel with NVIDIA A100 GPU, ARM-1 with NVIDIA V100 GPU and ARM-3 with NVIDIA A100 GPU. (Color figure online)

cases, for example for Intel machines the configuration with GPU has better performances that a four nodes CPU only platform. Moreover, the configurations with A100 GPU have little differences in performances: the Esterase model gives 634 ns/day for Intel case and 615 ns/day for ARM case, similar results for the Spike model obtaining 19 ns/day for Intel and 18.4 ns/day for ARM. Meaning that most of the effort is moved on the accelerator. No significant difference in behaviour is reported between the two models tested.

3.3 Energy Consumption

During the simulations the power consumption was monitored. We compute the instant power averaged over the execution time of the simulation, for the esterase case the measure is repeated four times while for the spike it was not necessary because the fluctuation was smaller due to a higher duration of the simulation. From the quantity of power used is possible to obtain a measure of the energy required to complete a simulation (energy to solution) computed as the amount of kWh required to compute a ns of the simulation. Systems with lower energy to solution values have a better compromise between performance and power consumption. Since machines used are in a different formfactors (twin-square for Intel and AMD servers, 2U for ARM ones) is necessary to scale idle consumption from measured power values. For a single node case AMD machines use the lowest average power while Intel, ARM-1 and ARM-2 gives similar results, when per-core power is compared ARM-2 machines have the best results with AMD and Intel case reported an increase respectively of four times and six times. The results in Fig. 3 show that ARM-1 machines are the most energy hungry while AMD and ARM-2 cases require the smallest energy amount to complete the simulation for a single node. When more than a node is considered Intel and AMD servers show a moderate incremental trend with the exception of Spike

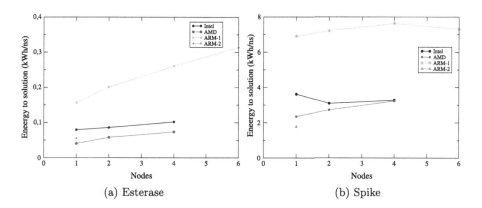

(a) Esterase   (b) Spike

Fig. 3. The Figure shows the energy usage, in terms of kWh/ns, for every architecture considered, without GPU acceleration. Values for Esterase model are in figure a, for Spike model are in figure b.

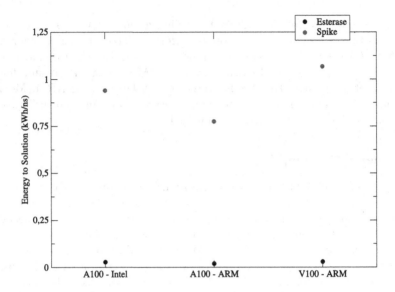

Fig. 4. In the figure are presented the results for energy consumption with GPU acceleration in terms of kWh/ns. The values are collected for every architecture considered, data for Esterase model are showed in black while for Spike model in red. For both cases there are values obtained in three different configurations: Intel with NVIDIA A100 GPU, ARM-1 with NVIDIA V100 GPU and ARM-3 with NVIDIA A100 GPU (Color figure online)

model on Intel machine where there is a slight decrease. The results highlight that Intel AMD energy consumption are comparable only in the worst case for AMD platform (i.e. four nodes configuration with Spike model).

When GPU acceleration is considered ARM nodes have the lowest power usage for both A100 and V100 cases respect to Intel one. Computing the energy to solution (Fig. 4) we obtain that NVIDIA V100 is not convenient respect to A100 and the most advantageous approach is to use NVIDIA A100 in combination with ARM CPUs. It is worth notice that comparing Intel single node cases that the use of a GPU accelerator could reduce the energy required by four times.

4 Conclusions

In this work, the performances of the Gromacs suite are tested on different CPU architectures, allowing the comparison of traditional Intel and AMD based machines with ARM based clusters. For this analysis we used two different protein models, one with small size (Esterase enzyme) and one with big size (Spike complex). The architectures considered were able to simulate both the models, reaching considerable performances even for large sized systems. Intel based server shows the highest absolute performance and the best scaling for both the models but comparing one node cases the best performance is obtained with an ARM based machine, although the scaling of ARM-2 server could not be tested because only one node was contemplated. In order to compare the platforms not only in terms of performance but also in terms of cost we have to consider system power consumption. Our results show that ARM-1 servers have the highest energy to solution values while ARM-2 and AMD clusters are the most suitable option to contain energy usage keeping reasonable performances. The use of GPU acceleration increases noticeably the performance reducing the energy consumption and the best configuration tested is an ARM based server combined with NVIDIA A100. Future works should explore a wider sample of GPU models, considering also different vendors, and test multiple GPU simulations.

References

1. McCammon, J.A., Gelin, B.R., Karplus, M.: Dynamics of folded proteins. Nature **267**(5612), 585–90 (1977)
2. Filipe, H.A.L., Loura, L.M.S.: Molecular dynamics simulations: advances and applications. Molecules **27**(7) (2022)
3. https://www.top500.org/
4. Kutzner, C., et al.: J. Chem. Inf. Model. **62**(7) (2022)
5. Hockney, R.W., Goel, S.P., Eastwood, J.: Quiet high resolution computer models of a plasma. J. Comp. Phys. **14**, 148–158 (1974)
6. Lindorff-Larsen, K., et al.: Improved side-chain torsion potentials for the Amber ff99SB protein force field. Proteins: Struct. Funct. Bioinform. **78**(8), 1950–1958 (2010)
7. Bussi, G., Donadio, D., Parrinello, M.: Canonical sampling through velocity rescaling. J. Chem. Phys. **126**(1) (2007)
8. Parrinello, M., Rahman, A.: Polymorphic transitions in single crystals: a new molecular dynamics method. J. Appl. Phys. **52**, 7182–7190 (1981)
9. Darden, T., York, D.M., Pedersen, L.G.: Particle mesh Ewald: an Nlog(N) method for Ewald sums in large systems. J. Chem. Phys. **98**(12), 10089–10092 (1993)
10. https://www.gromacs.org/
11. Abraham, M.J., et al.: GROMACS: high performance molecular simulations through multi-level parallelism from laptops to supercomputers. SoftwareX **1–2**, 19–25 (2015)
12. Grant, O.C., Montgomery, D., Ito, K., Woods, R.J.: Analysis of the Sars-Cov-2 spike protein glycan shield reveals implications for immune recognition. Sci. Rep. **10**, 14991 (2020)
13. Margesin, R., Schinner, F.: Properties of cold-adapted microorganisms and their potential role in biotechnology. J. Biotechnol. **33** (1994)
14. Brzuszkiewicz, A., et al.: Acta Cryst. **F65**, 862–865 (2009)

A Sentiment Analysis on Reviews of Italian Healthcare

Maria Chiara Martinis[1,2](✉) and Chiara Zucco[1,2]

[1] Data Analytics Research Center, University "Magna Græcia",
88100 Catanzaro, Italy
martinis@unicz.it, chiara.zucco@unicz.it
[2] Department of Medical and Surgical Sciences, University "Magna Græcia",
Catanzaro, Italy

Abstract. Sentiment analysis is gaining prominence in the evaluation of patient satisfaction within healthcare environments. The present article takes advantage of VADER-IT, an Italian language adaptation of the VADER sentiment analysis engine, to analyze patient reviews of hematological care facilities. In particular, 526 reviews published on the QSalute review platform (https://www.qsalute.it/), by patients who had received care in hospitals and medical facilities across Italy. Sentiment polarity was extracted from each Italian review using VADER-IT. To assess performance, the user's averaged rating was dichotomized and considered as ground truth. VADER-IT was further compared to the English version of VADER applied to reviews translated into English. The results reveal the superior performance of our method on Italian data compared to VADER applied to translated content.

Keywords: Text mining · Sentiment Analysis · VADER · VADER-IT

1 Introduction

The growing ubiquity of user-generated content in regard to different contexts, such as reviews, social media, forums, and online articles, has accentuated the interest registered in the last decade towards sentiment analysis, the research field that enables the comprehension and tracking of latent opinions. This, combined with the remarkable advancements in deep learning that have substantially contributed to enhance text classification performance and effectiveness, has made Sentiment analysis a pivotal and widely acknowledged tool in the realm of text analysis [2].

However, deep learning models are not exempt from disadvantages, including the need for vast amounts of data for training, the limited transparency of mechanisms leading to specific predictions, and the requirement to forget sensitive information about particular samples, a process called machine unlearning [5]. Additionally, of the five rules initially incorporated in the original VADER, the first two heuristics could be applied directly to the Italian language without requiring additional modifications. However, the last three rules needed manual adjustments to suit the Italian language. It's important to highlight that while VADER provides a polarity score within a continuous range of [-4, 4], VADER-IT also includes a method to predict a polarity class [9].

This article illustrates, as a first analysis, the use of VADER-IT in analyzing 526 reviews collected from the Qsalute platform. Through this Italian healthcare platform, patients were able to express their opinions and rate their experiences in hematology units at various Italian hospitals, covering aspects such as hospitals, doctors, medications, and more. Patients provided feedback through free text descriptions and assigned ratings on a scale ranging from 1 to 5 stars. The analysis achieved $Accuracy-score = 93.00\%$. As a second analysis, each review was automatically translated into English and the original version of VADER was applied to the translated data set, resulting in a $Accuracy-score = 87.07\%$.

The subsequent sections of the document are structured as follows: Sect. 2 provides basic information about Sentiment Analysis and an overview of polarity detection. Section 3 details data collection, the dataset, and the methodology employed in this application, while Sect. 4 delves into the analysis results. Finally, Sect. 5 presents the conclusions of the paper.

2 Background

Sentiment analysis is a method used in text mining to investigate the emotional or affective aspects present in textual content. This technique utilizes natural language processing to analyze and interpret the feelings and attitudes conveyed in different forms of communication, such as social media and related channels.

Additionally, it finds application in the healthcare sector, offering numerous benefits. This is particularly valuable in the healthcare field due to the vast amount of health-related information available online, including personal blogs, social media posts, and medical assessment websites, which are often not systematically collected but can significantly enhance the quality of healthcare [1].

In certain contexts, the use of a word lexicon method involves referencing a dictionary of words and their corresponding sentiment categories, indicating whether the word conveys a positive, negative, or neutral sentiment.

In different situations, dictionaries can also provide a sentiment score, where each word in the lexicon is assigned a score within a predefined range, usually [-1, 1], indicating the level of positivity linked to that word.

2.1 VADER

An example tool is VADER (Valence Aware Dictionary for sEntiment Reasoning), which is a sentiment analysis engine based on a lexicon which combines highly curated lexical resources with modeling based on five human-validated rules to predict a polarity from a text [8].

One of the most important advantages of VADER is that it does not need a training phase; it works well on short texts, which makes it usable for analyzing open-ended responses to questionnaires.

The VADER system begins with a widely applicable, valence-based sentiment lexicon, which is created using three established lexicons: LIWC, General Inquirer [10], and ANEW [6]. This lexicon is further enriched with several lexical elements commonly found in social media, including emojis, resulting in a total of 9,000 English terms. Subsequently, these terms are annotated on a scale ranging from $[-4, 4]$ through the crowd-sourcing service provided by Amazon Mechanical Turk.

Additionally, VADER incorporates five grammatical and syntactic rules to detect sentiment amplification or shifters in text. In particular, punctuation, capitalization, and a set of predefined nouns adjective and adverbs are considered amplifiers or downtoners of a sentence, while contrastive and negation particles are considered sentiment shifters, as they reverse polarity orientation.

3 Materials and Methods

3.1 VADER-IT

To adapt VADER for the Italian language, the approach described in [11] was followed. This involved utilizing Sentix [4], a lexicon that automatically extends the SentiWordNet annotation to Italian synsets found in MultiWordNet [7]. Furthermore, to create the Italian version of VADER's emoji list, a reference was made to the website to https://emojiterra.com/it/lista-it/. Given that the influence of word capitalization and the utilization of special punctuation marks as intensifiers remains consistent in the Italian language, only three out of VADER's five heuristics needed to be explicitly adapted to the Italian language.

The negation word set for VADER-IT was established by converting VADER's existing set of negation words into the Italian language. Furthermore, this set was expanded by including terms from the MultiWordNet syntagma for each word. The contrasting term 'but' was also translated to 'ma' in Italian, and its synonyms, such as 'per,' were considered. A similar approach was employed to adjust the set of word intensifiers and downtonings from the original VADER to Italian. Some additional linguistic rules tailored for the Italian language were used in conjunction with POS tagging features to identify new intensifier words in the dataset. However, identifying downtoning words proved more challenging.

While the original VADER also included certain idiomatic expressions in its sets of intensifiers, due to language variations, idioms were not initially included in the preliminary version of VADER-IT and will be explored in future developments.

Unlike VADER, which starts with a continuous polarity score discretization, VADER-IT also predicts a polarity class, categorizing text as positive, negative, and/or neutral.

3.2 The Qsalute Dataset

VADER-IT underwent testing using 526 reviews authored by patients dealing with Hematology-related illnesses.

These reviews were automatically gathered from the Italian platform QSalute (https://www.qsalute.it/), employing BeautifulSoup (https://beautiful-soup-4.readthedocs.io/en/latest/), a widely used Python library for web scraping.

The data collection process adhered to the methodology outlined in [3]. Alongside the review text, each review also furnished an average rating ranging from [1, 5]. This rating represented the mean of user-assigned scores across four distinct categories: Competence, Assistance, Cleaning, and Services.

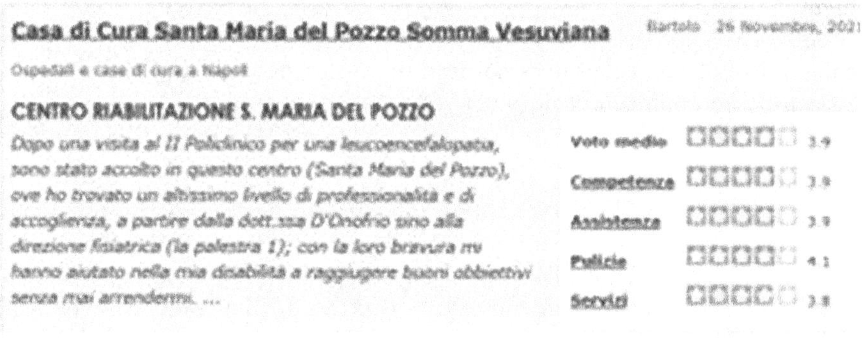

Fig. 1. A screenshot of Qsalute.

Figure 1. shows an example (in Italian) of a typical Qsalute review. Each review includes details such as the healthcare facility's name, the date, the review's content, a user-provided rating for the four aspects mentioned earlier (Competence, Assistance, Cleaning, and Services), and an overall score calculated by the system based on the aforementioned four ratings.

4 Results

In order to predict a sentiment class, the mean rating was discretized through the following approach: ratings in [1, 2] were associated with a negative polarity score, while the others were considered as positive. Then Vader-IT was used to predict polarity from each review. To compare the prediction accuracy, the review in the dataset were translated in the English language via the google translation API library for Python[1] and the original VADER was then apply to extract a polarity score that was further dichotomized to obtain a polarity class. Results, shown in Table 1.

---

[1] https://pypi.org/project/googletrans/.

Table 1. Accuracy and Jaccard score.

Model	Accuracy score	Jaccard micro score
VADER	87.07	77.10
VADER-IT	93.00	90.00

The dataset that has been divided into discrete categories is heavily skewed in favor of positive reviews. Through the use of VADER-IT applied to the dataset of 526 hematology reviews from different Italian hospitals, we denote a remarkable distribution toward the positive class with score equal to one compared to the negative score compared to the others. The results reported in Table 1, show that VADER-IT performs better than the original VADER applied to translated content.

Unlike the approaches based on machine learning, the VADER-IT tool on the dataset of 526 revisions does not lead to overfitting problems because, being a lexicon-based approach, it does not require a training phase.

5 Conclusion

The paper presents VADER-IT, a lexicon-based tool that adapts the popular VADER, or Valence Aware Dictionary for sEntiment Reasoning [8], to classify the polarity of Italian textual content. In contrast to the traditional VADER, which predicts a polarity score within a continuous range of [-4, 4], VADER-IT incorporates an additional approach to predict a polarity class. Out of the five rules present in the original VADER, only the first two heuristics could be applied to the Italian language without any modifications, while the remaining three had to be manually adjusted to suit the Italian language [9]. The results reported in Table 1, show that VADER-IT performs better than the original VADER applied to translated content.

References

1. Abualigah, L., Alfar, H.E., Shehab, M., Hussein, A.M.A.: Sentiment analysis in healthcare: a brief review. In: Recent Advances in NLP: The Case of Arabic Language, pp. 129–141 (2020)
2. Agarwal, A., Xie, B., Vovsha, I., Rambow, O., Passonneau, R.J.: Sentiment analysis of twitter data. In: Proceedings of the Workshop on Language in Social Media (LSM 2011), pp. 30–38 (2011)
3. Bacco, L., Cimino, A., Paulon, L., Merone, M., Dell'Orletta, F.: A machine learning approach for sentiment analysis for italian reviews in healthcare. Comput. Linguist. CLiC-it 2020 **630**(699), 16 (2020)
4. Basile, V., Nissim, M.: Sentiment analysis on Italian tweets. In: Proceedings of the 4th Workshop on Computational Approaches to Subjectivity, Sentiment and Social Media Analysis, pp. 100–107. Association for Computational Linguistics, Atlanta, Georgia (Jun 2013). https://www.aclweb.org/anthology/W13-1614

5. Bourtoule, L., et al.: Machine unlearning. In: 2021 IEEE Symposium on Security and Privacy (SP), pp. 141–159. IEEE (2021)
6. Bradley, M.M., Lang, P.J.: Affective norms for english words (anew): Instruction manual and affective ratings. Tech. rep., Technical report C-1, the center for research in psychophysiology (1999)
7. Catelli, R., Pelosi, S., Esposito, M.: Lexicon-based vs. bert-based sentiment analysis: a comparative study in italian. Electronics **11**, 374 (2022). https://doi.org/10.3390/electronics11030374
8. Hutto, C., Gilbert, E.: Vader: A parsimonious rule-based model for sentiment analysis of social media text. In: Proceedings of the International AAAI Conference on Web and Social Media, vol. 8, pp. 216–225 (2014)
9. Martinis, M.C., Zucco, C., Cannataro, M.: An Italian lexicon-based sentiment analysis approach for medical applications. In: Proceedings of the 13th ACM International Conference on Bioinformatics, Computational Biology and Health Informatics, pp. 1–4 (2022)
10. Pennebaker, J.W., Boyd, R.L., Jordan, K., Blackburn, K.: The development and psychometric properties of liwc2015. Tech. rep. (2015)
11. Zucco, C., Paglia, C., Graziano, S., Bella, S., Cannataro, M.: Sentiment analysis and text mining of questionnaires to support telemonitoring programs. Information **11**(12), 550 (2020)

A Numerical Approach to Basic Calculus

Layla Nasr[✉]

Lebanese University, Beirut, Lebanon
laylanasr1@hotmail.com

Abstract. A big body of research has been devoted to the complications faced by undergraduate students with basic calculus and the means of overcoming them. This situation imposes the urge to try educational approaches that are different from the traditional ones used in most curriculums worldwide.

In this paper, we deal with a non-traditional method in teaching basic calculus for undergraduate students. In particular, this study explores the efficiency of a numerical methodology for dealing with infinity and infinitesimals introduced by Yaroslav Sergeyev in the early 2000's. The Grossone methodology proposed by Sergeyev is a non-classical one yet it doesn't contradict the classical mathematics. Instead, it provides a simple and easy computational way for dealing with infinite quantities. It is noted that this methodology has been widely used in mathematics and applied sciences. Moreover, it has proved to be fruitful for educational purposes according to the results of several studies done recently.

The experimentation involves grade 11 and 12 students in Lebanon in addition to a small group of grade 12 students from an arbitrary secondary public school in Lebanon (ages 17–19). The aim of this study is to check students' intuitive reactions with the Grossone methodology and compare it to their performance after being subject to a teaching unit highlighting this methodology.

It is noted that this study follows the guidelines of a study done recently in Italy for the purpose of comparing the results of addressing Grossone methodology among different educational systems.

Keyword: infinity; Grossone methodology; calculus; mathematics education

1 Introduction

Calculus is considered a fundamental topic in mathematics curriculums worldwide for undergraduate students and higher education as well. However, numerous studies have reported complications faced by students in learning calculus [19–21].

A large body of research has embraced the inconsistencies in calculus education and has proposed several remedies for the latter. For instance, numerous studies suggested integrating software technologies like (GeoGebra or MATLAB) in teaching calculus [6], while others investigated the integration of STEM areas (science, technology, engineering and mathematics) in differential and integral calculus teaching [10].

In the same context, some recent studies have highlighted the psychological and conceptual aspects for boosting success in calculus courses. For instance, Hammoudi

A Numerical Approach to Basic Calculus 339

et al. [3] focused in their paper on enhancing students' intrinsic and extrinsic motivation. Bos et al. [2] focused on meaning-making in calculus and pre-calculus approaches.

In this study, we are interested in trying a new non-classical approach to basic calculus, which is the Grossone methodology [11]. Particularly talking, the study focus is: First, to investigate students' preliminary interactions with the Grossone-based system while having no previous knowledge of it. Second, to check students' interactions, performance and reflections on this new system after several teaching sessions involving Grossone-based system.

2 Grossone-Based System in Brief

The Grossone methodology is a new method for dealing with infinity and infinitesimals. It was proposed by Yaroslav Sergeyev in 2003. In the Grossone system, Sergeyev introduces a new infinite unit called *Grossone* and denoted by ①. ① represents the number of elements of set $\mathbb{N}$, in other words, for any $n \in \mathbb{N}$, $n \leq ①$. Hence, the set of natural numbers will be as follows:

$\mathbb{N} : \{1, 2, 3, \ldots, ①-2, ①-1, ①\}$, where ①-2, ①-1, ① are infinite numbers that belong to the set $\mathbb{N}$.

As a consequence, the infinitesimals would be the inverses of the Grossone-based infinite numbers.

An extended set of natural numbers is of the form:
$\hat{\mathbb{N}} : \{1, 2, 3, \ldots ①-1, ①, ①+1, ①+2, \ldots, 2①, 2①+1, \ldots, ①^2-1, ①^2, ①^2+1, \ldots\}$

In the Grossone system, infinite numbers and infinitesimals can be represented in the positional numeral system with infinite radix which is an extension of the system in base ten. Consequently, computations with infinite numbers and infinitesimals would be done in an analogous way to those done with finite numbers which makes it so easy and intuitive[1][11, 16, 18].

The Grossone-based system has wide applications in numerous fields of mathematics and computer sciences [12–17] in addition to its successful results in several pedagogical studies [1, 4, 5, 7–9].

3 Method

The study was done in three phases. In phase 1, a Google Form was distributed and filled by grade 11 (scientific section) and grade 12 students (general sciences, life sciences, economics and sociology section)[2]. Grade 11 and 12 students all have a basic background in traditional calculus. It is noted that in Lebanon most schools follow the same curriculum for secondary students (grades 10, 11, 12) who apply for unified official exams in grade 12.

---
[1] For more details, visit the website: http://www.theinfinitycomputer.com.
[2] In General sciences section mathematics is the basic subject learnt, in Life sciences section sciences are basic, Economics and sociology section learn finance and economics as major subject.

In phase 2, a small sample of 26 grade 12 students from an official secondary school in Mount Lebanon was randomly chosen. The sample consisted of two groups. The first group consisted of 13 students from the General Sciences class (GS) and the second consisted of 13 students from the Economics and Social Sciences class (SE). All students range from 17 to 19 years old. The mean grade of mathematics for the first term of the academic year 2022/2023 was 12/20 for GS class and 10.7/20 for SE class[3].

In this phase, the students performed a pre-test (similar to the Google Form but done on paper in class and including more items) followed by 6 teaching sessions (45 min each) for GS section and 2 sessions for SE section introducing the basics of Grossone methodology.

In phase 3, the two groups of students were subject to a post-test which is similar to the pre-test but more sophisticated.

This study was undergone during the academic year 2022/2023. The sessions were delivered starting from the second half of March and the post-test was administrated at the end of April.

3.1 Phase 1

As mentioned before, in the first phase of this study a Google Form was distributed among grade 11 and 12 students in Lebanon. The students, who learn in traditional methods haven't encountered the Grossone symbol before. Bearing in mind that the aim of this module was to check students' first impressions and interactions with Grossone (zero knowledge test), the students were just informed that Grossone represents the number of elements of set $\mathbb{N}$ without any further information.

The module consisted of 8 multiple choice/checklist questions. It was intended to be relatively short and with easy calculations to encourage a wide sample of students to fill it. The items included questions on computations and comparisons using Grossone. The last item quested points of view on introducing Grossone and its possible usage in mathematics[4].

3.2 Phase 2

The first step in this phase was performing a pre-test including 25 items. The items of the pre-test involved both multiple-choice questions and few open-ended questions. In addition to computations with Grossone, the students were asked about their expectations of the possible benefits of introducing Grossone in their opinion.

After performing the pre-test, six sessions (45 min each) were delivered to the GS students in 3 weeks. The first two sessions involved an introduction of the new methodology where students discussed and reflected on it. The third and fourth sessions were dedicated to arithmetic using Grossone-based framework. Students worked in groups on order and basic computations. The last two sessions constituted of a comparative study between the Grossone method and traditional methods in the context of limits and asymptotes[5].

---
[3] The passing grade in Lebanon is 10 out of 20.
[4] To view the Google Form: https://forms.gle/HUmZRQzgdE2FpZx58.
[5] For content of the sessions visit the site: http://www.numericalinfinities.com.

On the other hand, the SE students were subject to only 2 sessions (45 min each) introducing basics of Grossone-based system.

3.3 Phase 3

After the sessions being delivered in phase 2, a post-test was administrated to both groups of students. The test was similar to the pre-test with minor difference in difficulty in addition to questions of comparing infinity to Grossone in the context of limits and asymptotes. Below there are some proposed items:

- Simplify the following expressions:

  1) $2①-5①+1$
  2) $2(① - 3)(3① + 4)$
  3) $①(① - 1) + 3①$
  4) $\frac{①}{2} + \frac{①}{3}$
  5) $① \times \frac{1}{①}$
  6) $(-\frac{2}{3}① + 2)(\frac{①}{2} - 3) + 2$
  7) $3\frac{①}{2} + \frac{①}{5} - 1$

- Mark the following items as "true" or "false":

  1) $① < 3①$
  2) $① + 2 < ① \times ①$
  3) $① - 1 < ① < ① + 1$
  4) $\frac{①}{2} < \frac{①}{3}$
  5) $① < \infty$
  6) $① = \infty$
  7) $①$ and ∞ cannot be compared
  8) $\infty + 1 = \infty$
  9) $① + 1 = ①$
  10) $\frac{1}{①}$ is an indeterminate form

- Consider the function $f(x) = x + 1 + \frac{1}{x}$ defined for $x \in \mathbb{R}/\{0\}$

  1) Prove that the line $(d) : y = x + 1$ is an oblique asymptote to the graph of f.
  2) Calculate $f(①)$; $f(① + 1)$; $f(2①)$; $f(-①)$
  3) Calculate $f(①) - y_{(d)}(①)$; then $f(① + 1) - y_{(d)}(① + 1)$. Compare the results.

- Do you think that replacing ∞ by $①$ could be beneficial in mathematics. If yes, in what ways?

4 Results

4.1 Phase 1

In this section, we present the results of phase 1 of the study which is the Google Form questionnaire (Fig 1).

185 students of grades 11 and 12 had filled this module. The results are shown in the chart below:

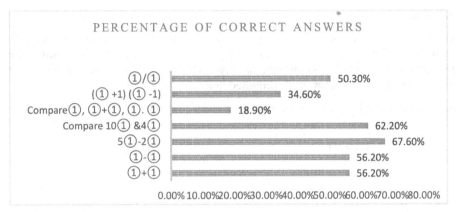

Fig.1. Percentage of correct answers for each of the items of the Google Module

The results shown in the chart above reveal that performing computations using Grossone-based system is highly intuitive. In most items, more than half of the participants were able to provide correct answers without having any prior knowledge of Grossone. Nevertheless, low percentages of correct answers show in items 2 and 3. The reason of those low percentages compared to the rest of the items could be related to the formulation of the questions. In particular, comparing ①, ① + ①, and ①. ① in the same item could have been confusing for the students. Maybe if each comparison was held in a separate question, the results would have been different. On the other hand, some students could have missed the second item due to calculation mistakes. People usually fill in forms using their mobile phones and do not give special attention to the accuracy of their answers since it is not a test and they stay anonymous.

In the last question, the students were asked what they think of introducing Grossone. They were given several options (useful in mathematics, not possible, makes sense, makes no sense) with the possibility of choosing more than one option (checklist question) and adding their own opinion. 55.1% of the participants responded that introducing Grossone could be useful in mathematics, 23.2% marked the option that Grossone makes sense, 25.4% chose that it doesn't make sense, and 22% responded that introducing Grossone is not possible.

Some students added their own answers:

"Introducing Grossone is the same as introducing infinity."
"Introducing Grossone seems useful in mathematics although it might be rare to use it considering it is a fixed and infinite number."
"Introducing ① may have implications in certain branches of mathematics, such as set theory or number theory."

Despite that it was the first time the participants encounter Grossone, yet expectations of its usefulness were high for the majority of them.

4.2 Phases 2 and 3

During the sessions, the students discussed and reflected on the new methodology, they also worked in groups performing computations in Grossone terms and comparing the traditional methods with Grossone methods in the context of limits.

The students showed acceptance and welcoming attitude to the Grossone method and seemed to be comfortable in dealing with it. Moreover, using it in the study of limits of functions provided a deeper insight of the behavior of functions at different infinite values.

Results of the pre-test are similar to that of the Google Form of phase 1. Even without having previous knowledge of Grossone, the students were able to cope with it. Nevertheless, after delivering some lectures, the results got even better as the table below shows. Results of the post-test for GS students (who took four more sessions) were better than those of SE students. The percentage of increase in students' performance between the pre- and the post-test was 43.4% for GS students and 32.4% for the SE students. The difference in performance between the two groups was clear in the questions related to limits and asymptotes. This result is no surprise, since GS students were exposed to such context during the sessions. This means that just few teaching sessions were enough to achieve remarkably high results!

It is noted that both the pre- and the post-test consisted of 25 graded items. For each correct answer the student collects one point and a minimum of 13 points is required to pass the test (Table 1).

Table 1. Results of pre-test and post-test of the two classes

Class	Pre-test			Post-test		
	Mean/25	Max score/25	Std Dev/25	Mean/25	Max score/25	Std Dev/25
GS	14.8	22	4.6	21.23	25	4.12
SE	14.23	21	4.24	18.6	23	3.64

The last item in the post-test was not graded since it reflects students' opinions. Most students didn't specify a yes/no answer. But most of them agreed that introducing Grossone could be useful in mathematics. Many students wrote that Grossone helps in providing more precise answers and dealing with it is easier than dealing with infinity.

Below there are some interesting answers:

"Grossone could be useful in solving equations or problems in mathematics that cannot be solved using infinity."
"Maybe Grossone could be used in engineering, specifically those related to artificial intelligence and robotics where high accuracy is needed."
"Taking in consideration that Grossone provides more accurate values, I think that it could be beneficial in astronomical studies to specify enormous distances, masses…"

Students' expectations and acceptance of Grossone are quite interesting. Their answers reveal that they are open to embracing new methods that are different from those they are used to.

5 Conclusions

The results of this study show the intuitive nature of the Grossone-based system. Results of the zero-knowledge test were very satisfying in both the Google Module and the pre-test of the class experiment. Moreover, providing the students with just few lectures lead to a remarkable elevation in their scores. Furthermore, the group of students who were subject to six teaching sessions performed better than those who took only two sessions. This means that the addition of just few sessions were enough to mark a noticeable increase in the results.

Our results are compatible with the results of previous studies done on this topic [1, 4, 5]. In particular, our results are parallel with [1] (the two studies follow similar research disciplines), regarding the remarkably positive results of the zero-knowledge test in addition to the elevation of students' performance after few teaching sessions. This indicates that using Grossone method yields to successful results in a relatively short teaching time and in different educational systems and cultures as well.

It would be interesting and necessary to carry on further educational experiments using Grossone-based system on a wider scale to check the efficiency of using this system in teaching/learning calculus.

References

1. Antoniotti, L., Caldarola, F., d'Atri, G., Pellegrini, M.: New approaches to basic calculus: an experimentation via numerical computation. In: Sergeyev, Y., Kvasov, D., (eds.) Numerical Computations: Theory and Algorithms. NUMTA 2019. Lecture Notes in Computer Science, vol 11973. Springer, Cham (2020)
2. Bos, R., Kouropatov, A., Swidan, O.: Editorial to the special issue on tools to support meaning-making in calculus and pre-calculus education. Teach. Math. Its Appl. 41, 87–91 (2022)
3. Hammoudi, M.M., Grira, S.: Improving students' motivation in calculus courses at institutions of higher education: evidence from graph-based visualization of two models. EURASIA J. Math. Sci. Tech. Educ., **19**(1), em2209 (2023)
4. Iannone, P., Rizza, D., Thoma, A.: Investigating secondary school students' epistemologies through a class activity concerning infinity. In: Bergqvist, E., et al. (eds.) Proceedings of the 42nd Conference of the International Group for the Psychology of Mathematics Education, Umeºa(Sweden), vol. 3, pp. 131–138. PME (2018)
5. Ingarozza, F., Adamo, M.T., Martino, M., Piscitelli, A.: A Grossone-Based Numerical Model for Computations with Infinity: A Case Study in an Italian High School. In: Sergeyev, Y.D., Kvasov, D.E. (eds.) NUMTA 2019. LNCS, vol. 11973, pp. 451–462. Springer, Cham (2020)
6. Listiani, Y., Aklimawati, A., Wulandari, W., Isfayani, E.: The effectiveness of the geogebra-assisted integral calculus module. Kalamatika: J. Pendidikan Matematika **7**(2), 177–190 (2022)
7. Mazzia, F.: A computational point of view on teaching derivatives. Inf. Educ. **37**(1), 79–86 (2022)

8. Nasr, L.: The effect of arithmetic of infinity methodology on students' beliefs of infinity. Med. J. Res. Math. Educ. **19**, 5–19 (2022)
9. Nasr, L.: Students' resolutions of some paradoxes of infinity in the lens of the grossone methodology. Inf. Educ. **38**(1), 83–91 (2023)
10. Pessoa da Silva, K.A., Borssoi, A.H., Ferruzzi, E.C.: Integration of STEM education in differential and integral calculus classes: aspects evidenced in a mathematical modelling activity. Acta Sci. (Canoas) **24**(7), 116–145 (2022)
11. Sergeyev, Y.D.: Arithmetic of Infinity. Edizioni Orizzonti Meridionali, Cosenza (2003)
12. Sergeyev, Y.D.: Evaluating the exact infinitesimal values of area of Sierpinski's carpet and volume of Menger's sponge. Chaos Solitons Fractals **42**(5), 3042–3046 (2009)
13. Sergeyev, Y.D.: Higher order numerical differentiation on the Infinity computer. Optim. Lett.. Lett. **5**(4), 575–585 (2011)
14. Sergeyev, Y.D.: Solving ordinary differential equations by working with infinitesimals numerically on the infinity computer. Appl. Math. Comput.Comput. **219**(22), 10668–10681 (2013)
15. Sergeyev, Y.D.: Using blinking fractals for mathematical modelling of processes of growth in biological systems. Informatica **22**(4), 559–576 (2011)
16. Sergeyev, Y.D.: Numerical infinities and infinitesimals: methodology, applications, and repercussions on two Hilbert problems. EMS Surv. Math. Sci. **4**(2), 219–320 (2017)
17. Sergeyev, Y.D., Mukhametzhanov, M.S., Mazzia, F., Iavernaro, F., Amodio, P.: Numerical methods for solving initial value problems on the Infinity Computer. Int. J. Unconv. Comput.Unconv. Comput. **12**(1), 55–66 (2016)
18. Sergeyev Ya.D., De Leone R. (eds.) Numerical infinities and infinitesimals in optimization. Springer, Cham (2022)
19. Tall, D.: Inconsistencies in the learning of calculus and analysis. Focus **12**(3&4), 49–63 (1990)
20. Tall, D.: Students' difficulties in calculus. In: Proceedings of Working Group 3 on Students' Difficulties in Calculus, ICME-7, Québec (Canada), pp. 13–28 (1993)
21. Tallman, M.A., Carlson, M.P., Bressoud, D.M., Pearson, M.: A characterization of calculus I final exams in US colleges and universities. Int. J. Res. Undergrad. Math. Educ. **2**, 105–133 (2016)

Modelling Hyperentanglement for Quantum Information Processes

Luca Salatino[1]([✉]), Luca Mariani[1], Carmine Attanasio[1,2],
Sergio Pagano[1,2,3], and Roberta Citro[1,2,3]

[1] Department of Physics "E. R. Caianiello", University of Salerno, via Giovanni Paolo II 132, 84014 Fisciano, (SA), Italy
lsalatino@unisa.it
[2] CNR - SPIN, c/o University of Salerno, via Giovanni Paolo II 132, 84014 Fisciano, (SA), Italy
[3] INFN, Gruppo collegato di Salerno, via Giovanni Paolo II 132, 84014 Fisciano, (SA), Italy

Abstract. Entanglement is a very useful phenomenon for quantum information processes. When more than one degree of freedom (DOF) is entangled, the hyperentanglement is realized. The main advantage of hyperentanglement is that of being able to encode a greater amount of information by exploiting the quantum coherence. This advantage is due to the fact that it is possible to extract information for each DOF. We propose to model a hyperentangled state through copies of a Bell state and evaluate its functionality transferring a quantum state through a classical channel. The goal is to simulate the transport of a hyperentangled state through a classical channel in the presence of a system-environment interaction. By virtue of an increased Hilbert space following the increase in the DOF, the entanglement quantifier, the so called concurrence, decays slower with the distance compared to the case of one DOF.

Keywords: Entanglement · Hyperentanglement · Quantum Protocol

1 Introduction

Entanglement is one of the most important resources in the context of quantum technologies, especially in the context of quantum information transfer processes [16]. In fact, there are many technologies related to quantum communication and quantum computing that exploit the phenomenon of entanglement, such as quantum teleportation [5], quantum key distribution [8], quantum state sharing [9,12], and other quantum communication protocols [6,15]. When we are in the presence of entanglement achieved with more than one DOF, we can refer to the concept of hyperentanglement. In recent years, research on hyperentanglement has produced numerous results [2,3,10,27], finding applications in the Bell-state analysis [1,4,13,19,21,24,28–30], entanglement purification [14,18,22,23,26] and entanglement concentration [17]. Once defined a hyperentangled quantum state understanding its transmission through a classical channel remains one of the most relevant problems for quantum information processes. On the other side, the simplest open quantum system model is a two-level system (a doublet) interacting with the modes of a harmonic oscillator. This model, known as Jaynes-Cummings (JC) model, was originally developed to study the interaction of atoms with the quantized electromagnetic field in order to investigate the phenomena of spontaneous emission and absorption of photons in a cavity [7,11]. In this article, using the JC model, the transfer of a hyperentangled quantum state through a classical channel is simulated. The goal is to verify, in terms of quantum bit error rate (QBER) and concurrence, the gain in using a hyperentangled state compared to an entangled state.

2 Transmission of an Hyperentangled States Through Classical Channel

We are going to study the transmission of an entangled quantum state of few DOF using the simulator of a quantum computer provided by the ©Qiskit platform. We report the quantum circuit that simulates the transmission of the quantum state between 2 nodes, evaluating their quantum bit error rate (QBER) and concurrence, that quantifies the degree of entanglement, as the number of DOF varies. Generally, we can identify a single entangled state as a special case of a more complex hyperentangled state, i.e., with only one DOF. In this case, the initially maximally entangled state corresponds to the Bell state

$$|\Phi^+\rangle = \frac{1}{\sqrt{2}} (|00\rangle + |11\rangle) \qquad (1)$$

where $|00\rangle$ and $|11\rangle$ are two-qubit states starting from the $(|0\rangle, |1\rangle)$ basis set. One of the possible ways of producing a Bell state is the one proposed in Fig. 1. It consists in creating a uniform superposition of states $|0\rangle$ and $|1\rangle$, through the Hadamard gate, followed by a CNOT. The circuit transforms input state $|00\rangle$ into $|\Phi^+\rangle$. Now, let us imagine we can expand the number of DOFs of the Bell state $|\Phi^+\rangle$. To do this we consider n replicas of the state $|\Phi^+\rangle$ which will correspond to the n DOFs of the system. The basic idea is therefore to use replicas of the system $|\Phi^+\rangle$ to increase the number of degrees of freedom of the system itself. In this way, by virtue of an increment of the Hilbert space, the number of states corresponds to 2^{2n}. We can define an hyperentangled Bell state as a tensor product of single qubit states

$$|\Phi^+\rangle^n = \frac{1}{\sqrt{2}} (|0\rangle_1 \otimes \ldots \otimes |0\rangle_n + |1\rangle_1 \otimes \ldots \otimes |1\rangle_n) \qquad (2)$$

Based on the previous scenario we consider the circuits that implement the transmission of the quantum state described in Eq. (2) as the DOF varies. After the measurement, the number of states that emerge is closely related to the

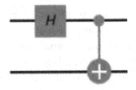

Fig. 1. Quantum circuit to generate Bell state $|\Phi^+\rangle$. (Adapted from IBM Quantum, https://quantum-computing.ibm.com/).

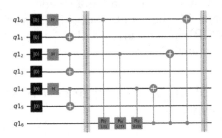

Fig. 2. Quantum circuits reproducing the transmission of a hyperentangled state with 3 DOF. (Adapted from IBM Quantum, https://quantum-computing.ibm.com/).

number of DOF n of the system. We define the QBER as the probability that the state obtained as a result of the measurement differs from the initial one. Let us first consider the case of two copies of a Bell state. Let us consider Alice (A) and Bob (B) the interpreters of the communication. For an initially maximally entangled state, we have

$$|\Phi^+\rangle = \frac{1}{\sqrt{2}}(|0\rangle_A \otimes |0\rangle_B + |1\rangle_A \otimes |1\rangle_B) \qquad (3)$$

Following the measurement, the system collapses with different probabilities in the states $|0\rangle_A|0\rangle_B$, $|1\rangle_A|0\rangle_B$, $|0\rangle_A|1\rangle_B$, $|0\rangle_A|0\rangle_B$. We consider the states $|0\rangle_A|0\rangle_B$ and $|1\rangle_A|1\rangle_B$ as not contributing to the QBER, as opposed to states $|1\rangle_A|0\rangle_B$, $|0\rangle_A|1\rangle_B$ because they are non-diagonal. In fact, when these states emerge, the density matrix that represents the system presents an increase in decoherence. With the same logic, for a system with a number of DOFs equal to 2, the number of useful states for information transmission rises to 4 out of a total of 16 states. Up to 6 DOFs were simulated. In order to simulate the external environment, the amplitude damping model was adopted to quantum circuit. It consists of a controlled rotation along the Y axis (R_y) and a CNOT. This operation is analogous to the unitary operation resulting from the Jaynes-Cummings interaction on the single-photon state.

The QBER calculated as described above is shown in Fig. 3 for different number if DOFs. As shown, it decreases at increasing the number of DOFs. The perturbation, in fact, being distributed over an increasing number of states, affects the system in a less invasive way. To study the advantage of hyperentan-

Modelling Hyperentanglement for Quantum Information Processes 349

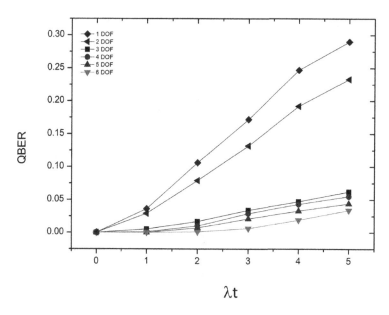

Fig. 3. QBER as a function of parameters of the Jaynes-Cummings model. The QBER decreases as the DOF increases.

gled states one can also study the degree of entanglement through the concurrence. For pure states of two qubits $|\Psi\rangle$ the concurrence $C(|\Psi\rangle)$, given by $C(|\Psi\rangle) := |\langle\Psi|\tilde{\Psi}\rangle|$ where $|\tilde{\Psi}\rangle = \sigma_y \otimes \sigma_y |\Psi\rangle^*$ is the temporally reversed state and σ_y is the Pauli spin matrix. Concurrence returns increasing values, as the number of DOF increases. In the adopted configuration, the starting state $|\Phi^+\rangle^n$ is the result of tensor products of separable Bell states. We denote the concurrence $f|\Phi_i\rangle$ as $C_i (i = 1, 2, \ldots, N)$. We can obtain the total concurrence of the hyperentangled state C_{hyper} as [25]

$$C_{hyper} = \sum_i^N C_N \qquad (4)$$

From above definition, if all the $|\Phi_i\rangle$ are maximally Bell states, we can get $C_{hyper} = N$.

Of course, when the coupling to the external environment is taken into account the equality is violated and due the interaction it starts to decrease. We have analyzed concurrence and QBER as a function of the Jaynes-Cummings model parameters. The abscissa axis λt represents the distance traveled by the qubit (see Ref. [20]) and lambda is the system-environment coupling frequency in the JC model. By virtue of an increased Hilbert space the concurrence, decays

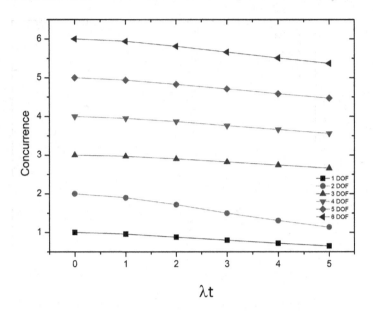

Fig. 4. Concurrence as a function of parameters of the Jaynes-Cummings model. The maximum value of concurrence varies in relation to the number of DOFs. The concurrence varies from 6 with 6 DOFs up to 1 with 1 DOF.

slower with the distance at increasing the number of DOFs, compared to the case of one DOF and remains more robust as the distance increases. This demonstrate that hyperentangled state are more robust to the coupling with the environment.

3 Conclusion

In this article, we modelled a hyperentangled qubit state through a quantum circuit. Starting from a maximally Bell state, n copies were considered in an attempt to simulate n DOFs corresponding to an increased dimension of the Hilbert space. To investigate the advantages related to hyperentanglement we studied, the transfer of the hyperentangled state through a classical channel, using the amplitude damping modelled whose operation is equivalent to that of the famous Jaynes-Cummings model for an open system. Indeed, it has been found that both the quantum bit error rate and the concurrence are improved by the hyperentanglement. The QBER increases slower at increasing the number of DOF and the concurrence is robust and decreases slower as a function of the distance. Our simulation indicates that hyperentanglement represents a promising resource to improve coherence and transmission in quantum information processing.

References

1. Barbieri, M., Vallone, G., Mataloni, P., De Martini, F.: Complete and deterministic discrimination of polarization bell states assisted by momentum entanglement. Phys. Rev. A **75**(4), 042317 (2007)
2. Barbieri, M., Cinelli, C., Mataloni, P., De Martini, F.: Polarization-momentum hyperentangled states: realization and characterization. Phys. Rev. A **72**(5), 052110 (2005)
3. Barreiro, J.T., Langford, N.K., Peters, N.A., Kwiat, P.G.: Generation of hyperentangled photon pairs. Phys. Rev. Lett. **95**(26), 260501 (2005)
4. Barreiro, J.T., Wei, T.C., Kwiat, P.G.: Beating the channel capacity limit for linear photonic superdense coding. Nat. Phys. **4**(4), 282–286 (2008)
5. Bennett, C.H., Brassard, G.: C. cr epeau, r. jozsa, a. peres and wk wootters. Phys. Rev. Lett. **70**, 1895 (1993)
6. Deng, F.G., Long, G.L., Liu, X.S.: Two-step quantum direct communication protocol using the einstein-podolsky-rosen pair block. Phys. Rev. A **68**(4), 042317 (2003)
7. Eberly, J.H., Narozhny, N., Sanchez-Mondragon, J.: Periodic spontaneous collapse and revival in a simple quantum model. Phys. Rev. Lett. **44**(20), 1323 (1980)
8. Ekert, A.K.: Quantum cryptography based on bell's theorem. Phys. Rev. Lett. **67**(6), 661 (1991)
9. Hillery, M., Bužek, V., Berthiaume, A.: Quantum secret sharing. Phys. Rev. A **59**(3), 1829 (1999)
10. Hu, B.L., Zhan, Y.B.: Generation of hyperentangled states between remote noninteracting atomic ions. Phys. Rev. A **82**(5), 054301 (2010)
11. Jaynes, E.T., Cummings, F.W.: Comparison of quantum and semiclassical radiation theories with application to the beam maser. Proc. IEEE **51**(1), 89–109 (1963)
12. Karlsson, A., Koashi, M., Imoto, N.: Quantum entanglement for secret sharing and secret splitting. Phys. Rev. A **59**(1), 162 (1999)
13. Kwiat, P.G., Weinfurter, H.: Embedded bell-state analysis. Phys. Rev. A **58**(4), R2623 (1998)
14. Li, X.H.: Deterministic polarization-entanglement purification using spatial entanglement. Phys. Rev. A **82**(4), 044304 (2010)
15. Long, G.L., Liu, X.S.: Theoretically efficient high-capacity quantum-key-distribution scheme. Phys. Rev. A **65**(3), 032302 (2002)
16. Nielsen, M.A., Chuang, I.L.: Quantum computation and quantum information. Cambridge University Press (2010)
17. Ren, B.C., Deng, F.G.: Hyperentanglement purification and concentration assisted by diamond nv centers inside photonic crystal cavities. Laser Phys. Lett. **10**(11), 115201 (2013)
18. Ren, B.C., Du, F.F., Deng, F.G.: Hyperentanglement concentration for two-photon four-qubit systems with linear optics. Phys. Rev. A **88**(1), 012302 (2013)
19. Ren, B.C., Wei, H.R., Hua, M., Li, T., Deng, F.G.: Complete hyperentangled-bell-state analysis for photon systems assisted by quantum-dot spins in optical microcavities. Opt. Express **20**(22), 24664–24677 (2012)
20. Salatino, L., Mariani, L., Attanasio, C., Pagano, S., Citro, R.: Dissipative dynamics in quantum key distribution. European Phys. J. Plus **138**(6), 517 (2023)
21. Schuck, C., Huber, G., Kurtsiefer, C., Weinfurter, H.: Complete deterministic linear optics bell state analysis. Phys. Rev. Lett. **96**(19), 190501 (2006)

22. Sheng, Y.B., Deng, F.G.: Deterministic entanglement purification and complete nonlocal bell-state analysis with hyperentanglement. Phys. Rev. A **81**(3), 032307 (2010)
23. Sheng, Y.B., Deng, F.G.: One-step deterministic polarization-entanglement purification using spatial entanglement. Phys. Rev. A **82**(4), 044305 (2010)
24. Sheng, Y.B., Deng, F.G., Long, G.L.: Complete hyperentangled-bell-state analysis for quantum communication. Phys. Rev. A **82**(3), 032318 (2010)
25. Sheng, Y.-B., Guo, R., Pan, J., Zhou, L., Wang, X.-F.: Two-step measurement of the concurrence for hyperentangled state. Quantum Inf. Process. **14**(3), 963–978 (2015). https://doi.org/10.1007/s11128-015-0916-1
26. Simon, C., Pan, J.W.: Polarization entanglement purification using spatial entanglement. Phys. Rev. Lett. **89**(25), 257901 (2002)
27. Vallone, G., Ceccarelli, R., De Martini, F., Mataloni, P.: Hyperentanglement of two photons in three degrees of freedom. Phys. Rev. A **79**(3), 030301 (2009)
28. Walborn, S., Pádua, S., Monken, C.: Hyperentanglement-assisted bell-state analysis. Phys. Rev. A **68**(4), 042313 (2003)
29. Wang, T.J., Lu, Y., Long, G.L.: Generation and complete analysis of the hyperentangled bell state for photons assisted by quantum-dot spins in optical microcavities. Phys. Rev. A **86**(4), 042337 (2012)
30. Wei, T.C., Barreiro, J.T., Kwiat, P.G.: Hyperentangled bell-state analysis. Phys. Rev. A **75**(6), 060305 (2007)

An Innovative Sentiment Analysis Model for COVID-19 Tweets

Areeba Umair[1(✉)] and Elio Masciari[1,2]

[1] University of Naples, Federico II, Naples, Italy
{areeba.umair,elio.masciari}@unina.it
[2] ICAR-CNR, Rende, Italy

Abstract. Worldwide health problems and feelings of worry and anxiety have been brought on by COVID-19, a terrible pandemic that the WHO has declared. Sentiment analysis is a vital technique for figuring out how people are responding to the pandemic. In order to find the most pertinent and instructive aspects of the embeddings, the research suggests a novel sentiment analysis model for COVID-19 tweets using CT-BERT as a base model and MAX Pooling function on the last four layers. The generated embeddings are joined with the classification (CLS) token before being sent to a classifier, which generates a probability distribution across all potential classes. The proposed technique acquired 91 and 92 % accuracy, 93 and 94% recall and 90 and 92% F-measure for positive and negative sentiment classification respectively. Insights into public sentiment and emotions have been gained during COVID-19 that have been useful for informing decision-making and communication tactics.

Keywords: Sentiment Analysis · COVID-19 · BERT Model

1 Introduction and Background

The technique of assessing and recognizing the feelings and opinions conveyed in a text, such as social media posts, product reviews, or news articles, is called sentiment analysis. To detect whether a text is positive, negative, or neutral in sentiment, natural language processing (NLP) methods are used.

One of the worst pandemics ever to hit the world was COVID-19, which the WHO deemed to be a pandemic. It has not only led to widespread health problems but has also made people feel scared and anxious, [1]. Sentiment analysis can be very helpful during the COVID-19 pandemic to comprehend how people are feeling and responding to the issue [5].

Twitter is a well-known social media site where users post brief text messages called tweets expressing their ideas and opinions. Twitter data is frequently subjected to sentiment analysis in order to better understand public sentiment toward a certain subject or event. Sentiment analysis processes text data using AI algorithms to extract the feelings and attitudes that the text expresses. To effectively classify text as positive, negative, or neutral, machine learning algorithms are trained on enormous datasets of labeled text data. Once trained, these algorithms can be used to examine fresh text data and forecast the sentiment that the text will convey. Sentiment analysis has undergone a revolutionary change because to artificial intelligence, which makes analyzing huge amounts of text data more precise and efficient. It is possible to evaluate enormous amounts of text data in real-time using AI-powered sentiment analysis, providing fast insights into people's attitudes and feelings. Additionally, AI-powered sentiment analysis may adjust to shifting linguistic and cultural nuances, increasing the analysis's accuracy.

The significance of this research lies in its innovative approach to sentiment analysis of COVID-19 tweets. By utilizing CT-BERT as the base model and implementing MAX Pooling on the last four layers, the study harnesses state-of-the-art natural language processing techniques. This method not only ensures a more in-depth understanding of the embedded data but also enhances the model's capability to provide more accurate and insightful sentiment analysis. By combining the generated embeddings with a classification token and employing a classifier, the research contributes to a more refined and efficient approach to understanding and categorizing sentiment within the context of the COVID-19 pandemic, which has significant implications for public health, communication, and decision-making.

The paper [8] presents CoQUAD, an efficient question-answering system for COVID-19 queries. It utilizes two datasets: a reference-standard dataset from CORD-19 and LitCOVID, and a gold-standard dataset created by public health experts. CoQUAD comprises a Retriever component employing the BM25 algorithm to retrieve relevant documents based on COVID-19 questions. It also features a Reader component using the MPNet Transformer model to extract answers from the retrieved documents. Notably, CoQUAD can address questions related to various stages of COVID-19, setting it apart from prior approaches. In another study [7] COVID-19 patient case reports are gathered from published literature, forming a corpus. Experts annotate a portion for ground truth labels, and a semi-supervised learning approach is employed for re-annotation. An NLP framework is created to extract SDOH from free texts, and a two-way evaluation method measures the method's quantity and quality.

The organization of the paper is stated as: Section 2 shows the overall methodology of the research. Section 3 explains the experiments and Sect. 4 shows the results of the experiments. Lastly, Sect. 5 concludes the overall research.

2 Methodology

2.1 Data Collection and Pre-processing

For our study, we obtained COVID-19 vaccine data from a freely available link "https://www.kaggle.com/datasets/gpreda/all-covid19-vaccines-tweets".

Pre-processing text data is a popular procedure, particularly when working with unstructured data like tweets. Pre-processing aids in preparing the raw data for analysis and modeling by cleaning and transforming it. According to the details you gave, it appears that the pre-processing processes listed below were completed:

2.2 Sentiment Generation

The polarity of text is an indication of the degree of positivity or negativity conveyed by the sentences. To determine the sentiment of the COVID-19 vaccine tweets, we utilized the TextBlob() function, which calculates the polarity values. TextBlob is a straightforward and user-friendly Python library for sentiment analysis, making it ideal for quick sentiment polarity assessments. Other approaches, like machine learning-based models, often require more data and computational resources to fine-tune and train, but they can provide higher accuracy and adaptability to various text types and languages. Choosing between TextBlob and other methods depends on the specific needs of the sentiment analysis task. Based on these polarity values, we classified the tweets into seven different sentiment classes, namely neutral, weakly positive, mild positive, strongly positive, weakly negative, mild negative, and strongly negative. This categorization allowed us to better understand the overall sentiment conveyed by the tweets in our dataset.

2.3 CT-BERT

Google created the pre-trained deep learning model BERT (Bidirectional Encoder Representations from Transformers) for natural language processing (NLP). BERT is a particular kind of neural network that can comprehend the context and meaning of words within a phrase, making it suitable for sentiment analysis applications [9,10].

The COVID-Twitter BERT model [6] is a machine learning model that examines tweets about COVID-19 using the BERT (Bidirectional Encoder Representations from Transformers) method. This model is intended to recognize and classify tweets according to their sentiment, topic, and other important characteristics [2].

2.4 Proposed Model

Multiple hidden layers in deep learning models, like neural networks, process and transform the input data before producing the output. These hidden layers might be conceptualized as a hierarchy of representations that take in progressively complicated and abstract input information. In addition, using the model's hidden layers can help modify the fundamental outcomes produced by pre-trained models and extract more complex information. A score expressing the overall sentiment of a piece of text (such as positive, negative, or neutral) might be the result of the final layer of a pre-trained BERT model in sentiment analysis, for instance. However, the model's hidden layers can be used to extract more precise sentiment data, such as the precise words or phrases that help to express the sentiment in the text [4].

The proposed method applies max pooling to each of the CT-BERT model's last four layers individually, choosing the highest value possible for each embedding dimension for each layer. Because only the highest value for each embedding dimension is chosen rather than the whole embedding, the output's dimensionality is minimized. The embeddings are layered and added using PyTorch routines after being pooled. To derive class-wise probabilities, the obtained embeddings are combined with the classification token (CLS) and fed into a classifier.

We proposed the Layer Fusion model shown in Fig. 1.

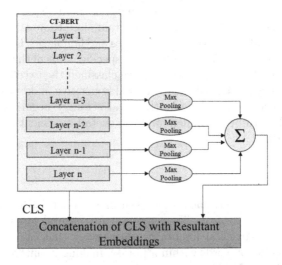

Fig. 1. Proposed Layer Fusion

3 Experiments

We used following 1 hyperparameters for fine tuning:

To test the effectiveness of our suggested paradigm, we ran experiments using cutting-edge models using 80% train and 20% test split of data. These models include:

1. BERT: A pre-trained model called BERT (Bidirectional Encoder Representations from Transformers) uses a transformer-based architecture to understand the contextual relationships between the words in a phrase.

Table 1. Hyperparameters for Sentiment Classification

Hyperparameter	Values
Model Name	CT-BERT
Max Sequence Length	128
Batch Size	32
Learning Rate	2e−6
Optimizer	AdamW
Number of Epochs	10
Warmup Proportion	0.1
Dropout Probability	0.1
Weight Decay	2e−4
Loss Function	Cross-Entropy

2. K Nearest Neighbors (KNN): KNN is a straightforward and efficient classification algorithm. It works by locating the k examples in a training set that are the closest to a given instance, then classifying the instance based on the class that is shared by its k closest neighbors. It's a non-parametric model, thus it doesn't assume anything about how the data are distributed in general [3].
3. Support Vector Machine (SVM): SVM is a well-known classifier that looks for a hyperplane that divides instances of various classes by the greatest margin. Given that it is a parametric model, it makes assumptions about the data's underlying distribution. Various NLP tasks, including sentiment analysis and text categorization, have seen success with the usage of SVM.
4. Naive Bayes: Naive Bayes is a straightforward probabilistic classification technique that employs the Bayes theorem to determine the likelihood of a class given an instance's features. It makes the assumption that features are independent of one another.

The matrices, we utilized for evaluation, are frequently employed in projects involving machine learning and natural language processing. The degree of accuracy in your model refers to the proportion of cases that are correctly categorised relative to the total number of occurrences. It is measured as the proportion of the model's accurate forecasts to all of its other predictions. Precision is the percentage of positive instances that were correctly classified out of all the positive instances that were classified as positive. It is determined by dividing the quantity of real positives—i.e., events that were correctly categorized as positive—by the total of true positives and false positives. Out of all the positive events in the dataset, recall is the percentage that were properly categorised. The ratio of true positives to the total of both true positives and false negatives, or instances that were mistakenly labeled as negative, is used to compute it.

Each of these matrices offers insightful data about how well your model is doing. It's crucial to remember that based on the precise task and project goals, certain metrics could be more or less useful.

4 Results and Discussion

Figure 2 presents the classification findings for positive sentiment. The plot demonstrates that our model surpasses all cutting-edge models by achieving maximum recall, maximum accuracy, and maximum accuracy. The proposed model has the greatest accuracy of 0.91%, which indicates that it classified 91% of the tweets correctly. The suggested model has good precision and recall values of 0.92 and 0.93, respectively. This shows that the model accurately categorized the tweets and avoided incorrectly identifying tweets as positive or negative.

When compared to other models, our model had the second-highest accuracy, however SVM's precision and recall values were considerably lower than those of the suggested model. In comparison to the other models, KNN and Naive Bayes display lesser accuracy and precision values, but Classical BERT also demonstrates a comparatively high accuracy.

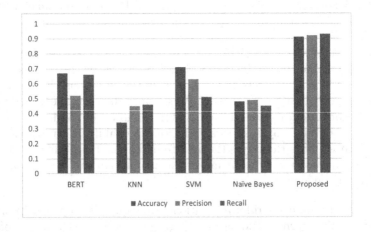

Fig. 2. Results of Positive sentiment classification

Figure 3 compares our model against the state-of-the-art for classifying negative sentiment using accuracy, precision, and recall. The results demonstrate that our model continues to beat all other models by achieving the highest accuracy, precision, and recall for the categorization of negative sentiment.

It is clear from the data in Fig. 4 that our model surpasses cutting-edge algorithms for neutral sentiment categorization in terms of accuracy, precision, and recall. The use of BERT as a base model and the utilization of the final four layers of BERT to carry out the sentiment analysis task ensure a fine-tuning

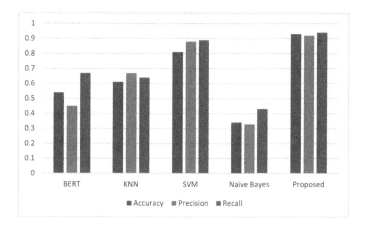

Fig. 3. Results of Negative sentiment classification

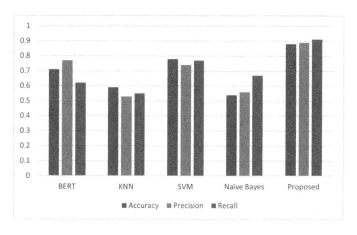

Fig. 4. Results of Neutral sentiment classification

that is not carried out by rival approaches, which helps explain why our model performs better than others.

More specifically, when compared to the proposed model (BERT-based), KNN, Naive Bayes, and SVM perform worse in terms of accuracy, precision, and recall. These findings' primary drivers are: 1) Limited feature representation: Compared to BERT, a pre-trained language model that can handle vast volumes of data and complicated word associations, these models may not be able to capture the complex relationships and patterns in the data as well. 2) Model complexity: Compared to BERT, models like KNN, Naive Bayes, and SVM are much simpler, and they might not be able to handle the complexity of the sentiment analysis problem. BERT, in comparison, can catch more nuances in the data due to its deeper and more complicated architecture. 3. Hyperparameter tuning: By selecting the best values for k in KNN or an acceptable kernel function in SVM, for example, the performance of these models may be enhanced. However, this can be a laborious and costly computer operation.

5 Conclusion

The COVID-19 epidemic has had a profound effect on people's views and emotions about numerous facets of life. Sentiment analysis has been a crucial technique for monitoring public opinion and sentiment regarding the virus, governmental policies, healthcare systems, and similar topics as social media has grown in popularity. In this study, utilizing BERT as a basis model, we suggested a unique model for sentiment analysis. Our model seeks to classify tweets using our suggested architecture into positive, negative, and neutral attitudes. This study offers perceptions on how people feel about the COVID-19 pandemic, which might assist decision-makers in their deliberations. On Twitter dataset connected to COVID-19, proposed model has been demonstrated to be a highly successful strategy for sentiment categorization, yielding cutting-edge results. In conclusion, using deep models' hidden layers can assist extract more complex and in-depth information from text input and can result in better performance on NLP tasks like sentiment analysis. For future work, it is crucial to explore the scalability of the CT-BERT-based sentiment analysis model to handle larger datasets, as this can enhance its applicability and performance. Additionally, further investigation into optimizing the classification task, such as exploring different classifiers and fine-tuning strategies, can improve the model's predictive accuracy. Lastly, the research could extend its application to other domains or languages to assess its adaptability and effectiveness in broader contexts.

References

1. Adday, B.N., Shaban, F.A.J., Jawad, M.R., Jaleel, R.A., Zahra, M.M.A.: Enhanced vaccine recommender system to prevent covid-19 based on clustering and classification. In: 2021 International Conference on Engineering and Emerging Technologies (ICEET), pp. 1–6. IEEE (2021)
2. Chakraborty, T., Shu, K., Bernard, H.R., Liu, H., Akhtar, M.S. (eds.): CONSTRAINT 2021. CCIS, vol. 1402. Springer, Cham (2021). https://doi.org/10.1007/978-3-030-73696-5
3. Hota, S., Pathak, S.: KNN classifier based approach for multi-class sentiment analysis of twitter data. Int. J. Eng. Technol. **7**(3), 1372–1375 (2018)
4. Karimi, A., Rossi, L., Prati, A.: Improving BERT performance for aspect-based sentiment analysis. arXiv preprint arXiv:2010.11731 (2020). http://arxiv.org/abs/2010.11731
5. Masciari, E.: RFID data management for effective objects tracking. In: Proceedings of the 2007 ACM Symposium on Applied Computing, pp. 457–461 (2007)
6. Müller, M., Salathé, M., Kummervold, P.E.: Covid-Twitter-BERT: a natural language processing model to analyse covid-19 content on twitter. arXiv:2005.07503 (2020). http://arxiv.org/abs/2005.07503

7. Raza, S., Dolatabadi, E., Ondrusek, N., Rosella, L., Schwartz, B.: Discovering social determinants of health from case reports using natural language processing: algorithmic development and validation. BMC Digit. Health **1**(1), 35 (2023)
8. Raza, S., Schwartz, B., Rosella, L.C.: CoQUAD: a covid-19 question answering dataset system, facilitating research, benchmarking, and practice. BMC Bioinf. **23**(1), 1–28 (2022)
9. Umair, A., Masciari, E.: Sentimental and spatial analysis of covid-19 vaccines tweets. J. Intell. Inf. Syst., 1–21 (2022)
10. Umair, A., Masciari, E.: Using high performance approaches to covid-19 vaccines sentiment analysis. In: 2022 30th Euromicro International Conference on Parallel, Distributed and Network-based Processing (PDP), pp. 197–204. IEEE (2022)

Exploring Hierarchical MPI Reduction Collective Algorithms Targeted to Multicore Node Clusters

Gladys Utrera[1](✉)[iD], Marisa Gil[1][iD], Xavier Martorell[1][iD], William Spataro[2][iD], and Andrea Giordano[3][iD]

[1] Universitat Politécnica de Catalunya. BarcelonaTECH, Barcelona 08034, Spain
{gladys.utrera,marisa.gil,xavier.martorell}@upc.edu
[2] Department of Mathematics and Computer Science (DeMACS), University of Calabria, Rende, (CS) 87036, Italy
william.spataro@unical.it
[3] Institute of High Performance Computing (ICAR), National Research Counsil (CNR) Ponte Bucci, Rende, (CS) 87036, Italy
andrea.giordano@icar.cnr.it

Abstract. High-performance computing applications heavily rely on message-passing mechanisms for data sharing in cluster environments, and the MPI library stands as the default communication library for parallel applications. Significant efforts have been directed toward optimizing data distribution and buffering based on size. This optimization aims to enhance communication performance and prevent issues such as running out of memory on the target node.

Furthermore, the emergence of multicore clusters with larger node sizes has stimulated the investigation of hierarchical collective algorithms that consider the placement of processes within the cluster and the memory hierarchy.

This paper studies and compares the performance of the algorithm of the reduction collective from the literature, specifically several implementations that do not form part of the current MPI standard, which tackle this issue. We implement the algorithms on top of Intel MPI and OpenMPI libraries using the MPI profiling interface.

Experimental results with the Intel MPI Benchmarks on a multicore cluster, Intel Platinum processor-based and OmniPath interconnection network show much room for improvement in the performance of collectives depending on the message sizes.

Keywords: MPI · HPC · multicore cluster · collective algorithms

1 Introduction

The Message Passing Interface (MPI) library is a widely used standard for communication in high-performance computing (HPC) systems based on the message-passing model. It provides a portable and efficient way for multiple processes running on different computing nodes to communicate and coordinate in parallel computing environments.

In this sense, ongoing research focuses on optimizing collective communication performance due to their potential to create performance bottlenecks in HPC applications. Moreover, in Chunduri et al. [4], the authors identify the reduction and broadcasting collective operations as the most time-consuming MPI operations in a production system. On the other hand, the increasing size of nodes in multicore clusters has led to the exploration of hierarchical collective algorithms, leveraging shared-memory architectures and faster interconnection networks within individual nodes.

In this work, we analyze different implementation proposals for the reduction algorithm (MPI_Allreduce) from the literature [1,2,11], not included yet in the MPI standard. The hierarchical design of these algorithms considers the placement of processes within the cluster, aiming to leverage the benefits of shared memory and exploit the available parallelism within each computing node, especially in the context of the growing number of node-level cores. In this work, we seek to identify the parameters that impact the algorithm when executed on different platforms and MPI library implementations. We have already analyzed in a previous work [12], the MPI_Bcast function, which is an asymmetric operation, i.e. not all the processes have the same role. On the contrary, the MPI_Allreduce is a symmetric operation, where all the processes contribute equally to the final solution of the operation. To that aim, we develop the implementations on top of the Intel MPI library [8] and OpenMPI library [9], taking advantage of the MPI profiling interface and leveraging the shared-memory functionality offered by MPI-3 within nodes.

In summary, the main contributions of this work are:

- Implementation, evaluation and comparison of the three hierarchical Allreduce collective algorithms: Multi-leader, Multi-lane and Hier. The comparisons are made against the native reduction algorithm from the Intel MPI library and OpenMPI library. In both cases, we have selected from all the native available algorithms, the one in each library, that obtained the best performance in our platform. We consider it relevant to explore different MPI implementations as the hierarchical implementations are built on top of other MPI operations which serve native implementations.
- Implementation of the Multi-leader hierarchical algorithm using the MPIWin facility from the MPI-3 standard. We have also built the best configuration for this algorithm by evaluating different parameters like the number of leaders and the size of the groups.
- Evaluation of the different reduction algorithms in conjunction with Intel MPI and OpenMPI libraries, varying the message size and other parameters like group size and the number of leaders for the specific algorithms.

As already stated in our previous work [12], with an asymmetric collective, in this work, we affirm, also with a symmetric collective, the necessity for hierarchical algorithms that exhibit awareness of the underlying platform characteristics. We observe the significance of how different implementations of the MPI library exploit platform characteristics such as shared memory utilization using RDMA. Consequently, a hierarchical algorithm should be designed to adapt to parameters beyond node size, encompassing considerations such as memory hierarchy and characteristics of interconnection networks.

The subsequent sections of this document are structured as follows: initially; we provide an overview of related works concerning the design and assessment of collective algorithms. Subsequently, we delineate our methodology, describe the implemented and evaluated algorithms, and introduce our proposed. Following this, we detail the conducted evaluations and present our observed findings. Ultimately, we encapsulate our insights and delineate avenues for future research in the concluding section.

2 Related Work

Being MPI the de-facto library for cluster applications, there is continuous research work to analyse, evaluate and improve MPI primitives. This section shows some of them, chosen by similarity with the primitive to study MPI_Allreduce, the methodology - improving the memory performance- or the library. In [6], the authors aim to improve the collectives' local (in-core) behaviour when working in a multicore. To do this, they analyse several collectives, in addition to MPI_Allreduce: they study the behaviour of the collective algorithms for shared memory and compare it to that of point-to-point implementations.

In [10], the goal is to achieve the best possible communication throughput, and they do so from process-to-processor mapping. To do this, they work with the collective rank assignment to processes.

The work in [3] focuses on neural networks; specifically, they develop the study in the TensorFlow framework and how to improve the training phase. To do this, the authors study the behaviour of the different implementations of the MPI_Allreduce collective in three libraries, MPICH, OpenMPI and IntelMPI, then select the best option based on the number of nodes/message size for each library. The investigation leads them to detect an improvement of between 1.2x and 2.8x when they work with a specific model and dataset in addition to the most suitable algorithm of the three libraries.

In [12], the authors present their first work evaluating and comparing the hierarchical collective algorithms for the broadcasting operation on an ARM platform. There were noticeable improvements over the native Broadcast from the OpenMPI library depending on the message size, number of groups and number of leaders. In this work, we also compare the Intel MPI library in a supercomputer platform and with a different collective, symmetric for the processes working in contrast to broadcast.

3 Methodology

We explore recently proposed algorithms that consider at least one of the characteristics: large node sizes, shared-memory facility, and use of available parallelism. The selected algorithms are:

- **Multi-leader**: The proposed algorithm by Bayatpour et al. [7,8] introduces a multi-leader approach wherein collectives are organized into groups of processes, with a designated leader responsible for inter-group communication. Their study suggests adapting the group size to align with the system's socket architecture. In this algorithm, intra-group communication is facilitated through shared memory. Figure 1 illustrates a graphical representation of the algorithm, featuring two groups, each comprising four processes. The message is initially copied to the shared-memory region, and the leader process performs a local reduction followed by a reduction operation with the other leaders. Finally, the leader copies the resultant value, while non-leader processes retrieve the outcome from the shared-memory region. The choice of an appropriate configuration for the number of leaders per node is contingent upon the specific collective operation being performed.

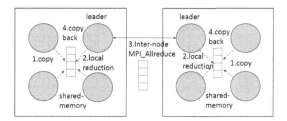

Fig. 1. Multi-leader scheme

- **Multi-lane**: This algorithm proposed by J. Traff et al. [11] suggests having multiple lanes per group. In this scheme, each group member takes responsibility for initially executing a local reduction on the assigned partition of the original vector within its respective group. Subsequently, group members collaborate in performing reductions across nodes with their counterparts in other groups. Finally, the original message is reconstructed through an allgatherv operation in each group. The operational mechanism is illustrated in Fig. 2.
- **Hier**: This algorithm, proposed by J. Traff et al. [11], offers an alternative to Multi-lane. It closely resembles Multileader but without explicitly utilizing shared memory, which ultimately would depend on the MPI library implementation. The collective operation is disassembled into a scatter operation, allocating each vector partition to a group member, followed by inter-node reductions, and concluding with reconstructing the original message through an allgatherv operation. Figure 3 illustrates the operational process of the algorithm.

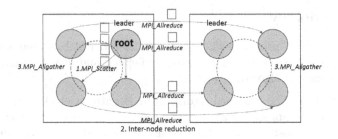

Fig. 2. Multi-lane scheme

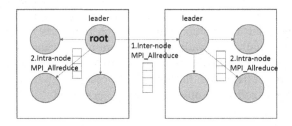

Fig. 3. Hier scheme

After exploring different configurations for group sizes and number of leaders, we selected one group per socket (24 processes per group), which means two groups per node. This was set for both MPI libraries. We present only the experimental results for this best configuration.

The experiments are carried out on the MarenostrumIV Supercomputer, a multicore-cluster based on 2x Intel Xeon Platinum 8160 24C, a 2.1 GHz processor with 100Gb Intel Omni-Path Full-Fat Tree interconnection network.

The operating system is SUSE Linux Enterprise Server 12 SP2, and for the experiments, we have used Intel MPI library version 2018.4 and OpenMPI library version 4.1.3. To test the MPI_Allreduce collective algorithms, we use the Intel MPI Benchmarks version 2019 with the option "cache off" to avoid distortion of results from repeated executions. We repeat at least four times each experiment.

4 Experimental Results

This section presents the performance results after evaluating the different reduction algorithms described in Sect. 3.

In the plots in Figs. 4 and 5, we can observe the average execution times of the different reduction algorithms when varying the message size in values power of two. To appreciate the difference in performance between the evaluated algorithms, we categorize the results in small, medium and large message sizes, which correspond to 8Bytes-4KBytes, 8KBytes-128KBytes and 256KBytes-4MBytes, respectively.

We appreciate that the Hier approach overperforms the rest of the algorithms in both MPI libraries for small message sizes. In contrast, the original algorithm (Intel and OpenMPI) performs quite similarly to the best in medium message sizes. However, the Multi-lane approach best serves large message sizes under both MPI libraries.

Even though the Multi-leader approach has an algorithm similar to Hier, its performance shows the worst results, especially as message size increases. The intra-node shared-memory usage in this centralized algorithm does not contribute to any improvement and results in a worse performance. In the case of the Hier algorithm, at least in the OpenMPI library, which is open source, the version used does not exploit intra-node shared memory for this operation.

The Multi-lane approach does not benefit small message sizes, as the initial message size, before the partitioning, is already small enough to accomplish small latencies and fit in memory caches. The centralized approach can take advantage of fast intra-node communication. However, as the message sizes increase, the Multi-lane algorithm can exploit the bandwidth better than the centralized and native algorithms, leading to better performance. In the case of the IMPI library, this advantage is enhanced because of the RDMA usage, which helps to improve significantly the performance of the memory copies at the intra-node level.

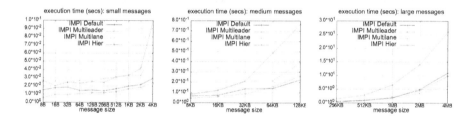

Fig. 4. Performance results for MPI_Allreduce collective using Intel MPI library

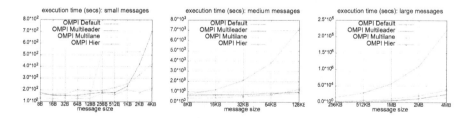

Fig. 5. Performance results for MPI_Allreduce collective using OpenMPI library

5 Conclusions

The research on the performance of MPI collective operations demonstrates that they are the bottleneck of HPC applications in high-performance computing systems.

On the other hand, the increasing number of cores per node in multicore clusters deserves to have a unique role in the hierarchy of collective algorithms. The availability of shared memory and faster interconnection networks within a node claim to adapt those existing algorithms.

Recently, there have been many efforts in that direction. In this work, we focused on the reduction operation. To that aim, we have selected three hierarchical approaches, which we implemented, evaluated and compared with default algorithms from the OpenMPI and Intel MPI libraries. They are Multi-leader, Multi-lane and Hier collective algorithms. In particular, in the Multi-leader approach, at the intra-node level, the communication is performed through shared memory, which we implemented using the MPIWin facility from MPI-3. The other two decompose the collective into other collectives, differentiating between intra- and inter-node communication.

Our assessments on a supercomputer equipped with Intel Platinum processors, utilizing two different MPI library implementations, revealed that Hier is suitable for small message sizes. In contrast, Multi-lane is preferable for large message sizes. This algorithm exploits the bandwidth better as the intra/inter-node communications are performed concurrently on partitions of the original message.

We intend to expand our evaluations to encompass many-core architectures such as GPGPU clusters. Concurrently, our ongoing research involves the examination of real applications and libraries capable of managing data ranging from small to large quantities [5,7].

Acknowledgements. The authors acknowledge the support of the Spanish Ministry of Education (PID2019-107255GB-C22) and the Generalitat de Catalunya (2021-SGR-01007).

References

1. Bayatpour, M., Maqbool Hashmi, J., Chakraborty, S., Subramoni, H., Kousha, P., Panda, D.K.: Salar: scalable and adaptive designs for large message reduction collectives. In: 2018 IEEE International Conference on Cluster Computing (CLUSTER), pp. 12–23 (2018). https://doi.org/10.1109/CLUSTER.2018.00014
2. Bayatpour, M., Chakraborty, S., Subramoni, H., Lu, X., Panda, D.K.D.: Scalable reduction collectives with data partitioning-based multi-leader design. In: Proceedings of the International Conference for High Performance Computing, Networking, Storage and Analysis, SC 2017, Association for Computing Machinery, New York (2017). https://doi.org/10.1145/3126908.3126954, https://doi.org/10.1145/3126908.3126954

3. Castelló, A., Catalán, M., Dolz, M.F., Quintana-Ortí, E.S., Duato, J.: Analyzing the impact of the MPI allreduce in distributed training of convolutional neural networks. Computing , 1–19 (2021). https://doi.org/10.1007/s00607-021-01029-2
4. Chunduri, S., Parker, S., Balaji, P., Harms, K., Kumaran, K.: Characterization of mpi usage on a production supercomputer. In: SC18: International Conference for High Performance Computing, Networking, Storage and Analysis, pp. 386–400 (Nov 2018). https://doi.org/10.1109/SC.2018.00033
5. Giordano, A., De Rango, A., Rongo, R., D'Ambrosio, D., Spataro, W.: Dynamic load balancing in parallel execution of cellular automata. IEEE Trans. Parallel Distrib. Syst. **32**(2), 470–484 (2021). https://doi.org/10.1109/TPDS.2020.3025102
6. Graham, R.L., Shipman, G.: MPI support for multi-core architectures: optimized shared memory collectives. In: Lastovetsky, A., Kechadi, T., Dongarra, J. (eds.) EuroPVM/MPI 2008. LNCS, vol. 5205, pp. 130–140. Springer, Heidelberg (2008). https://doi.org/10.1007/978-3-540-87475-1_21
7. Heroux, M.A., et al.: Improving Performance via Mini-applications. Tech. Rep. SAND2009-5574, Sandia National Laboratories (2009)
8. Intel MPI library: Intel MPI Library. https://www.intel.com/content/www/us/en/developer/tools/oneapi/mpi-library.html
9. OpenMPI Open Source HPC: OpenMPI Open Source HPC. http://www.open-mpi.org/
10. Rico-Gallego, J., Díaz-Martin, J.: Improving the performance of the mpi_allreduce collective operation through rank renaming. In: In Proceedings of First International Workshop on Sustainable Ultrascale Computing Systems (August 2014). https://core.ac.uk/download/pdf/30277086.pdf
11. Träff, J.L., Hunold, S.: Decomposing mpi collectives for exploiting multi-lane communication. In: 2020 IEEE International Conference on Cluster Computing (CLUSTER), pp. 270–280 (2020). https://doi.org/10.1109/CLUSTER49012.2020.00037
12. Utrera, G., Gil, M., Martorell, X.: Analyzing the performance of hierarchical collective algorithms on arm-based multicore clusters. In: 2022 30th Euromicro International Conference on Parallel, Distributed and Network-based Processing (PDP), pp. 230–233 (2022). https://doi.org/10.1109/PDP55904.2022.00043

Author Index

A
Acuto, Alberto III-166
Addawe, Rizavel C. II-78
Agapito, Giuseppe III-15
Akopov, Andranik S. I-273
Alì, Giuseppe III-3
Alvelos, Filipe I-351
Amisano, Maristella II-184
Amodio, Pierluigi II-167
Antoniotti, Luigi III-223, III-232
Archetti, Francesco I-49, I-151
Astorino, Annabella I-281, III-223
Attanasio, Carmine III-346

B
Bagdasaryan, Armen II-175
Bajārs, Jānis II-3
Barilla, David I-290
Barillà, Paola III-166
Barillaro, Luca III-15
Barkalov, Konstantin I-3
Basta, Daniele II-19
Battiti, Roberto I-19
Becciu, Gianfranco II-298
Beklaryan, Armen L. I-273
Bertacchini, Arrigo III-240
Bertacchini, Francesca III-3, III-262
Bertolazzi, Enrico II-31, II-47
Bevacqua, Marida II-184
Bilotta, Eleonora III-3, III-262
Biral, Francesco II-47
Boada Gutierrez, Maria Gabriela II-94
Boronina, Marina III-254
Bottino, Lorella III-26
Bozzolo, Ludovico III-166
Brunetti, Guglielmo Federico Antonio II-184
Brusco, Anna Chiara II-331
Buonaiuto, Giuseppe III-98

C
Caldarola, Fabio II-192, III-223
Candelieri, Antonio I-34, I-49
Cannataro, Mario III-26
Capano, Gilda II-19
Carini, Manuela II-192, II-315
Caristi, Giuseppe I-290
Castagna, Jessica II-200
Caterina, Gianluca III-84
Cavallaro, Lucia I-296
Cavoretto, Roberto II-207, II-215
Cesario, Eugenio III-248
Chebak, Ahmed I-343
Chernoshtanov, Ivan III-254
Chernov, Alexey I-257
Chernykh, Igor III-254
Citro, Roberta III-346
Colace, Simone III-207
Comito, Carmela III-40
Consiglio, Mirko III-56
Conterno, Matteo III-166
Corazza, Marco I-304
Cordero, Elena II-63
Costa, Lino A. I-351
Curcio, Efrem II-19
Curcio, Giulia Maria II-19
Curulli, Giuseppe II-110

D
D'Ambrosio, Claudia I-210
D'Ambrosio, Donato II-223, II-346
d'Atri, Gianfranco III-113, III-301
De Luca, Pasquale III-71
De Pietro, Giuseppe III-98
De Rango, Alessio II-223, II-346
De Rango, Floriano II-290
De Rossi, Alessandra II-207, II-231
De Stefano, Angelafrancesca II-281
Dell'Amico, Mauro I-108

Demarco, Francesco III-3, III-262
Di Francesco, Massimo I-281
Diveev, Askhat I-64, I-383

E

Efimova, Anna III-254
Egidi, Nadaniela II-238, II-246
Esposito, Massimo III-98
Evangelista, Davide I-360

F

Falco, Salvatore II-331
Falcone, Alberto III-270
Falcone, Deborah III-40
Fasano, Giovanni I-304
Fatone, Lorella II-254
Faydaoğlu, Şerife II-261
Filatovas, Ernestas III-278
Fiorentino, Michele Giuliano III-154
Fioretto, Simona III-286
Flerova, Anna I-257
Folino, Francesco I-79
Folino, Gianluigi I-79, III-293
Forestiero, Agostino III-207, III-293
Franchini, Giorgia I-94, I-108
Funaro, Daniele II-254
Furnari, Luca II-268, II-346
Fusaro, Pierpaolo Antonio III-240

G

Galletti, Ardelio III-71
Gangle, Rocco III-84
Garajová, Elif I-312
Garro, Alfredo III-270
Gaudio, Roberto II-110
Gaudioso, Manlio I-281
Germano, Daniele III-207
Giacchi, Gianluca II-63
Giacomini, Josephin II-238, II-246
Giallombardo, Giovanni I-320
Gil, Amparo II-275
Gil, Marisa III-362
Giordano, Andrea III-362
Gonçalves, Rui I-351
Gorgone, Enrico I-281
Gregori, Daniele III-324
Grishagin, Vladimir A. I-3, I-336
Grossi, Giovanna II-298
Guarasci, Raffaele III-98

Guarascio, Massimo I-79
Guerreiro Lopes, Luiz I-165, I-182

H

Hladík, Milan I-312

I

Iembo, Rosanna III-113
Ingarozza, Francesco III-113, III-127
Ivano, Mària III-316

J

Jafariani, Zahra I-290
Javadi Nejad, Hana II-281

K

Kalampakas, Antonios II-175
Kamach, Oulaid I-343
Kanzi, Nader I-290
Karamzin, Dmitry I-328
Kozinov, Evgeny I-3
Kozyriev, Anton I-136
Kulikov, Igor III-254
Kumam, Poom II-134
Kvasov, Dmitri E. I-336

L

La Regina, Francesco III-301
Lancellotti, Sandro II-215, II-231
Larcher, Matteo II-149
Laurita, Sara III-207
Lazaga, Junley L. II-78
Lazzaro, Ilaria III-309
Le Thi, Hoai An I-368
Libatique, Criselda P. II-78
Lindia, Paolo III-248
Liotta, Antonio I-296
Loli Piccolomini, Elena I-360
Luzza, Francesco III-180

M

Magnani, Matteo I-108
Maiolo, Maria Rita II-184, III-316
Maiolo, Mario II-19, II-192
Mallah, Sara I-343
Manca, Benedetto I-281
Maponi, Pierluigi II-238, II-246
Marcellino, Livia III-71
Marchetti, Filippo III-324

Author Index

Mariani, Luca   III-346
Marigmen, Joseph Ludwin D. C.   II-78
Marotta, Corrado Mariano   III-139, III-232
Martínez, Ángeles   I-121
Martinis, Maria Chiara   III-332
Martorell, Xavier   III-362
Masciari, Elio   III-286, III-353
Masmoudi, Malek   I-343
Matos, Marina A.   I-351
Melicchio, Andrea   III-139, III-232
Mendicino, Giuseppe   II-200, II-223
Miglionico, Giovanna   I-320
Milano, Marianna   III-309
Montone, Antonella   III-154
Morales Escalante, José A.   II-94
Morotti, Elena   I-360

N
Naghib, Seyed Navid   II-331
Napolitano, Enea Vincenzo   III-286
Nasr, Layla   III-338
Nguyen, Thi Tuyet Trinh   I-368
Norkin, Vladimir   I-136

O
Orts, Francisco   III-278

P
Pagano, Sergio   III-346
Palermo, Stefania Anna   II-281, II-290
Pantano, Pietro   III-3, III-262
Papaleo, Maria Anastasia   III-232
Papuzzo, Giuseppe   III-207, III-293
Paulavičius, Remigijus   III-278
Penna, Nadia   II-110
Pichler, Alois   I-136
Piro, Patrizia   II-121, II-290
Pirouz, Behrouz   II-121, II-298
Pirouz, Behzad   II-281
Piscitelli, Aldo   III-127
Pizzuti, Clara   I-197
Poleksic, Aleksandar   I-376
Policicchio, Antonio   III-166
Ponti, Andrea   I-49, I-151
Pontieri, Luigi   I-79
Porta, Federica   I-94
Presta, Ludovica   II-298

Prokopiev, Igor   I-383
Promenzio, Luigi   III-26
Punzalan, Jean Milnard S.   II-78

R
Repaci, Vincenzo   II-19
Ricciardiello, Giuditta   III-154
Rocha, Ana Maria A. C.   I-351
Romaniello, Federico   II-215, II-231
Ruggiero, Valeria   I-94
Ruiz-Antolín, Diego   II-275
Rukavishnikov, Viktor A.   II-306

S
Saburov, Mansoor   II-175
Sakhno, M.Yu.   I-241
Salatino, Luca   III-346
Saleh, Mohammed M.   II-290
Salisu, Sani   II-134
Sammarra, Marcello   I-320
Sanfilippo, Umberto   II-298
Santoro, Sergio   II-19
Scarcella, Maria Antonietta   II-315
Scarpino, Ileana   III-180
Scuro, Carmelo   III-262
Segura, Javier   II-275
Senatore, Alfonso   II-268, II-346
Serafin, Tommaso   I-296
Sergeyev, Yaroslav D.   II-207, III-270
Serpe, Annarosa   III-193
Settino, Marzia   III-26
Shmalko, Elizaveta   I-64, I-383
Siciliano, Alessio   II-19
Silva, Bruno   I-165, I-182
Socievole, Annalisa   I-197
Spataro, William   III-362
Spencer Trindade, Renan   I-210
Sriwongsa, Songpon   II-134
Stocco, Davide   II-149
Straface, Salvatore   II-19
Stripinis, Linas   III-278

T
Takahashi, Daisuke   II-323
Temme, Nico M.   II-275
Tohmé, Fernando   III-84
Trombini, Ilaria   I-94

Tropea, Mauro II-290
Tseveendorj, Ider I-391
Tubera-Panes, Donnabel II-78
Turco, Michele II-281, II-331

U
Umair, Areeba III-353
Utrera, Gladys III-362

V
Vallelunga, Rosarina III-180
Vena, Stefano III-3
Viernes, Jhunas Paul T. II-78
Vinci, Andrea III-248
Viola, Marco I-121
Viviani, Mariangela III-207
Vshivkov, Vitaly III-254

W
Wang, Xia II-339
Weins, Dmitry III-254

Y
Yamshanov, Konstantin I-383
Yousaf, Umair II-346
Yousefi, Mahsa I-121

Z
Zakharov, Aleksey I-226
Zakharova, Yulia V. I-241, I-400
Zampoli, Vittorio II-315
Zanni, Luca I-94, I-108
Zhukova, Aleksandra I-257
Zou, Qinmeng II-339
Zucco, Chiara III-332
Zupi, Massimo III-240

Printed in the United States
by Baker & Taylor Publisher Services